The Earthscan Reader on Gender and Forests

There is currently much interest in mainstreaming gender in natural resource management, including forestry. This reader provides a collection of key articles on gender and forests published over the last 30 years. Including an editorial introduction and overview, it provides an accessible collection of excellent forestry-relevant social science within an overarching analytical framework and demonstrates the leading debates in the field.

The book will be of great value to both biophysical science and social science students and to professionals in training. It focuses on people and forest interactions, providing a range of studies from both developed and developing countries. It includes theoretical analyses, methodological pieces, case studies, and cross-country comparisons, and it forms a companion volume to *Gender and Forests: Climate Change, Tenure, Value Chains and Emerging Issues* (2016).

Carol J. Pierce Colfer is a Senior Associate at the Center for International Forestry Research (CIFOR) and Visiting Scholar at Cornell University's Southeast Asia Program, Ithaca, New York, USA.

Marlène Elias is a Gender Specialist at Bioversity International, Rome, Italy.

Bimbika Sijapati Basnett is Gender Coordinator at CIFOR in Bogor, Indonesia.

Susan Stevens Hummel is a Research Forester at the Pacific Northwest (PNW) Research Station, US Department of Agriculture Forest Service, Portland, Oregon, USA.

The Earthscan Reader Series

The Earthscan Reader in Environmental Values
Edited by Linda Kalof and Terre Satterfield

The Earthscan Reader in Forestry and Development
Edited by Jeffrey Sayer

The Earthscan Reader in Sustainable Agriculture
Edited by Jules Pretty

The Earthscan Reader in Rural-Urban Linkages
Edited by Cecilia Tacoli

The Earthscan Reader on Sustainable Consumption
Edited by Tim Jackson

The Earthscan Reader on Adaptation to Climate Change
Edited by E. Lisa F. Schipper and Ian Burton

The Earthscan Reader on Risk
Edited by Ragnar E. Lofstedt and Asa Boholm

The Earthscan Reader on Poverty and Biodiversity Conservation
Edited by Dilys Roe and Joanna Elliot

The Earthscan Reader on Gender and Forests
Edited by Carol J. Pierce Colfer, Marlène Elias, Bimbika Sijapati Basnett and Susan Stevens Hummel

For more information please visit https://www.routledge.com/Earthscan-Reader-Series/book-series/ECRES

The Earthscan Reader on Gender and Forests

Edited by Carol J. Pierce Colfer, Marlène Elias, Bimbika Sijapati Basnett and Susan Stevens Hummel

Routledge
Taylor & Francis Group
LONDON AND NEW YORK

earthscan
from Routledge

First published 2017
by Routledge
2 Park Square, Milton Park, Abingdon, Oxon OX14 4RN

and by Routledge
711 Third Avenue, New York, NY 10017

Routledge is an imprint of the Taylor & Francis Group, an informa business

British Library Cataloguing-in-Publication Data
A catalogue record for this book is available from the British Library

Library of Congress Cataloging in Publication Data
A catalog record for this book has been requested.

ISBN: 978-1-138-23158-0 (hbk)
ISBN: 978-1-138-23159-7 (pbk)

Typeset in Bembo
by Keystroke, Neville Lodge, Tettenhall, Wolverhampton

The Center for International Forestry Research (CIFOR) envisions a more equitable world where forestry and landscapes enhance the environment and well-being for all. CIFOR is a non-profit, scientific institution that conducts research on the most pressing challenges of forest and landscape management around the world. Using a global, multidisciplinary approach, we aim to improve human well-being, protect the environment, and increase equity. To do so, we conduct innovative research, develop partners' capacity, and actively engage in dialogue with all stakeholders to inform policies and practices that affect forests and people. CIFOR is a CGIAR Research Center, and leads the CGIAR Research Program on Forests, Trees and Agroforestry (FTA). Our headquarters are in Bogor, Indonesia, with offices in Nairobi, Kenya, Yaounde, Cameroon, and Lima, Peru.

We dedicate this book to researchers, practitioners, policymakers, advocates and local people everywhere who are striving for an equitable and verdant planet.

"This is a valuable compilation of the conceptual and theoretical frameworks regarding gender and forests and their evolution in diverse contexts since the 1970s. For all those working on equitably strengthening the interface between communities and forests, this reader will be an excellent resource for better understanding the often invisible power relations determining whether women gain or lose from well-intentioned interventions."

Madhu Sarin, The Campaign for Survival and Dignity, India

"Forestry is ultimately a social science if our aim is to benefit people. Often gender issues in forestry are addressed superficially, without adequate thought to underlying assumptions and theoretical frameworks or consideration of case examples. This volume—and its 2016 companion—thus fills a gap for planners, researchers and managers who care about improving how gender is represented in forestry."

Cynthia Mackie, US Forest Service International Programs, Washington DC, USA

"'What can we do about gender?' ask many research and development actors working to sustain forest landscapes globally. This book makes a compelling case that by pursuing gender-focused strategies and opportunities, all kinds of people, along with forests, trees and the environment, will be better off."

Patricia Kristjanson, World Agroforestry Center, Kenya

"This volume will educate and inspire those wishing to conserve forested landscapes and make the people living in and around forests better off. It thoughtfully analyzes critical gender issues related to forests and what we know about what works and what doesn't. This will be valuable for researchers, development agencies, governments and civil society organizations interested in capitalizing on the positive role forests can play for all kinds of people and their environments."

Werner L Kornexl, Program on Forests, World Bank Group, Washington DC, USA

Contents

List of illustrations

Figures

Tables

List of contributors

Agarwal, Bina, Professor, Development Economics and Environment, University of Manchester, Manchester, UK

Arora-Jonsson, Seema, Associate Professor Department of Urban and Rural Development, Swedish University of Agricultural Sciences, Uppsala, Sweden

Carney, Judith, Professor, Department of Geography, University of California, Los Angeles, USA

Colfer, Carol J. Pierce, Senior Associate, Center for International Forestry Research, Bogor, Indonesia; Visiting Scholar, Southeast Asia Program, Cornell University, Ithaca, New York, USA

Elias, Marlène, Gender Specialist, Bioversity International, Rome, Italy

Elmhirst, Rebecca, Reader in Human Geography, School of Environment and Technology, University of Brighton, UK

Fortmann, Louise, Professor, Natural Resource Sociology, College of Natural Resources; Rudy Grah Chair in Forestry and Sustainable Development, University of California, Berkeley, California, USA

Hummel, Susan Stevens, Research Forester, Portland Forestry Sciences Lab, US Forest Service Pacific Northwest Research Station, Portland, Oregon, USA

Kroma, Margaret, Coordinator, Gender Integration Team, Forests, Trees and Agroforestry Collaborative Research Program of the CGIAR; Assistant Director General, Partnerships and Impact, World Agroforestry Center, Washington, DC, USA

Leach, Melissa, Director, Institute of Development Studies, University of Sussex, UK

Li, Tania Murray, Professor, Canada Research Chair in the Political-Economy and Culture of Asia; Director, Centre for Southeast Asian Studies, University of Toronto, Canada

Nightingale, Andrea, Professor, Chair of Rural Development in the Global South, Swedish University of Agricultural Sciences, Uppsala, Sweden

Norgaard, Kari Marie, Professor, Department of Sociology, University of Oregon, Eugene, Oregon, USA

Paulson, Susan, Professor, Center for Latin American Studies, University of Florida, Gainesville, Florida, USA

Reed, Maureen G., Professor and Assistant Director, School of Environment and Sustainability, University of Saskatchewan, Saskatoon, Saskatchewan, Canada

Rocheleau, Dianne, Professor, Graduate School of Geography, Clark University, Worcester, Massachusetts, USA

Ross, Laurie, Associate Professor, Community Development and Planning, Department of International Development, Community, and Environment, Clark University, Worcester, Massachusetts, USA

Schroeder, Richard A., Professor of Geography, Dean of Social and Behavioral Sciences, School of Arts and Sciences, Rutgers University, Rutgers, New Jersey, USA

Sijapati Basnett, Bimbika, Gender Focal Point, Forest, Trees and Agroforestry Program, Center for International Forestry Research, Bogor, Indonesia

Foreword

The management of natural resources, including forests and tree resources, requires the involvement of men and women, often in differentiated roles and with responsibilities that uniquely mediate their understandings and knowledge of the natural resource base. But the gendered nature of knowledge in everyday practice has historically been subjugated to the putatively objective scientific knowledge. It took critical Feminist scholarship beginning in the 1970s to throw into sharp relief the partial and subjective nature of knowledge produced through conventional science. This opened a floodgate of conceptual discourses and articulation of critical gender theories of knowledge that helped inform and guide professionals in their work.

It is important to mention that other strands in Feminist scholarship in the same period offered interesting glimpses into cognitive and psychoanalytical aspects in relation to gendered effects on behavior. However, none in published literature establish any direct connections to neuroplasticity – changes in the structure and functions of the human brain – in relation to male/female-specific knowledge and attitudes in natural resource management. The theoretical discourses in this reader locate gender as structural, relational, and dynamic, drawing on relevant literature, thus deftly avoiding the uncomfortable path of biological reductionism and essentialist thought around the female/male nature and environment conundrum.

There has been a growing recognition in recent years that notwithstanding the awareness of the relevance of gender in the work of natural resource and forest researchers and conservation communities, many do not have the backgrounds or theoretical lens they need to robustly address such issues. The purpose of this volume is to respond to the lacunae in theoretical grounding of analyses on gender and forests. In this effort, the editors have roundly succeeded.

Through a thorough revisit of the historical literature, the editors have seamlessly pulled together a compelling mosaic of seminal work on gender theory from anthropological, geographic, economic, and sociological perspectives. These theoretical frameworks have undergirded a wide range of multidisciplinary empirical analyses in the 70s and 80s across forests and tree landscape niches. Empirical studies from Asia, sub-Saharan Africa, and North America illuminate the gendered nature of knowledge, underlying power relations, and social structures that limit women forest users' opportunities to develop viable economic livelihoods or derive benefits from management of forests or tree resources.

Encapsulating a wide range of scholarship on gender and forests across different geographies and historical moments, the studies reaffirm the enduring relevance and explanatory power of gender constructs, and the commonalities of the "gender problematique" in the global South and global North. In the process, the possibilities of methodological eclecticism and pluralism that could help to develop a gendered and holistic account of the world are also brought to light.

This reader is a timely and much-needed contribution to research and practice in the broad arena of natural resources management, including forestry, agroforestry, and environmental conservation. While the methodological treatment of gender is nascent and relatively unexplored in forest and conservation work, the reader provides brief insights that should help professionals recognize recurrent elements in gender analyses and aid in conducting or facilitating pertinent studies. This notwithstanding, a methodological gap remains. This flags an urgency that should hopefully stimulate thought and contribution to the discourse and scholarship in the social science and forestry/conservation/agroforestry community.

It is noted that while the reader has not addressed "risks" and "trust" as important social relational dimensions of gender in forest management, these elements are flagged as needing significant priority, and rightly so. The causes and underlying drivers of gender inequality in natural resource/forest management are deeply interlocking. Uncoupling and illuminating the interlocking elements for necessary action could come with social risks to women who are often the marginalized and vulnerable in natural resource and forest management. This underscores the importance of theoretically grounded methods and robust analyses to help professionals navigate unequal power relations and gender inequities, identifying the social risks to women in access and benefits sharing. In the contested contexts of natural resource and forest management, norms of trust and reciprocity become significant dimensions in evolving sustainable futures and should be treated as important conceptual and methodological elements.

Margaret Kroma
Coordinator, Gender Integration Team
Forests, Trees and Agroforestry Collaborative Research Program of the CGIAR
Assistant Director General, Partnerships & Impact
World Agroforestry Center

Acknowledgements

We are grateful to the Center for International Forestry Research for its key role in bringing this project to fruition. The Center provided additional funding support from the CGIAR Fund as the book has been developed within the framework of the CGIAR Research Consortium on Forests, Trees and Agroforestry (CRP FTA) and the UKAID Knowledge for Forestry Programme. The US Forest Service and Bioversity International both provided additional funding, including from the CGIAR donor fund and CGIAR's Research Program on Forests, Trees and Agroforestry. Cornell University made their extensive library and other facilities available to the team for developing this collection; and the Southeast Asia Program in particular hosted Colfer as a visiting scholar. Thanks also go to Tim Hardwick, Ashley Wright, and Alanna Donaldson from Routledge for their consistently valuable advice and cooperation throughout this endeavour, and to Maggie Reid for copy-editing new material. Without Tim's 'care and feeding' related to the process of obtaining permission to reprint and publish these materials online, we would have had far more trouble than we did. We thank the various authors and publishers (see Publication details) for granting permission to use their materials as well as the people everywhere who made the contributions possible. And we thank Maria Brockhaus, Lee Cerveny, Jessica Clendenning, Shelley Feldman, Tim Forsyth, Anne Larson, Melissa Leach, Rebakah Daro Minarchek, and Marianne Schmink for their thoughtful and full critiques during the review process, as well as Rebecca McLain and Gun Lidestav for substantive aid in finding materials on the global North. Although none bears any responsibility for the final product, we are confident that the book has benefited from their inputs.

We are grateful to the CGIAR donors and UK Aid for supporting this book.

The views expressed in this collection are solely those of the authors and do not reflect the views of the supporting organizations.

Publication details

Chapter 2: Melissa Leach, "Earth Mother Myths and Other Ecofeminist Fables: How a Strategic Notion Rose and Fell," *Development and Change* 38 (1), 67–85.

Chapter 3: Louise Fortmann, "In Theory and in Practice: Women Creating Better Accounts of the World," in *Women and Gender Equity in Development Theory and Practice*, Edited by Jane S. Jaquette, Gale Summerfield, 191–201.

Chapter 4: Carol J. Pierce Colfer, "On Communication among 'Unequals,'" *International Journal of Intercultural Relations* 7 (3), 263–283.

Chapter 5: Seema Arora-Jonsson, "Discordant Connections: Discourses on Gender and Grassroots Activism in Two Forest Communities in India and Sweden," *Signs* 35 (1), 213–240.

Chapter 6: Andrea Nightingale, "A Feminist in the Forest: Situated Knowledges and Mixing Methods in Natural Resource Management," *ACME: An International E-Journal for Critical Geographies* 2 (1).

Chapter 7: Bina Agarwal, "The Hidden Side of Group Behaviour: A Gender Analysis of Community Forestry in South Asia," Chapter 9 in *Group Behaviour and Development: Is the Market Destroying Cooperation?* Edited by Heyer, J., Stewart, F. & Thorp, R., 186–208.

Chapter 8: Richard A. Schroeder, "Shady Practice: Gender and the Political Ecology of Resource Stabilization in Gambian Garden/Orchards," *Economic*

Geography 69 (4), 349–365. Copyright 1993 Clark University. Reproduced with permission.

Chapter 9: Marlène Elias and Judith Carney, "African Shea Butter: A Feminized Subsidy from Nature," *Africa* 77 (1), 37–62. Copyright 2007 The International African Institute. Published by Cambridge University Press. Reproduced with permission.

Chapter 10: Tania Murray Li, "Working Separately but Eating Together: Personhood, Property, and Power in Conjugal Relations," *American Ethnologist* 25 (4), 675–694. Copyright 1998 American Anthropological Association. Published by Wiley. Reproduced with permission.

Chapter 11: Rebecca Elmhirst, "Migrant Pathways to Resource Access in Lampung's Political Forest: Gender, Citizenship and Creative Conjugality," *Geoforum* 42 (2), 173–183. Copyright 2011. Elsevier Ltd. All rights reserved. Reproduced with permission.

Chapter 12: Susan Paulson, "Changing Masculinities and Femininities in the (Re)production of Andean Silvopasture Systems," Chapter 12 in the *Earthscan Reader on Gender and Forests* (this volume), Edited by Carol J. Pierce Colfer, Marlène Elias, Bimbika Sijapati Basnett, and Susan Stevens Hummel. Copyright 2017 Center for International Forestry Research. Published by Routledge.

Chapter 13: Dianne Rocheleau and Laurie Ross, "Trees as Tools, Trees as Text: Struggles over Resources in Zambrana-Chacuey, Dominican Republic," *Antipode* 27 (4), 407–428. Copyright 1995, Editorial Board of *Antipode*. Published by Wiley Blackwell. Reproduced with permission.

Chapter 14: Maureen G. Reed, "Marginality and Gender at Work in Forestry Communities of British Columbia, Canada," *Journal of Rural Studies* 19 (3), 373–389. Copyright 2003. Elsevier Science Ltd. All rights reserved. Reproduced with permission.

Chapter 15: Kari Marie Norgaard, "The Politics of Invasive Weed Management: Gender, Race, and Risk Perception in Rural California," *Rural Sociology* 72 (3), 450–477. Copyright 2007 the Rural Sociological Society. Published by Wiley Blackwell. Reproduced with permission.

Glossary

agency the capacity of individuals to act independently and to make their own free choices

bilateral systems that trace descent and inheritance through both mother and father

conjugal contract terms on which husbands and wives exchange goods, incomes, and services, including labour, within the household

customary rights rights allocated to people according to their own traditions, which in many forest areas, differ from their legal rights

dependency and world systems theories of social change that interpret phenomena in relation to their role in the global political economic system as well as consequences for relevant parts – highlighting dynamics of unequal exchange through which core countries fuel their own economic growth and development via exploitation of natural and human resources from peripheral countries

discourse ways of constituting knowledge, together with the social practices, forms of subjectivity, and power relations inherent in such knowledges and relations between them

do[ing] gender individuals complying with their culture's definition of appropriate roles for men and women – involving perceptions, interactions, and micropolitics that define masculine and feminine 'natures' in that group

elite capture the tendency for those with more power to be able to gain illicit and/or unfair access to new assets and opportunities more readily than those with less

embeddedness here, the extent to which women's labour can or should be deemed an extension of their conjugal role or the action of autonomous individuals

embodied the ways people incorporate biologically the social and material world in which they live

enculturation the process by which people acquire the mental representations (e.g., beliefs, knowledge) and patterns of behaviour necessary to function as a member of a culture

equity justice; or the allocation of resources, programmes, and decision-making fairly without discrimination, addressing any imbalances in the benefits available

essentialist a view that links natures and dispositions intrinsically with specific categories of people (such as women)

GAD Gender and Development

GED Gender, Environment and Development

gender a sociocultural system that organizes the identities, practices, and relationships that play out among humans, and between humans and their environment, infusing them with power and meaning that refers symbolically to sex and sexuality

gendered knowledge knowledge that differs by gender

governmentality how to govern ourselves, be governed, accept governing by whomever, and being the best governor possible

hegemonic masculinity patterns of cultural practice and meaning that establish certain forms of masculinity as natural and superior to all forms of femininity – also supporting inequalities between more privileged men in position to fulfill hegemonic norms and poorer, darker, queerer men who are positioned in subordinate masculinities

heteronormative denoting or relating to a world view that promotes heterosexuality as the normal and preferred sexual orientation

individuation (degree of) how far a person is treated as an individual distinct from the social relations that form them and link them to others

intersectionality overlapping or intersecting social identities and related systems of oppression, domination, or discrimination

interstices small gaps (e.g., within forests or between fields)

justice fairness, equity

masculine hegemonies see 'hegemonic masculinity'

masculinities constellations of qualities, behaviours, attitudes, and accomplishments that – within particular communities of interpretation – are associated with the category or subcategories of 'man'

narrative stories that guide given communities to structure and assign meaning to past and present phenomena. Narratives are developed and expressed through cultural and religious traditions, popular culture, media, and science as well as politics and development work. Narratives backed by different degrees and types of power compete to instill ideas about causal relations as well as moral values and ethical behaviour.

particularist an orientation that honours personal relationships and cultural context above following rules intended for all. Mottos: 'relationships evolve'; 'things change'; 'people see things differently'.

patriarchal bargain a term used to explain why women collude in gender subordination on the understanding that to do otherwise could be harmful and that, in colluding, this will bring other kinds of rewards later on

patrilineal a system that allocates membership in a kinship group through the male line (father to son); it also refers to inheritance patterns where goods are transferred from father to sons and not to daughters

performative the capacity of speech and communication not simply to communicate but, rather, to act or consummate an action, or to construct and perform an identity – typically with symbolic implications

personhood the status of being a person with distinct rights and capacities

political forest both political land-use zones meant to remain in permanent forest and species defined as forest. Their management encapsulates ideas, practices, and institutions that seek to regulate people's access to resources, providing recognition and legitimacy to some, whilst excluding and criminalizing others.

positionality consideration of the metaphorical location where an actor or group stands in relation to others distinguished by ethnoracial, gender, class, geographical, and other terms. The 'position' of an individual or group within intersecting systems of opportunity and adversity relates to their 'strategic interests' in relations of difference and power involving decision-making or control over resources.

positivism viewpoint that factual knowledge based on the observable and verifiable is trustworthy

reproduction (social) those activities involved in reproducing the species, from the narrow biological meaning of reproduction to the maintenance of a household or a particular cultural or other system (including capitalism): child- or eldercare, care of the sick, cleaning, collecting water and firewood, and shopping have been seen as part of reproduction

sex segregated a division of labour in which genders have their own separate spheres, such as home gardens for women and larger fields for men

sex sequential a division of labour in which the labour of one gender follows the labour of another in time, such as when women weed the crops men plant

situated knowledges the idea that there is no single truth to be discovered and thus all knowledge is partial, linked to contexts and perspectives

subjectivities the set of processes by which a subject or self is constituted, usually in relation to others, including attitudes, values, expectations, memories, and dispositions

subsistence economy one that relies on natural resources for basic needs (hunting, gathering, subsistence agriculture) with no or minimal use of cash

tenure regime a system by which landownership is organized, allocated, and maintained in a given society

territorialization governance of space; often related to decentralization and/or implying recognition of varying definitions of a group's legitimate territory

universalist an orientation that honours rules that apply to all, irrespective of social position and relationships. Mottos: 'a rule is a rule'; 'there is one truth'.

usufruct rights rights to use resources (short of ownership)

WED Women, Environment and Development

WID Women in Development

1 Intoduction to gender and forests

Themes, contents, and gaps

Carol J. Pierce Colfer, Marlène Elias, Susan Stevens Hummel, and Bimbika Sijapati Basnett

This volume follows close on the heels of *Gender and Forests: Climate Change, Tenure, Value Chains, and Emerging Issues* (Colfer *et al.* 2016). The process of producing that volume unearthed a wealth of materials on gender and forests, key material that is not easily accessible to the forestry community. Pulling these classics together, as we do here, illuminates the broad strokes and trajectories of research into gender and forestry, thus consolidating it as a field of its own. Knowing where we've been and sharing that history with both junior and more seasoned scholars and practitioners as well as non-specialists can help orient future research and action.

We are acutely aware of the growing need to understand how gender and forests are interrelated and how this knowledge can contribute to an equitable and environmentally sound world. Scholars and practitioners in research, development, conservation, and management now recognize the relevance of gender in their work; but many lack the necessary backgrounds to address it. Initially, we imagined that the need for such materials might be greatest in the global South[1] due to the paucity of academic literature there (see Canagarajah 1996, for a still timely explication of the barriers southern researchers can face). As we looked into the material on the global North, however, seeking to compare and contrast key regional issues, we discovered, to our dismay, that the topic of gender and forests is poorly represented in the North – despite the fact that access to literature and other resources is so much easier there.

In selecting the papers for this volume,[2] we were cognizant that progress in any field involves 'building on the shoulders of giants' and thus begins with the work of previous researchers. We envision this book contributing to the usual incremental progression in sophistication (in methods, knowledge, conceptual frameworks) and serving as a base from which current-day researchers can build and expand. We sought works that reach across a broad range of complementary issues and geographical contexts. We value the variety of approaches taken by different disciplines and are drawn to triangulating on various forms of 'truth', examining gender and forests through different methodological and analytical lenses. In these pages, biologists and ecologists will recognize the concepts of forest structure and composition being affected by disturbance and are likely familiar with theories of landscape patterns emerging from ecological processes

(Turner 1989; Connell and Slatyer 1977). But these disturbances also affect women and men (see Eriksen 2013 on gender and wildfire in Australia). Many may be unfamiliar with social processes involving human knowledge and power relations and how these, too, affect landscape patterns directly and indirectly in forested and wooded ecosystems (e.g. Agarwal; Colfer; Nightingale; all this volume). In social scientific terms, these studies show the 'embeddedness' of knowledge, its inevitable partiality (discussed further below; see Glossary).

In interaction with biological and physical scientists in natural resource management, social scientists may be seen as 'handmaidens', expected to do assessments on demand rather than to build on relevant social sciences as an end in itself (Cernea and Kassam 2006).[3] Here, we offer examples of excellent social science on gender and forests and explain its value. We hope, thereby, both to avoid this pitfall and to assert the importance of social sciences (and gender research specifically) for understanding and improving natural resource management globally.

Why focus on gender and forestry?

While works on gender in agriculture have proliferated (e.g. World Bank *et al.* 2009; FAO 2011; Meinzen-Dick *et al.* 2011), considerably less research attention has focused on gender in forests. Scholarship on gender in agriculture can usefully inform gender analyses in forest settings – and a holistic understanding of rural livelihoods and agro-ecological systems must consider the connections and interdependency of what happens across the farm-to-forest continuum. Here, we highlight six arenas that are particularly relevant in relation to forests and substantiate the need to deepen understanding of the gender and forests nexus. Although much of the research reported here was conducted in the global South, those familiar with conditions in the North will recognize numerous parallels. Studies focused on the global North are indicated with an 'N' after the reference. Arora-Jonsson's article in this volume provides one interpretation of why this topic is so nearly invisible in the North.

Tenure and lands

A critical distinction between agricultural and forest settings relates to the tenure regimes that influence access to and governance of lands under crop or tree cover. Whereas farming typically occurs on (*de facto* or *de jure*) 'private' landholdings, forests are typically state held (86 per cent) with a much smaller proportion falling under private or communal ownership (FAO 2006).[4] In the field, these differentiations tend to be extremely muddled due to overlaps and differing perceptions about traditional, legally private, and publicly held lands (Marfo *et al.* 2010). As forest governance is progressively decentralized, however, forest lands that are technically state owned are increasingly being managed as common property by local communities and organizations or as private timber concessions by logging companies (Agarwal *et al.* 2008).

These systems of governance, and shifts therein, raise questions related to the use of power, the nature of institutions governing the use and management of communal resources, and related benefit-sharing. In particular, one has to question whether such institutions exhibit specific processes or systematic biases associated with gender.

Research attention has focused on women's (and men's) participation (or more often lack thereof) in decision-making with respect to forests generally (e.g. Arora-Jonsson, this volume, N; Larson *et al.* 2016; Reed 2010, N) and participation in forest user groups, such as joint forest management committees, more specifically (e.g. Agarwal this volume and 2001, 2009, 2010; Das 2011; Mwangi *et al.* 2011; Nightingale, this volume; Pinchot Institute of Conservation 2006, N; Redmore and Tynon 2010, N; Strong *et al.* 2013, N). Also related to forest governance are questions about the ways collective action and social movements to safeguard forests and access to their products are gender specific (e.g. Agarwal 1997; Asher 2016; Escobar *et al.* 2002; Kennedy 2016, N; Shanley *et al.* 2011).

Tenure and trees

Tenure issues can be particularly complex in the case of trees, especially in the global South. As in agriculture, gender-specific use and management of tree resources and associated knowledge systems are partly a product of the gender norms that structure customary and formal rights (Elias 2016). For millennia, humans have taken advantage of the regenerative capacities of trees to harvest their wood, branches, bark, resin, roots, leaves, and buds without killing the trees themselves. Living trees with evidence of past use can be found worldwide. Such 'culturally modified trees' are classified by Turner *et al.* (2009, N) by the alteration purposes, including: 1) incidental harvest results (e.g. collecting inner bark tissues for food, fibre for weaving); 2) intentional tending for future production (e.g. pruning and coppicing); and 3) intentional tending for marking (e.g. boundary, tenure/ownership, or as witness trees).

In patrilineal systems, women's rights to natural resources are mediated by their relationship with their male counterparts (Meinzen-Dick *et al.* 1997).[5] Access rights to trees are frequently 'layered' and distinct from land tenure such that many individuals can have usufruct rights to a single tree (Osborn 1989). Control of trees and their products is generally determined by the user's relationship to the formal resource 'owner' and is subject to negotiation (Fortmann and Nabane 1992; Rocheleau and Edmunds 1997; Schroeder, this volume). Differences exist among genders over the right to plant, harvest, and cull trees (Kevane and Gray 1999). These are complicated by the intersection of gender with other social distinctions, such as residence status (as autochthon or migrant), marital status, household structure, age, religion, socio-economic status, and more, that endow different women or men with distinct tenure and access rights (Elias and Arora-Jonsson 2016; Rocheleau and Ross, this volume).

Different rules also regulate access to planted versus wild trees, to exotic versus native species, to different taxa, and to tree products used for subsistence versus commercial purposes (Fortmann and Bruce 1988). Men (particularly those from wealthy families or dominant ethnic groups) often have stronger claims to trees that carry economic value (Osborn 1989; Rocheleau and Edmunds 1997). Differentiations are also observed in the types of trees owned and managed, as in the case of the Ibo where women tend to own fruit trees and men, timber species (Obi 1988). Differences also stem from the parts of the tree that are used; for example, women harvest a species for its fruit, green manure, or fodder and men, for timber or poles (Bonnard and Scherr 1994). Multiple-purpose trees differ from single-purpose agricultural plants. This can cause conflicts over product uses that vary by gender and other social differentiations. Competing claims, exclusions, and negotiations in relation to access to trees and their products call for detailed analyses that explicitly include gender and other locally relevant social distinctions.

Gender norms, taboos, and space

Gender norms and taboos also mediate the geographic spaces women and men access in the forest, which creates gender-specific constraints and opportunities in forest use (see Reed, N, this volume). These vary by age, caste, occupation, socio-economic status, and other factors, causing women and men from the same community or even the same household to access trees located in different spaces. Spatial variations related to gender may be due to different means of transportation (e.g. typically greater male access to bicycles and motorcycles, allowing them to canvass greater forest areas), to cultural norms that sanction women's entry into certain forest zones, to women-specific security concerns, or to household responsibilities that keep women closer to the homestead, limiting their ability to venture deep into the forest (Paumgarten 2007). These distinctions are manifest in foraging activities that affect forest structure and composition. For example, in some rainforest societies, men gather tree products from primary forest areas and women gather from secondary forests and interstices between fields closer to home (Coe and Anderson 1996; Kainer and Duryea 1992; Nyamweru 2003). Spatial collection patterns are context specific (Ireson 1997; Rocheleau 1988). In West Kalimantan in 1992–1993, Colfer (Dudley and Colfer 1993) found ethnicity linked with different tree species: locals agreed that the Melayu lived in areas where *Fragraea fragrans* grows[6] and the neighbouring Iban, where *Eusideroxylon zwageri* grows. These spatial patterns relate directly to the tenure, labour, and differentiated knowledge that capture the attention of scholars working on gender and forests.

Gender-differentiated access to and use of forest benefits

Tenure patterns associated with gender ('gendered patterns') also influence provisioning practices, consumption, and income generation from the sale of forest products. Rural women's often limited access to land, varying along with other social differentiations, can make them highly dependent on common property forest resources for their livelihoods (Agarwal 2010). The availability of certain forest products during the lean season, before the harvest, helps women fulfill their food-provisioning responsibilities and plays an important role in household food security. Forest product-derived income is often particularly significant for women (Kaimowitz 2003; Shackleton *et al.* 2007; Shackleton and Cobban 2016; Sunderland *et al.* 2014, for a global comparison). This partly owes to the fact that local trade in these products entails lower barriers to entry in terms of land than agricultural products.

Gendered division of labour

Flexibility in terms of labour also facilitates women's participation in forest product collection, processing, and sales. This flexibility often entails disadvantages, including invisibility, temporary and/or low-wage employment, and devaluation of women's work. On the positive side, tree products may require no or very limited labour to grow, and their collection often happens as women multitask on their way to and from their fields. Processing is commonly done at non-peak labour hours, at the homestead, and can thus be integrated with other domestic responsibilities (Shackleton *et al.* 2007). Despite the low returns non-timber forest products often reap in local markets, these conditions present a different set of gender-specific opportunities than those offered to women in agriculture. Scholars and practitioners have therefore focused on enhancing the prospects for women in forest product value chains as a way to improve gender equity and women's livelihoods (Ingram *et al.* 2016; Purnomo *et al.* 2011; Shackleton *et al.* 2011).

Gender-differentiated knowledge

Norms, (tenure) rights, and (labour) responsibilities result in knowledge of forests and tree-based systems that varies with gender because myriad life forms and ecosystem services are accessed, harvested, and managed differently by women, men, boys, and girls (further nuanced by other social characteristics; Bonnard and Scherr 1994; Pfeiffer and Butz 2005). Throughout Latin America, Asia, and Africa, women and girls collect up to 80 per cent of wild vegetables and hold specialized and localized knowledge of wild plants used for fodder and medicine (FAO 1999). Gender-specific interactions with the forest and related knowledge differences are linked to the gender division of labour. For example, women play an important role in the provisioning of forest foods, fuelwood, and medicinals to satisfy their care responsibilities, and men play a dominant role in timber harvesting, though regional variations exist (Agarwal 2010; Mai *et al.* 2011; Pfeiffer and Butz 2005). Use and knowledge may be linked to life form (e.g. annuals, short-lived perennials, long-lived perennials), as in Eastern Tanzania where Kwere and Zigua men work with arborescent plants and women, with herbaceous species (Luoga *et al.* 2000). Gender specialization is manifest in the gathering, management, and processing of most forest products (Sunderland *et al.* 2014), imbuing women and men with different sets of forest-related knowledge(s). Scholars have examined these differences as well as the overlaps and complementarities between women's and men's environmental knowledge, their role in guiding resource management decision-making and practices, and the implications these have for the sustainable management of forest or agroforest ecosystems (e.g. Elias 2015). Although governments have expressed some interest in integrating indigenous knowledge into management practices, it is highly probable that most of that knowledge so far has come from men.

These six arenas summarize a vast range of variation from place to place. In this short volume, we can only provide a 'taste' of what's available in the literature (which in turn provides only a taste of what's 'out there' in the world). As our understanding of the ubiquity of these gendered differences has grown, the urgency of learning more about them has increased. The next section clarifies choices we have made about what to include.

What is included in this collection?

Most fundamentally, we have sought 'classic' articles. But we have also sought balance across regions (Figure 1.1),[7] years of publication/research, methodological and topical variety, attention to men as well as women,[8] relevance beyond the case in question, and readability for non-gender specialists. We include older and newer articles that capture both the foundations and more recent developments in thinking about gender and forests. Hence, publication dates range from 1983 to one newly published article.

Here, we expand on four factors in our choice of articles, concluding by presenting the book's organization.

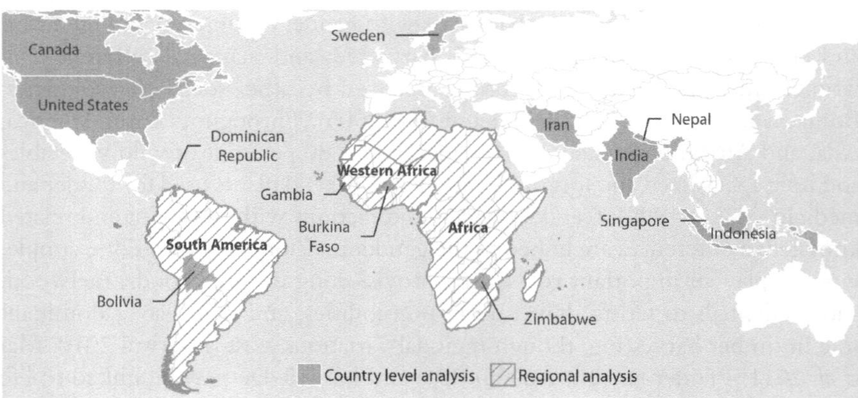

Figure 1.1 Map of countries where research for this book was conducted.

The North versus the South

The attempt to include contributions from a variety of geographic areas led to our recognition that gender and forests have been under-studied in the North. The number of concerns shared by the global South and the global North, we realized, is far greater than typically recognized. Fortmann (this volume) reports the most typical attitude: participants in a conference on community forestry could imagine nothing of interest about gender in the North; a sentiment echoed in Sweden (Arora-Jonsson, this volume). The introductions to each of the seven Parts in this volume provide additional discussion of the significant thematic overlap between the North and the South.

Contributors and their theoretical orientations

Our conviction – based partly on recognition of the partiality of all knowledge – that multiple disciplines have significant contributions to make in gender analysis led to the great variety represented herein (see the list of contributors). Contributors' orientations vary accordingly; but all recognize that dealing well with gender and forests requires ideas, conceptual frameworks, and evidence from multiple disciplines. Human ecology has roots that vary from home economics to anthropology and ecology – linking human behaviour and beliefs with ecological processes. Feminist political ecology draws on and integrates ideas from feminism, Marxism, political science/economics, human ecology, and geography – one of its key strengths is its attention to power in human relations. Discourse or narrative analysis, taken up by many fields, draws on anthropology, political science, linguistics, and literature – these analyses bring to bear the power of ideology in influencing human behaviour and policy. Gender analysis pertaining to forests typically draws on any or all of these orientations as will become clear in the selections to follow.

The fact that only 16 authors are represented here reflects the earlier trend – more common in the social sciences vis-à-vis the biophysical sciences – for single authorship of papers. Other than the contributions by the book's editors, only two chapters were co-authored. One serious failing of the field of gender studies that we readily and regretfully acknowledge is the minority participation of men[9]; this imbalance is reflected in this collection.

Quantitative versus qualitative approaches

The field of gender and forests addresses behavioural and perceptual phenomena, which requires combining qualitative and quantitative approaches (e.g. Behrman *et al.* 2014). Although some behavioural elements lend themselves more easily to quantification, when one enters the realm of perception and knowledge, additional complexities emerge. In this book, qualitative approaches dominate because many of the phenomena studied are better captured qualitatively. These include, for instance, value systems, social norms, emotional aspects of caretaking, 'doing gender',[10] and spiritual elements, among others. Furthermore, one of the biggest contributions of feminist and gender studies to the epistemological foundations of the social sciences is precisely in pointing out that knowledge is inevitably partial rather than objective and verifiable. Knowledge is a product of a number of factors including a 'mutually interpretive relationship' between researchers and subjects of research as well as the researchers' 'positionality' in the sociopolitical systems of which they are a part (Feldman and Welsh 1995; Haraway 1988; Harding 1986).

Intra- and inter-community variation in experience also call for interdisciplinary collaboration. People (whether forest dwellers or researchers) with direct, tactile experience of plant tending and harvest, for instance (e.g. Dobkins *et al.* 2016; Hummel and Lake 2015), can be expected to learn about their environment differently from people without such experience, given what we now know about the human brain (Buonomano and Merzenich 1998). These differences can produce and reinforce different types of knowledge. These multiple ways of learning and types of knowledge mean that not all information is accessible to everyone; it is derived from, and reinforced by, different experiences. Individuals think to ask certain questions or are disposed to observe certain relationships and responses based on a cultural lens that is reinforced cognitively. Qualitative approaches are needed to gain access to these differences. Recognizing the value and credibility of qualitative as well as quantitative research – and the overlaps between them – is an important step for those seeking to integrate gender in their work with forests. We hope this book will contribute to this process.

Women, men, and 'others' in analysis

Many contributions deal comparatively equitably with women and men in this collection (though there is very little in the way of recognition of the LGBTQ worlds).[11] Gender studies arose because of the near absence of material on women's lives and because of sociopolitical and economic systems that variably disadvantage women in relation to men. This has meant that the classics in the field focus attention more on women than men. There is strong current interest in strengthening attention to the 'other half of gender' as a legitimate part of the gender and forests topic (see, e.g. Bannon and Correia 2006; Cleaver 2002; Cornwall *et al.* 2011; de Mel *et al.* 2013, for analyses of men's lives more generally) – a concern we share (see Paulson, this volume). Attention to non-conventional sexuality has lagged far behind, however.

Organization of the book

This book is composed of seven Parts. Part I begins with four chapters – on Zimbabwe, the USA, Iran, and India and Sweden – that set out some of the fundamental conceptual, epistemological, and methodological issues that underpin other works in the field. It includes a critique of ecofeminism (Leach), the utility of sharing information and knowledge among divergent groups (Fortmann), the significance and nature of culturally influenced, cognitive, gender (and other) differences (Colfer), and discourses as they vary across geographies (Arora-Jonsson). The next five Parts (II–VI) are divided geographically, with each indicating one or more dominant themes in scholarship on gender in the region's forests. The topics were determined inductively, based on the particular materials we found pertaining to that region.

In Part II, Nightingale analyzes the differences between indigenous mapping knowledges and those of scientists; and Agarwal examines the ways community groups function and disadvantage particular groups of women in South Asia. In the chapters on Africa (Part III), Schroeder explores the unintended gender implications of externally driven change in the Gambia, emphasizing tenure and gendered negotiations; and Elias and Carney look at gendered knowledges, management, and marketing of shea trees and their butter. In Part IV, Li and Elmhirst move inward, looking at family relations. Li's chapter compares the household-scale inequities in access to assets (forest and otherwise) in Singapore and Sulawesi, places where comparative gender equity is usually proclaimed; and Elmhirst looks at the household and community implications of national governmental narratives about heteronormativity. Part V takes us to the New World, where Paulson's chapter links men and masculinities, ecology, history, and broader scales explicitly with local conditions in Bolivia. Rocheleau and Ross trace the multi-scalar influences and effects, both good and bad, of the introduction of *Acacia mangium* into communities in the Dominican Republic. Reed's article on how women in Canadian forest communities are marginalized and marginalize themselves begins Part VI. Next, Norgaard's chapter emphasizes the significance of differing perspectives of risk among the Karuk Native American group, a non-Indian group, and the US Forest Service in northern California. In conclusion, Part VII brings these thoughts together, with some ideas about ways forward.

We present fuller introductions to the papers themselves in the introductions to Parts II–VI. Here, we highlight four topics that have both conceptual and data-gathering implications for scholars as well as policymakers, forest managers, and development/conservation practitioners wishing to incorporate gender in their research. Although within the classic chapters presented here, one can find hints about these topics, they are not dealt with in a comprehensive way. We encourage gender scholars to fill these gaps.[12]

Gaps in gender studies today

We prioritize four key gaps here, beginning with intersectionality. This topic stems from a critique of a binary view of gender and a failure to recognize its fluidity, dynamism, pluralism, and relational- and context-specificity (see also Vinyeta *et al.* 2015). Second, we critically address the common differentiation between production and reproduction. As part of this discussion, we call for additional empirical data to unpack the gendered division of labour and associated negotiations/power dynamics, including the full spectrum of human behaviour in forests. Third, we look briefly at the logic of men's greater involvement in the domestic world. The last topic is population and reproductive health (overlapping with the 'production–reproduction' question). We argue that population has received too little attention in the gender and forests literature given its centrality to women's and men's well-being in forest (and other) contexts.

Intersectionality

Intersectionality refers to overlapping or intersecting social identities and related systems of oppression, domination, or discrimination (Crenshaw 1989). Gender power relations intersect with other social relations, thereby producing varying experiences that cannot be captured in a binary analysis of women and men alone. Even as gender relations across societies serve to disadvantage 'women' at an aggregate level more than their male counterparts, the constraints and opportunities that 'women' (and 'men') face vastly differ depending on their relative position in various sociopolitical hierarchies. These relate to life cycle processes, ethnicity, caste, class, sexual orientation, and ethnicity, amongst others. We cannot focus on gender alone but, rather, should examine gender as it intersects and interacts with other social relations (see Paulson, this volume).

We know that being, for instance, an illiterate, animist, gay, pygmy woman is likely to be disadvantageous in many ways. The ways such intersections function deeply affect all social groups, including people living in or near forests.[13] Yet mainstream feminism has also failed to address intersectionality fully (Vinyeta *et al.* 2015). Moreover, while gender norms, structures, and practices constrain people's freedoms, such constraints are constantly subject to alteration through political struggles, negotiations, and social and psychological change. Gender analyses need to reflect these complexities in more nuanced ways than we currently do (including designing policies and interventions accordingly). Our ability to effectively incorporate such awareness smoothly into forest management or conservation remains underdeveloped.

The following are four clear ways[14] that individuals who carry multiple, intersecting, disadvantaged identities experience exclusion in relation to forests and their benefits. The first two are internal to the individual, the second two external:

- *Cognitive* – A person who is raised among others who marginalize him or her has to develop greater skills than do elites at navigating the multiple cultural systems, including those of elites, their own, and other marginalized groups (see Colfer, this volume). This complexity can interfere with people's abilities to participate effectively in forest management; it requires them to struggle to mesh with and function within differing systems that conflict in a variety of ways (cf. Colfer 1974, for an analysis of such difficulties in a Bureau of Indian Affairs boarding school in the USA).

- *Emotional* – The internal experience of recurrent or constant discrimination by others has adverse effects on a person's sense of self, reducing their self-confidence, increasing their sense of risk, and making positive action or change in forests more difficult (see Lachapelle *et al.* 2004, on marginalized forest users in Nepal for adverse effects; and see Kusumanto *et al.* 2005, on Indonesia, or various contributions in Colfer 2005, on improvements in these considerations in collaborative management).
- *Social* – Negative narratives, discrimination, and stereotypes by others create concrete barriers to the marginalized, keeping them out of valuable forest-related networks, reducing both their access to forest-related information and ability to gain skills needed for forest-related work. The knowledge and perspectives of highly marginalized people are not valued or heard by elites, further hindering participation of marginalized individuals in forest governance (Agarwal and Reed, both this volume; Bose 2011).
- *Economic* – Inequitable rules of inheritance and reduced access to jobs and education can result in more limited access to assets. This can reduce people's access to the forest and options and motivation for participating in good forest management. See, for example, Bandiaky-Badji *et al.* (2016) on Liberia and Cameroon, Bhalla (2016) on India, or Pedulla (2016) on the USA.

Social scientists need to address intersectionality in a more sophisticated manner, and forest managers need to slot it into their social awareness. Collaboration among researchers, practitioners, and the state is required to devise effective ways to address compounding systems of discrimination on the ground.

Production versus reproduction – avoiding a pseudo-dichotomy

Chronological changes in global intellectual trends represent one example of the time dimension put forth in Colfer and Minarchek's Gender Box (2013; see Part VII). The tendency to differentiate production[15] and reproduction[16] so distinctly dates back to the Industrial Revolution. Prior to that time (and in many parts of the forested world today), men and women typically lived interdependently and with close connection to land. As industry developed in Europe, Marx and Engels observed the development of a clear differentiation between the world of production (associated with male-dominated wage work) and reproduction (associated with women 'reproducing' the workforce). Concurrently the worlds of production and reproduction moved apart.

These differentiations fit smoothly with other conceptual dualities common in forest-related research (public–private, modern–backward, male–female, rich–poor, paid–unpaid, production–consumption) that simplify the world's complexity. These binaries – better thought of as sliding scales or fluidities (Wieringa 2015) – are supported and reinforced by patriarchy. Use of these binaries solidifies and reifies them – though the tyranny of language compels us to use them (Lakoff *et al.* 2004).[17]

Most fundamentally, those activities that have been divided into production and reproduction equally constitute *work*.[18] Yet, when work is categorized and divided in this way, the concepts reinforce gendered ideologies (like 'hegemonic masculinity'[19] or women's inferiority). Such ideologies disadvantage women and overly determine men's roles as well.

Research related to forests – even some on gender – is not alone in overemphasizing the productive end of this continuum, to the near exclusion of the reproductive. This focus on production evolved partially in response to Sanday's (1974) influential article that showed productive activity as "necessary but not sufficient" for the development of high female status. The emphasis on production was also strategic: feminist scholars in the 1980s[20] recognized that focusing on women's economic contributions could serve as a wedge to insert women's concerns into broader global development narratives. In the long run, though, this has been directly disadvantageous for women, and indirectly problematic for all, as it has maintained the invisibility of women's work and ignored the trade-offs women face when participating in income-generating activities.

Nonetheless, researchers have made good progress in understanding women's productive roles, which had been comparatively invisible to scholars and development practitioners. There is now broad understanding that women have valuable productive roles and that we need to take those into account. However, the world of forestry has lagged behind: major international studies of non-timber forest products (NTFPs) by the Center for International Forestry Research (CIFOR), for instance, explicitly ruled out any subsistence-related use or management of NTFPs – exactly the most common ways in which women deal productively with forests.

Additionally, the emphasis on women's work in the productive sphere has resulted in a paucity of information on the unremunerated work and cultural activities that take place at home (the 'reproductive sphere'). Some methods, such as observational time allocation studies (proposed by Johnson 1975, who studied the forest-dwelling Machiguenga), have been explicitly designed to provide information on the full range of human activity and can help fill this research gap.

The shortcomings of a rigid differentiation between production and reproduction have become increasingly obvious, particularly in subsistence, unpaid, and barter economies. Is a woman's unpaid processing of rattan while taking care of children, or a man transporting his elderly mother to market to sell her forest-fibre mats, production or reproduction?

As we seek to understand existing gender roles and expectations, we struggle with difficult questions: How do we account accurately for multitasking, so common in women's lives? How do we deal with the 'care economy'? (See Folbre 2006; or Singh 2015, for care of the environment.) We still lack the methodological tools for adequately capturing the elements in the reproductive sphere (Van Esterik 1999, 229, calls for a "vocabulary of care") – a sphere that has been generally deemed far less important despite its essential role in sustaining the household and propping up 'productive' activities.

In similar and related fashion, the less quantifiable elements of social/cultural systems have been difficult to capture. Chung (2017), addressing the gendered effects of land-grabbing in Tanzania, notes that "It uproots [the people] from socio-ecological knowledges, cultural practices, and historical memories, which are rooted on the land and articulated through gender". These gendered elements are important components of enculturation processes, many of which are in women's hands. Studies of these cultural elements suffer the same fate as does gender in the 'reproductive realm'. We need to put serious effort into developing methods that can capture – and communicate – these key and ignored aspects of life. Chung (2017) suggests the use of 'social reproduction' rather than simply 'reproduction', to avoid the oversimplifications that carry over from the original Marxist analyses.[21]

However, the global scholarly community has acquired sufficient information, often acknowledged *in passing*, to realize that women's productive work supplements a whole range of activities at home and that calls for greater involvement in projects or programmes can increase women's burdens. Efforts to empower women – through education, trade and employment, and collective action in forest management and conservation, among other areas – have confronted a 'brick wall': women's time is limited. There are 24 hours in a day. And the tasks they (we) perform at home are vital to human existence. Children and others must be fed; clothing must be washed; houses must be cleaned; the sick must be cared for. In a less clearly/purely 'domestic' sense, crops must be cleaned, sorted, and stored; water and fuelwood must be collected; handicrafts must be made; cash crops must be taken to market and sold.

Rigid sex roles – strengthened and reinforced by the production–reproduction[22] dichotomy – make women's involvement in education, employment, community activism, and thus forest use and conservation difficult (see e.g. Colfer *et al.* 2008, on these interactions) and can preclude men from participating in activities at home that they might well enjoy, such as child- or eldercare, and to which they might contribute beneficially. Continued social scientific use of the production–reproduction differentiation simply reinforces social processes that more and more thinkers would like to change.

Men in the domestic world

Simple logic should have told us that if women no longer have time for such vital tasks (productive, reproductive, or community service), we would need to look elsewhere to accomplish them. Considerable evidence has been collected showing that men tend to have more leisure time than women (e.g. Sigman *et al.* 1989, in Sumatra; discussed more broadly in Kabeer 1998); this is commonly noted anecdotally by female and male researchers and community members alike. Gender studies in Africa have bemoaned the lack of paid work opportunities for men (e.g. Amuyunzu-Nyamongo and Francis 2006, on Kenya; Kandirikirira 2002, on Namibia; Silberschmidt 2001, on East Africa) as opportunities for women increase. Although most observers have noted the continued lack of involvement of men in domestic tasks, Kelbert and Hossain (2014) find emancipatory potential in these disruptions of conventional sex roles. Overall though, so far, such disruptions have resulted in further burdens for women and sometimes an increase in violence against them (e.g. Silberschmidt 2001; Correia and Bannon 2006).

Men's lack of involvement in domestic tasks was already recognized as problematic for women's opportunities by Van Esterik and Greiner as long ago as 1981. Yet our own gender expectations – and perhaps also recognition of the difficulty of changing behaviour – have blinded us to an obvious solution: involving men more meaningfully in work beyond the technically 'productive' (Barker 2014; Mies 2007; Razavi 2002).

Silberschmidt (2011) asks a pertinent question: "What would make men interested in gender equality?" This is not the place to answer this question exhaustively, but the following advantages have been perceived by some men (see e.g. Correia and Bannon 2006, on fatherhood, or Jacobsen 2006, whose analysis decries the current situation, the obverse of these advantages)[23]:

- less pressure to be the sole provider;
- more opportunity to spend time with their children and in a less structured environment (home);
- advantages to their children of interaction with and learning from their father;
- better relations with a grateful wife;
- learning new skills;
- a sense of justice and a desire to be equitable.

Barker (2014, 85), a student of 'men and development', concludes that

> men and boys doing gender justice and achieving richer and fuller (including healthier and less violent) lives – and women and girls achieving their full potential in political, social and work spheres – requires nothing less than a radical redistribution of care work.

We agree. And to bring that about, more attention needs to be paid to those 'reproductive' elements of life, both in the senses of the work that needs to be done and of the cultural elements in need of protection and sustenance.

Population and reproductive health issues[24] for empowerment

For decades now, the relevance of human population size to forest maintenance has been a taboo subject (lessening in recent years). This taboo emerged partly in response to draconian steps taken by India and China in the mid and late 1970s, respectively, to limit population growth. Feminists and others in the South also decried the emphasis on stopping southern population growth without a call for a reduction in northern consumption.

These legitimate concerns that led to the aforementioned taboo do not obliterate the further need to address the growing size of the Earth's population (now approaching 7.5 billion). Such concern is particularly pertinent for forests, which cannot thrive on any significant scale in cities or other densely populated areas. At a macro scale, the forests so essential to the livelihoods of forest communities[25] and beloved by ecologists, foresters, and nature lovers evolved in the context of low human population density. Some continue to do so (Amazon, Congo Basin, Borneo), though they are under threat – further exacerbated by large-scale resource extraction, land-grabs, and governmentally planned migration programmes.

If we shift to the micro scale, the relevance of the ability to control one's fertility and preserve one's reproductive health emerge as crucial elements for women's empowerment (see Norgaard, this volume, N). Going to school, getting and keeping a job, having time for community action or for oneself, and maintaining good health with age are all greatly simplified by the right to make decisions over one's own fertility and the availability of birth control technologies that enable women and men to implement their family planning decisions. A woman's ability to space childbearing is advantageous, health-wise, for the entire family. She and indeed her partner have more time for child- and eldercare, for preparing nutritious meals, and for contributing to family income and subsistence.[26] They also may have more time for each other and greater ability to finance their (fewer) children's schooling.

It is time to begin discussing and planning viable strategies for making birth control technologies available to women and men who want them.[27] Time and again in the field, women have privately approached us enquiring about methods of birth control and expressing a desire to plan their family. There is a significant and totally underserved population in remote forest areas. In sharing their concerns here, we urge responding to a felt need, working towards enhancing the affordability of such technologies, and informing people of options they may not know exist. The payoff is greater opportunities for women's empowerment, sustainable forest management, and biodiversity conservation – a genuine win–win–win opportunity.

In this section, we have tried to bring together some of the interconnections among production, social reproduction (to borrow from Chung 2017), and population. Women's involvement in production often – though not always – serves to reduce their willingness and propensity to reproduce (in the biological sense; see Colfer *et al.* 2008). We see a shift in men's and women's collaborative involvement in social reproduction (in the broadest sense) and production serving to enhance people's lives generally.

Conclusions

Here, we summarize the chapter and highlight some of the ways attention to gender and forests can improve forest management.

Summary

We began this chapter by introducing our decision-making related to the inclusion of chapters in this book. We then discussed the rationale for addressing gender and forests specifically (particularly vis-à-vis agriculture). The bulk of the chapter was devoted to discussing four substantive topics: 1) intersectionality, 2) the pseudo-dichotomy of production/reproduction, 3) men's roles in domestic work, and 4) population and the role of reproductive health in empowering people, particularly women. Subsequently, each Part introduction will introduce the chapters in that Part, highlighting issues of particular interest and relevance for gender and forests.

Highlights for forest managers

Early in this chapter, we introduced six key themes pertaining to gender and forests. Many of these issues affect forests; and even more affect those who live in forests. Uncertain or inequitable access to forests, trees, and their benefits discourages forest stewardship and reduces women's and men's willingness to develop or follow rules. Existing norms and taboos regarding forest use and management – whether those of communities or of the forest services and ministries – reduce formal forest managers' abilities to catalyze the energies of those marginalized (often women). Men and women have differential knowledge about and use of forests. Current practice often results in women's knowledge and use being unavailable to formal forest managers – which in turn means those managers cannot make decisions 'with a full deck'. These issues are explored more fully in the chapters to follow.

We also discussed four themes that need more attention. Noting the ways in which individual forest dwellers are multiply marginalized can help forestry professionals to ensure greater equity and gain access to such people's forest-relevant knowledge. Taking into account both reproductive and productive realms can strengthen women's effective and beneficial participation in forest management. Recognizing that women's and men's interests in forests vary from place to place and that stereotypes about sex roles unfairly inhibit women from participating are valuable first steps. Viewing the division of labour with a more open mind (women can deal with trees; men can be involved in domestic work) can stimulate new approaches to management. Attending to reproductive health can similarly open up new paths for desirable community involvement in forest management. Opportunities for education, paid work, and community action emerge for women who can control their own fertility.

Ultimately we hope that revisiting these classic texts can guide future scholarship and praxis to support greater equity, enhanced well-being, and improved resource management in forested landscapes.

Notes

1 By the 'global South' we refer to what some call 'developing countries' (e.g. Indonesia, Cameroon, Mexico). The 'global North' refers to 'developed countries' (e.g. USA, Canada, Europe, Australia, Japan).

2 We had to rule out – with considerable regret – a number of excellent papers we originally intended to include, due to financial stringency.

3 Three of this book's four editors are affiliated with CGIAR (formerly the Consultative Group on International Agricultural Research), where such perspectives are common. Hummel reported similar sentiments expressed by social scientists in the US Forest Service; and Colfer has seen them in government forest agencies throughout the tropics.

4 The Rights and Resources Initiative has been monitoring land tenure issues since about 2005 increase space between full point and they. They find 2013 'statutory land tenure' to be ~73 per cent. http://rightsandresources.org/en/resources/tenure-data/#.V2LUSI-cGyI.

5 As Li (this volume) shows, this can also be the case, if less explicitly, in bilateral systems (see also Lidestav 2010, N). Even in matrilineal systems, a woman's brothers and uncles have strong, sometimes determining, voices in land-related decision-making (e.g. Blackwood 1995).

6 Harwell (2010), working nearby, was told that the Melayu area was characterized by *Dryobalanops abnormis.*

7 Two of the chapters deal with broad regions in Africa; country-specific analyses come from Bolivia, Burkina Faso, Canada, Dominican Republic, Gambia, India, Indonesia, Iran, Nepal, Singapore, the USA, and Zimbabwe.

8 We recognize that a binary distribution of gender misrepresents the extant diversity. However, unfortunately, the gender and forestry literature has not yet incorporated this understanding very significantly. See Vinyeta *et al.* (2015) for a good (and rare) discussion of this.

9 This lack is particularly problematic if one acknowledges the embeddedness and partiality of knowledge.

10 'Doing gender' was proposed initially by West and Zimmerman (1987, 126) who state:

> We contend that the "doing" of gender is undertaken by women and men whose competence as members of society is hostage to its production. Doing gender involves a complex of socially guided perceptual, interactional, and micropolitical activities that cast particular pursuits as expressions of masculine and feminine "natures".

11 The body of queer theory – with relevance also beyond the Lesbian, Gay, Bisexual, Transsexual, Queer (LGBTQ) communities – has not been tapped signficantly in the natural resource literature, though there are surely gay people in forests! See Butler (2004) or Keller (2015) for good discussions of such theory (with a 'northern' emphasis). Vinyeta *et al.* (2015) provide the most thorough forest-related discussion, noting for instance that some Native American groups accept people with 'two-spirits' (non-conventional sexuality) far better than current-day North Americans (or see Graham-Davies 2004, on similar views in southern Sulawesi – where five genders are recognized).

12 See Edholm *et al.* (1978) for a prescient, if more theoretical, examination of these issues.

13 With our recent interest in gender in the global North, we wonder if our insights and findings apply as well to forest visitors, as are common in US and other 'northern' forests.

14 These titles (cognitive, emotional, social, and economic) are simply shorthand; we do not intend to encompass the full meaning of those terms here.

15 'Production', here, refers to paid or subsistence agricultural production, jobs, or other income-generating activities.

16 'Reproduction', here, refers to those activities involved in reproducing the species, from the narrow biological meaning of reproduction to the maintenance of a household or a

particular cultural or other system (including capitalism). Activities like child- or eldercare, care of the sick, cleaning, collecting water and firewood, and shopping have been seen as part of reproduction.

17 The existence of this differentiation forces its language upon us. The work that has been ignored and devalued has been called domestic or reproductive work. To redress that imbalance, we have to discuss it (further reinforcing it). Feldman (2012, 1), commenting on her own writing, says: "posing production and reproduction as a relation actually reproduces ... a separation between these sites of labor, and reinforces a public–private distinction that often is presumed to be a natural rather than a naturalized relation".

18 Mies (2007) notes with dismay the tendency to see only paid labour as work. She compares labour (women's, but also increasingly men's) to an iceberg, with ever rarer full-time, paid labour as the part above water; and housework, the informal sector, and nature, the much larger part below (see also Paulson, this volume); Mies links this to capitalism and patriarchy.

19 Hegemonic masculinity is variously defined. Initially it was the idea that 'real men' have to demonstrate reaching manhood by being successful providers, powerful, strong, and in control of women and children (see e.g. collections by Bannon and Correia 2006; Cornwall *et al.* 2011; Inhorn *et al.* 2009). More recently, it has been seen as patterns of cultural practice and meaning that establish certain forms of masculinity as natural and as superior to all forms of femininity. Hegemonic masculinities also support inequalities between more privileged men in position to fulfill hegemonic norms and poorer, darker, queerer men who are positioned in subordinate masculinities (Paulson, this volume).

20 The WID, or Women in Development, contingent, of which Colfer was a part. See Glossary for other relevant acronyms (e.g. GAD, WED).

21 See Edholm *et al.* (1978) for a thorough discussion of these concepts.

22 The links among production, reproduction, and consumption are also relevant, but not addressed here.

23 Such advantages were often proposed in the 1970s in women's 'consciousness-raising groups' in the USA (in which Colfer participated). See Little (2015) or Pini *et al.* (2015, 198), who state:

> Some of the men [newly unemployed with a sawmill closure] embraced broader and more flexible understandings of rural masculinity as encompassing involved fatherhood, the sharing of domestic labor, the nurturing of relationships with their wives and children, and a greater sense of life satisfaction despite their job loss.

24 Women's health generally has important forest-related elements (Allotey *et al.* 2008), but very little has been written about them – another important topic for future research.

25 The Worldwide Fund for Nature estimates that 300 million people live in forests and 1.6 billion depend on them for their livelihoods (www.worldwildlife.org/habitats/forest-habitat; accessed 16 May 2016).

26 We are by no means suggesting here that in terms of population policy, one size fits all. It definitely does not. In some places, a woman's or man's place in the community and family is closely linked to bearing many children; in other areas, not so. These issues are also tied to a preference for sons over daughters (due to various norms, systems of land tenure, and cultural and religious practices) that encourages spouses to have many children, to a lack of social security for ageing parents, and to high levels of child mortality in certain contexts. Such differences require different approaches to the provision of family planning and birth control technology.

27 This also has positive implications for controlling the spread of devastating sexually transmitted diseases, such as HIV – which also affect forest management (Lopez 2008).

References

Agarwal, B. 1997. Environmental Action, Gender Equity and Women's Participation. *Development and Change* 28 (1): 1–44.

Agarwal, B. 2001. Participatory Exclusions, Community Forestry, and Gender: An Analysis for South Asia and a Conceptual Framework. *World Development* 29 (10): 1623–1648.

Agarwal, B. 2009. Gender and Forest Conservation: The Impact of Women's Participation in Community Forest Governance. *Ecological Economics* 68 (11): 2785–2799.

Agarwal, B. 2010. *Gender and Green Governance: The Political Economy of Women's Presence Within and Beyond Community Forestry*. Oxford, UK: Oxford University Press.

Agarwal, A., Chhatre, A., and Hardin, R. 2008. Changing Governance of the World's Forests. *Science* 320 (5882): 1460–1462.

Allotey, P., Gyapong, M., and Colfer, C. J. P. 2008. The Gender Agenda and Tropical Forest Diseases. In C. J. P. Colfer, ed., *Human Health and Forests: A Global Overview of Issues, Practice and Policy*. London: Earthscan, 135–160.

Amuyunzu-Nyamongo, M., and Francis, P. 2006. Collapsing Livelihoods and the Crisis of Masculinity in Rural Kenya. In I. Bannon and M. Correia, eds, *The Other Half of Gender: Men's Issues in Development*. Washington, DC: The World Bank, 219–244.

Asher, K. 2016. Tenure vs. Territory: Black Women's Struggles in the Pacific Lowlands of Colombia. In C. J. P. Colfer, B. Sijapati Basnett, and M. Elias, eds, *Gender and Forests: Climate Change, Tenure, Value Chains and Emerging Issues*. London: Earthscan/Routledge, 206–218.

Bandiaky-Badji, S., Ntamag-Ndjebet, C. N., Weah, J. T. B., and Meyers. J. 2016. Women and Tenure in Liberia and Cameroon. In C. J. P. Colfer, B. Sijapati Basnett, M. Elias, eds, *Gender and Forests: Climate Change, Tenure, Value Chains and Emerging Issues*. London: Earthscan/CIFOR, 171–185.

Bannon, I., and Correia, M. C., eds. 2006. *The Other Half of Gender: Men's Issues in Development*. Washington, DC: The World Bank.

Barker, G. 2014. A Radical Agenda for Men's Caregiving. *IDS Bulletin* 45 (1): 85–90.

Behrman, J., Meinzen-Dick, R., and Quisumbing, A. 2014. Understanding Gender and Culture in Agriculture: The Role of Qualitative and Quantitative Approaches. In A. Quisumbing, R. Meinzen-Dick, T. Raney, A. Croppenstedt, J. A. Behrman, and A. Peterman, eds, *Gender in Agriculture and Food Security: Closing the Knowledge Gap*. Dordrecht, Netherlands: Springer, 31–54.

Bhalla, P. 2016. Gender Dynamics in Odisha's Forest Rights Act. In C. J. P. Colfer, B. Sijapati Basnett, and M. Elias, eds, *Gender and Forests: Climate Change, Tenure, Value Chains, and Emerging Issues*. London: Earthscan/Routledge, 186–205.

Blackwood, E. 1995. Senior Women, Model Mothers, and Dutiful Wives: Managing Gender Contradictions in a Minangkabau Village. In A. Ong and M. Peletz, eds, *Bewitching Women, Pious Men: Gender and Body Politics in Southeast Asia*. Berkeley, CA: University of California Press, 124–158.

Bonnard, P., and Scherr, S. 1994. Within Gender Differences in Tree Management: Is Gender Distinction a Reliable Concept? *Agroforestry Systems* 25 (2): 71–93.

Bose, P. 2011. Forest Tenure Reform: Exclusion of Tribal Women's Rights in Semi-Arid Rajasthan, India. *International Forestry Review* 13 (2): 220–232.

Buonomano, D. V., and Merzenich, M. M. 1998. Cortical Maps: From Synapse to Maps. *Annual Review of Neuroscience* 21: 149–186.

Butler, J. 2004. *Undoing Gender*. New York: Routledge.

Canagarajah, A. S. 1996. "Non-discursive" Elements in Academic Publishing, Material Resources of Periphery Scholars, and the Politics of Knowledge Production. *Written Communication* 13 (4): 435–472.

Cernea, M. M., and Kassam, A. H., eds. 2006. *Researching the Culture in Agri-Culture: Social Research for International Development*. Cambridge, MA: CABI Publishers.

Chung, Y. B. 2017. Engendering the New Enclosures: Development, Involuntary Resettlement and the Struggles for Social Reproduction in Coastal Tanzania. *Development and Change* 48 (1): 98–120.

Cleaver, F., ed. 2002. *Masculinities Matter! Men, Gender and Development*. London: Zed Books.

Coe, F. G., and Anderson, G. J. 1996. Ethnobotany of the Garífuna of Eastern Nicaragua. *Economic Botany* 50 (1): 71–107.

Colfer, C. J. P. 1974. *An Ethnography of Leaderlong Indian School*, Thesis, Anthropology, University of Washington, Seattle.

Colfer, C. J. P., ed. 2005. *The Equitable Forest: Diversity, Community and Natural Resources*. Washington, DC: RFF/CIFOR.

Colfer, C. J. P., and Minarchek, R. D. 2013. Introducing "the Gender Box": A Framework for Analysing Gender Roles in Forest Management. *International Forestry Review* 15 (4): 1–16.

Colfer, C. J. P., Dudley, R. G., and Gardner, R. 2008. Forest Women, Health and Childbearing. In C. J. P. Colfer, ed., *Human Health and Forests: A Global Overview of Issues, Practice and Policy*. London: Earthscan, 113–134.

Colfer, C. J. P., Sijapati Basnett, B., and Elias, M. eds. 2016. *Gender and Forests: Climate Change, Tenure, Value Chains and Emerging Issues*. London: Earthscan/Routledge.

Connell, J. H., and Slatyer, R. O. 1977. Mechanisms of Succession in Natural Communities and Their Role in Community Stability and Organization. *The American Naturalist* 111 (982): 1119–1144.

Cornwall, A., Edström, J., and Greig, A., eds. 2011. *Men and Development: Politicizing Masculinities*. London: Zed Books.

Correia, M. C., and Bannon, I. 2006. Gender and its Discontents: Moving to Men-Streaming in Development. In I. Bannon and M. Correia, eds, *The Other Half of Gender: Men's Issues in Development*. Washington, DC: The World Bank, 245–260.

Crenshaw, K. W. 1989. Demarginalizing the Intersection of Race and Sex: A Black Feminist Critique of Antidiscrimination Doctrine, Feminist Theory and Antiracist Politics. *University of Chicago Legal Forum* 1: 139–167.

Das, N. 2011. Women's Dependence on Forest and Participation in Forestry: A Case Study of Joint Forest Management Programme in West Bengal. *Journal of Forest Economics* 17 (1): 67–89.

de Mel, N., Peiris, P., and Gomez, S. 2013. *Broadening Gender: Why Masculinities Matter (Attitudes, Practices and Gender-Based Violence in Four Districts in Sri Lanka)*. Colombo, Sri Lanka: Care.

Dobkins, R., Lewis, C., Hummel, S., and Dickey, E. 2016. *Cultural Plant Harvests on Federal Lands: Perspectives from Members of the Northwest Native American Basketweavers Association*, Research Paper PNW-RP-608. Portland, OR: US Department of Agriculture, Forest Service, Pacific Northwest Research Station.

Dudley, R. G., and Colfer, C. J. P. 1993. *Conservation Sub-Project Quarterly Report and Attachments, Indonesia Tropical Forestry Management Project* (June, unpublished report), Danau Sentarum Wildlife Refuge, West Kalimantan, Indonesia.

Edholm, F., Harris, O., and Young, K. 1978. Conceptualising Women. *Critique of Anthropology* 3 (9–10): 101–130.

Elias, M. 2015. Gender, Knowledge-Sharing and Management of Shea (*Vitellaria paradoxa*) Parklands in Central-West Burkina Faso. *Journal of Rural Studies* 38: 27–38.

Elias, M. 2016. Gendered Knowledge Sharing and Management of Shea (*Vitellaria paradoxa*) in Central-West Burkina Faso. In C. Colfer, B. Sijapati Basnett, and M. Elias, eds, *Gender and Forests: Climate Change, Tenure, Value Chains and Emerging Issues*. Abingdon: Routledge, 263–282.

Elias, M., and Arora-Jonsson, S. 2016. Negotiating Across Difference: Gendered Exclusions and Cooperation in the Shea Value Chain. *Environment and Planning D: Society and Space*, online first. doi: 10.1177/0263775816657084

Eriksen, C. 2013. *Gender and Wildfire: Landscapes of Uncertainty*. New York: Routledge.

Escobar, A., Rocheleau, D., and Kothari, A. S. 2002. Environmental Social Movements and the Politics of Place. *Society for International Development* 44 (1): 28–53.

Feldman, S. 2012. The Intersection of Production and Reproduction: A Politics of Work in a Reorganizing Capitalism. Paper presented at *Political Economy of the World System XXXVI: Labor, Democracy and Global Capitalism*, Clark University, Worster, MA, 20–22 April.

Feldman, S., and Welsh, R. 1995. Feminist Knowledge Claims, Local Knowledge, and Gender Divisions of Agricultural Labor. *Rural Sociology* 60 (1): 23–43.

Folbre, N. 2006. Measuring Care: Gender, Empowerment, and the Care Economy. *Journal of Human Development* 7 (2): 183–199.

Food and Agriculture Organization (FAO). 1999. *Women: Users, Preservers and Managers of Agrobiodiversity*. Rome: FAO.

Food and Agriculture Organization (FAO). 2006. *Global Forest Resources Assessment 2005: Progress Towards Sustainable Forest Management*. Rome: FAO.

Food and Agriculture Organization (FAO). 2011. *The State of Food and Agriculture. Women in Agriculture: Closing the Gender Gap for Development*. Rome: FAO.

Fortmann, L., and Bruce, J. W. 1988. *In Whose Trees? Proprietary Dimensions of Forestry*. Boulder, CO: Westview Press.

Fortmann, L., and Nabane, N. 1992. *The Fruits of Their Labour: Gender, Property and Trees in Mhondoro District*, CASS Publications Paper Series 6/1992. Harare: Centre for Applied Social Sciences, University of Zimbabwe.

Graham-Davies, S. 2004. *Hunters, Wedding Mothers, and Androgynous Priests: Conceptualising Gender among Bugis in South Sulawesi, Indonesia*. PhD thesis, University of Western Australia.

Haraway, D. J. 1988. Situated Knowledges: The Science Question in Feminism and the Privilege of Partial Perspective. *Feminist Studies* 14 (3): 575–599.

Harding, S. 1986. *The Science Question in Feminism*. Milton Keynes, UK: Open University Press.

Harwell, E. 2010. Fluid Landscape and Contested Boundaries in Danau Sentarum. *Borneo Research Bulletin* 41: 36–62.

Hummel, S., and Lake, F. K. 2015. Forest Site Classification for Cultural Plant Harvest by Tribal Weavers Can Inform Management. *Journal of Forestry* 113 (1): 30–39.

Ingram, V., Haverhals, M., Petersen, S., Elias, M., Sijapati Basnett, B., and Phosiso, S. 2016. Gender and Forest, Tree and Agroforestry Value Chains: Evidence from literature. In C. J. P. Colfer, B. Sijapati Basnett, and M. Elias, eds, *Gender and Forests: Climate Change, Tenure, Value Chains and Emerging Issues*. London: Earthscan/Routledge, 221–242.

Inhorn, M. C., Tjornhoj-Thomsen, T., Goldbert, H., and Mosegaard, M. I. C. (eds). 2009. *Reconceiving the Second Sex: Men, Masculinity and Reproduction*. New York: Berghahn Books.

Ireson, C. 1997. Women's Forest Work in Laos. In C. E. Sachs, ed., *Women Working in the Environment*. Washington, DC and London: Taylor & Francis, 15–29.

Jacobsen, J. P. 2006. Men's Issues in Development. In I. Bannon and M. Correia, eds, *The Other Half of Gender: Men's Issues in Development*. Washington, DC: The World Bank, 1–28.

Johnson, A. 1975. Time Allocation in a Machiguenga Community. *Ethnology* 14 (3): 301–310.

Kabeer, N. 1998. Jumping to Conclusions: Struggles over Meaning and Method in the Study of Household Economics. In C. Jackson and R. Pearson, eds, *Feminist Visions of Development: Gender Analysis and Policy*. London: Routledge, 91–107.

Kaimowitz, D. 2003. Not By Bread Alone . . . Forests and Rural Livelihoods in Sub-Saharan Africa. In T. Oksanen, B. Pajari, and T. Toumasjukka, eds, *Forestry in Poverty Reduction Strategies: Capturing the Potential*, EFI Proceedings No. 47. Joensuu, Finland: European Forest Institute, 45–64.

Kainer, K. A. and M. L. Duryea (1992). "Tapping Women's Knowledge: Plant Resource Use in Extractive Reserves, Acre, Brazil." *Economic Botany* 46(4): 408–425.

Kandirikirira, N. 2002. Deconstructing Domination: Gender Disempowerment and the Legacy of Colonialism and Apartheid in Omaheke, Namibia. In F. Cleaver, ed., *Masculinities Matter! Men, Gender and Development*. London: Zed Books, 112–137.

Kelbert, A., and Hossain, N. 2014. Poor Man's Patriarchy: Gender Roles and Global Crises. *IDS Bulletin* 14 (1): 20–28.

Keller, J. 2015. Rural Queer Theory. In B. Pini, B. Brandth, and J. Little, eds, *Feminisms and Ruralities*. New York: Lexington Books, 155–166.

Kennedy, V. 2016. Living Conservation Values: Women and Conservation Easement Protection in Central New York. In C. J. P. Colfer, B. Sijapati Basnett, and M. Elias, eds, *Gender and Forests: Climate Change, Tenure, Value Chains, and Emerging Issues*. London: Earthscan/Routledge, 35–52.

Kevane, M., and Gray, L. 1999. A Woman's Field is Made at Night: Gendered Land Rights and Norms in Burkina Faso. *Feminist Economics* 5 (3): 1–26.

Kusumanto, T., Yuliani, L., Macoun, P., Indriatmoko, Y., and Adnan, H. 2005. *Learning to Adapt: Managing Forests Together in Indonesia*. Bogor, Indonesia: CIFOR.

Lachapelle, P. R., Smith, P. D., and McCool, S. F. 2004. Access to Power or Genuine Empowerment? An Analysis of Three Community Forest Groups in Nepal. *Human Ecology Review* 11 (1): 1–12.

Lakoff, G., Dean, H., and Hazen, D. 2004. *Don't Think of an Elephant! Know Your Values and Frame the Debate – The Essential Guide for Progressives*. White River Junction, VT: Chelsea Green Publishing.

Larson, A. M., Dokken, T., Duchelle, A. E., Atmadja, S., Resosudarmo, I. A. P., Cronkleton, P., Cromberg, M., Sunderlin, W., Awono, A., and Selaya, G. 2016. Gender gaps in REDD+: Women's Participation is Not Enough. In C. J. P. Colfer, B. Sijapati Basnett, and M. Elias, eds, *Gender and Forests: Climate Change, Tenure, Value Chains and Emerging Issues*. London: Earthscan/Routledge, 68–88.

Lidestav, G. 2010. In Competition with a Brother: Women's Inheritance Positions in Contemporary Swedish Family Forestry. *Scandinavian Journal of Forest Research*, 25 (S9): 14–24.

Little, J. 2015. The Development of Feminist Prespectives in Rural Gender Studies. In B. Pini, B. Brandth, and J. Little, eds, *Feminisms and Realities*. New York: Lexington Books, 107–118.

Lopez, P. 2008. The Subversive Links between HIV/AIDS and the Forest Sector. In C. J. P. Colfer, ed., *Human Health and Forests: A Global Overview of Issues, Practice and Policy*. London: Earthscan, 221–238.

Luoga, E. J., Witkowski, E. T. F., and Balkwill, K. 2000. Differential Utilization and Ethnobotany of Trees in Kitanghalo Forest Reserve and Surrounding Communal Lands, Eastern Tanzania. *Economic Botany* 54 (3): 328–43.

Mai, Y. H., Mwangi, E., and Wan, M. 2011. Gender Analysis in Forestry Research: Looking Back and Thinking Ahead. *International Forestry Review* 13 (2): 245–258.

Marfo, E., Colfer, C. J. P., Kante, B., and Elias, S. 2010. From Discourse to Policy: The Practical Interface of Statutory and Customary Land and Forest Rights. In A. M. Larson, D. Barry, G. Dahal, and C. J. P. Colfer, eds, *Forests for People: Community Rights and Forest Tenure Reform*. London: Earthscan, 69–90.

Meinzen-Dick, R., Brown, L. R., Feldstein, H. S., and Quisumbing, A. R. 1997. Gender, Property Rights, and Natural Resources. *World Development* 25 (8): 1303–1315.

Meinzen-Dick, R., Johnson, N., Quisumbing, A., Njuki, J., Behrman, J., Rubin, D., Peterman, A., and Waitanji, E. 2011. *Gender, Assets, and Agricultural Development Programs: A Conceptual Framework*, CAPRi Working Paper No. 99. Washington, DC: International Food Policy Research Institute.

Mies, M. 2007. Patriarchy and Accumulation on a World Scale – Revisited. *International Journal of Green Economics* 1 (3/4): 268–275.

Mwangi, E., Meinzen-Dick, R., and Sun, Y. 2011. Gender and Sustainable Forest Management in East Africa and Latin America. *Ecology and Society* 16 (1), article 17. www.ecologyandsociety.org/vol16/iss1/art17/

Nyamweru, Celia. 2003. Women and sacred groves in coastal Kenya: a contribution to the ecofeminist debate. In *Ecofeminism and Globalization: Exploring Culture, Context, and Religion*, eds. H. Eaton and L.A. Lorentzen, 41–56. Roman and Littlefield Publishers, Lanham.

Obi, C. S. N. 1988. Women's Rights and Interests in Trees. In L. Fortmann and J. W. Bruce, eds, In *Whose Trees? Proprietary Dimensions of Forestry*. Boulder, CO: Westview Press, 240–242.

Osborn, E. 1989. *Tree Tenure: The Distribution of Rights and Responsibilities in Two Mandinka Villages*. MS thesis, University of California, Berkeley.

Paumgarten, F. 2007. *The Significance of the Safety-Net Role of NTFPs in Rural Livelihoods, South Africa*. MSc thesis, Rhodes University, Grahamstown.

Pedulla, D. S. 2016. How Race and Unemployment Shape Labor Market Opportunities: Additive, Amplified, or Muted Effects? Presentation at Cornell University, Ithaca, NY.

Pfeiffer, J. M. and R. J. Butz. 2005. "Assessing Cultural and Ecological Variation in Ethnobiological Research: The Importance of Gender." *Journal of Ethnobiology* 25(2): 240–278.

Pinchot Institute of Conservation. 2006. *Understanding the Role of Women in Forestry: A General Overview and a Closer Look at Female Forest Landowners in the US*. Portland, OR: Western Regional Office.

Pini, B., Little, J., and Brandth, B. 2015. Conclusion. In B. Pini, B. Brandth, and J. Little, eds, *Feminisms and Ruralities*. New York: Lexington Books, 195–204.

Purnomo, H., et al. 2011. "Scenario-Based Actions to Upgrade Small-Scale Furniture Producers and their Impacts on Women in Central Java, Indonesia." *International Forestry Review* 13(2): 163–176.

Razavi, S. 2002. Introduction. In S. Razavi, ed., *Shifting Burdens: Gender and Agrarian Change under Neoliberalism*. Bloomfield, CT: Kumarian Press, 1–34.

Redmore, L. E., and Tynon, J. F. 2010. WOWnet: A Communication and Networking Model for Women. *Journal of Extension* 48 (5). https://joe.org/joe/2010october/a2.php

Reed, M. G. 2010. Guess Who's (Not) Coming for Dinner: Expanding the Terms of Public Involvement in Sustainable Forest Management. *Scandinavian Journal of Forest Research* 25 (S9): 45–54.

Rocheleau, D. 1988. Women, Trees and Tenure: Implications for Agroforestry. In L. Fortmann and J. W. Bruce, eds, In *Whose Trees? Proprietary Dimensions of Forestry*. Boulder, CO: Westview Press, 254–272.

Rocheleau, D., and Edmunds, D. 1997. Women, Men and Trees: Gender, Power and Property in Forest and Agrarian Landscapes. *World Development* 25 (8): 1351–1371.

Sanday, P. 1974. Female Status in the Public Domain. In M. Rosaldo and L. Lamphere, eds, *Woman, Culture and Society*. Stanford, CA: Stanford University Press, 189–206.

Shackleton, S. and L. Cobban. 2016. Gender and vulnerability to multiple stressors, including climate change, in rural South Africa. In *Gender and Forests: Climate Change, Tenure, Value Chains, and Emerging Issues*. C. J. P. Colfer, B. Sijapati Basnett and M. Elias. London: Earthscan/Routledge, 126–149.

Shackleton, S., Shanley, P., and Ndoye, O. 2007. Invisible but Viable: Recognising Local Markets for Non-Timber Forest Products. *International Forestry Review* 9 (3): 697–712.

Shackleton, S. et al. 2011. "Opportunities for Enhancing Poor Women's Socio-Economic Empowerment in the Value Chains of Three African Non-Timber Forest Products (NTFPs)." *International Forestry Review* 13(2): 136–151.

Shanley, P., Da Silva, F. C., and Macdonald, T. 2011. Brazil's Social Movement, Women and Forests: A Case Study from the National Council of Rubber Tappers. *International Forestry Review* 13 (2): 233–244.

Sigman, V. A., Colfer, C. J. P., Wilson, K., and Yost, R. 1989. Indonesia: Farm-Based Research in the Tropsoils Project, Sitiung. In H. Feldstein and S. Poats, eds, *Working Together: Gender Analysis in Agriculture*. Harford, CT: Kumarian Press, 135–166.

Silberschmidt, M. 2001. Disempowerment of Men in Rural and Urban East Africa: Implications for Male Identity and Sexual Behaviour. *World Development* 29 (4): 657–671.

Silberschmidt, M. 2011. What Would Make Men Interested in Gender Equality? Reflections from East Africa. In A. Cornwall, J. Edström, and A. Greig, eds, *Men and Development: Politicizing Masculinities*. London: Zed Books, 98–110.

Singh, N. M. 2015. Rethinking Payments for Ecosystem Services Using the Gift Paradigm: Sharing the Burden and Joy of Environmental Care. *Ecological Economics* 117 C: 53–61.

Strong, N., Sagor, E., Muth, A., Subjin, A., Walkingstick, T., Gupta, A., and Grotta, A. 2013. Women Owning Woodlands in the United States: Status and Trends in Extension Outreach and Education. Paper presented at *Future Direction of Small-Scale and Community-Based Forestry*, International Union of Forestry Research Organizations, Fukuoka, Japan, 8–13 September.

Sunderland, T., Achdiawan, R., Angelsen, A., Babifumira, R., Ickowitz, A., Paumgarten, F., Reyes-Garcia, V., and Shively, G. 2014. Challenging Perceptions about Men, Women, and Forest Product Use: A Global Comparative Study. *World Development* 64 (S1): S56–S66.

Turner, M. G. 1989. Landscape Ecology: The Effect of Pattern on Process. *Annual Review of Ecology and Systematics* 20: 171–197.

Turner, N. J., Ari, Y., Berkes, F., Davidson-Hunt, I., Fusun Ertug, Z., and Miller, A. 2009. Cultural Management of Living Trees: An International Perspective. *Journal of Ethnobiology* 29 (2): 237–270.

Van Esterik, P. 1999. Right to Food; Right to Feed; Right to be Fed. The Intersection of Women's Rights and the Right to Food. *Agriculture and Human Values* 16 (2): 225–232.

Van Esterik, P., and Greiner, T. 1981. Breastfeeding and Women's Work: Constraints and Opportunities. *Studies in Family Planning* 12 (4): 184–197.

Vinyeta, K., Whyte, K. P., and Lynn, K. 2015. *Climate Change through an Intersectional Lens: Gendered Vulnerability and Resilience in Indigenous Communities in the United States*, PNW-GTR-923. Portland, OR: US Department of Agriculture, Forest Service, Pacific Northwest Research Station.

West, C., and Zimmerman, D. H. 1987. Doing Gender. *Gender and Society* 1 (2): 125–151.

Wieringa, S. E. 2015. *Heteronormativity, Passionate Aesthetics and Symbolic Subversion in Asia*. Brighton, UK: Sussex Academic Press.

World Bank, Food and Agriculture Organization, and International Fund for Agricultural Development. 2009. *Gender in Agriculture Sourcebook*. Washington, DC: The International Bank for Reconstruction and Development/The World Bank.

Part I

Global conceptual, epistemological and methodological analyses

The contributions in Part I are designed to acquaint readers with some outstanding conceptual frameworks, theoretical orientations and methodological issues of use in gender studies. Here, we provide brief summaries of key insights from each of the four papers in this Part.

We begin this Part with Leach's (2007) article, which begins with a brief introduction to the history of gender approaches over the last four decades (from WID to WED to GAD and GED; see Glossary). She critically describes the part ecofeminism has played in these approaches, pointing out the widespread interest it has spawned – particularly in relation to the management of forests and other natural resources. Ecofeminism – an orientation that links women with the environment in what many consider an essentialist manner – was most famously promulgated by Shiva (1989), expanded by others, e.g., Diamond and Orenstein (1990) and Roszak *et al.* (1995), and incorporated into much development and conservation discourse and the practice it informed.

Much of Leach's paper is devoted to a clear-minded critique of ecofeminism. She finds particularly problematic the idea that women are closer to nature biologically than men, taking issue with the essentialist and universal qualities that authors like Shiva attribute to women. Mallory (2006, 2010) and Moore (2008) have more recently defended slightly adapted ecofeminist views, recognizing the value of Leach's scepticism, but also convincingly arguing that we should not 'throw out the baby with the bathwater'. As Leach would readily agree, there are certainly cultures in which women tend to act in more environmentally aware ways and show more concern for their surroundings. But the authors represented in this volume argue that this arises out of social – not biological – factors/processes. The links between women and environments remain a researchable question.

In this article Leach also raises other important issues: that of women's time and labour, and its usurpation in many development/conservation projects; and the tendency to see women as a homogeneous group when in fact they have interests, capabilities, and goals that differ one from another.

Fortmann's (2006) chapter also raises a number of issues that recur throughout this book:

- the question of the long-standing and widespread invisibility of women;
- the adverse implications for the environment of gender inequality (using an example from Zimbabwe);
- the inevitable partiality of knowledge and the usefulness of looking at things through various lenses, from divergent perspectives;
- the utility of recognizing the value of local knowledge, including women's knowledge, and linking it with what we normally consider 'scientific knowledge' in what she calls 'interdependent science'.

But she also raises the question of the value of cross-fertilization of understandings between the global South and the global North – an issue that has also interested us. We see many parallels in the North and South (highlighted in subsequent Parts), and we hope gender researchers will delve into these parallels for better understanding of gender in both regions – particularly making use of her ideas about collaboration and mutual exchange of knowledge between researchers and community members.

In recent years, particularly in gender studies, recognition of the importance of interactions across scales has grown. Colfer's (1983) piece, the oldest in this collection, looks at interpersonal interactions that occur similarly from the very micro scale (within a North American forest household) to the meso scale (village interactions between Qashqa'i tribal people, American researchers, and urban Persians) to the macro scale (interactions among scholars from around the world at a research centre in Hawaii). Her work, which includes but is not limited to gender, suggests an approach to dealing with what we now call intersectionality (see Chapter 1). The people she studies behave differently in different contexts and are seen differently by people across contexts as well (reflecting the 'situated' nature of what she observes). Another recurring issue is that of the 'inarticulateness' of normally articulate but marginalized group members (Ardener 1975) in the company of their 'superiors', and the inability of modern science and managers to capture the models of reality used by the marginalized. These are critical points for those wishing to engage with individuals through research and practice within the context of inequitable power relations – particularly germane for the gender contexts discussed in this book. Her conclusions also suggest a need to devote greater attention to individual and contextual differences in forests and their management on the one hand and the ways power flows or operates on the other.

Arora-Jonsson's (2009) article, besides providing glimpses of how gender functions in the forests of India and Sweden, elaborates on the topics introduced above by Fortmann (similarities between the North and the South) and Colfer (how power differentials play out). Arora-Jonsson looks at women organizing in the two divergent settings: India, where women were more vocal about discrimination against them and where outside interventions in gender issues were considered acceptable; and Sweden, where the rhetoric of gender equality masked gender inequalities and made efforts to challenge the supposed neutrality of forest management structures and processes more difficult. She demonstrates how the dominant discourses that pervade policy narratives framed women's activism and influenced people 'on the ground'. One of the newest articles in this collection, this article clearly focuses on gender *relations* (see also in this volume: Leach; Paulson). It includes a focus on women in groups, another popular topic of relevance for forest management; and it clarifies differing views of what capacity to act (agency) and empowerment mean in different contexts.

Arora-Jonsson discusses a crucial methodological orientation: that of action research (also raised in this volume by Fortmann and by Reed) combined with triangulation (the use of multiple methods to address the same question). One element that will be alien to most people working on forestry is the call for reflexivity, or the need to examine one's own assumptions about gender when doing research on the subject. We are all enmeshed in gender systems of our own, rendering 'objectivity' even more elusive than with other topics. Getting outside our own assumptions about gender is unusually difficult given how integral our gender identity is to who we are.

References

Ardener, E. 1975. Belief and the Problem of Women. In S. Ardener, ed., *Perceiving Women*. London: Malaby Press, 1–17.

Arora-Jonsson, S. 2009. Discordant Connections: Discourses on Gender and Grassroots Activism in Two Forest Communities in India and Sweden. *Signs* 35 (1): 213–240.

Colfer, C. J. P. 1983. On Communication among 'Unequals'. *International Journal of Intercultural Communication* 7 (3): 263–283.

Diamond, I., and Orenstein, G., eds. 1990. *Reweaving the World: The Emergence of Ecofeminism*. San Francisco, CA: Sierra Club Books.

Fortmann, L. 2006. In Theory and in Practice: Women Creating Better Accounts of the World. In J. S. Jaquette and G. Summerfield, eds, *Women and Gender Equity in Development Theory and Practice*. Durham, NC: Duke University Press, 191–201.

Leach, M. 2007. Earth Mother Myths and Other Ecofeminist Fables: How a Strategic Notion Rose and Fell. *Development and Change* 38 (1): 67–85.

Mallory, C. 2006. Ecofeminism and Forest Defense in Cascadia: Gender, Theory and Radical Activism. *Capitalism Nature Socialism* 17 (1): 32–49.

Mallory, C. 2010. The Spiritual is Political: Gender, Spirituality, and Essentialism in Forest Defense. *Journal for the Study of Religion, Nature and Culture* [Equinoxonline] 4 (1): 48–71.

Moore, N. 2008. The Rise and Rise of Ecofeminism as a Development Fable: A Response to Melissa Leach's 'Earth Mothers and Other Ecofeminist Fables: How a Strategic Notion Rose and Fell'. *Development and Change* 39 (3): 461–475.

Roszak, T., Gomes, M. E., and Kanner, A. D., eds. 1995. *Ecopsychology*. San Francisco, CA: Sierra Club Books.

Shiva, V. 1989. *Staying Alive: Women, Ecology and Development*. London: Zed Books.

2 Earth mother myths and other ecofeminist fables

How a strategic notion rose and fell

Melissa Leach

Introduction

The woman carrying firewood on her head across a barren landscape has become an environment and development icon. Reproduced in policy reports, NGO glossies and academic books alike, her image encapsulates powerful and appealing messages. For a time in the 1980s, the message was that women have a special relationship with the environment. They are deeply reliant on land and trees in their day-to-day work; they are so purely as 'women' (the image is uncomplicated by men, kin, differences or relationships); this is a timeless, perhaps even natural role; subsistence, domesticity and environment are entwined as a female domain; women are victims of environmental degradation (walking ever further for that wood) but also environmental carers, and key fixers of environmental problems.

Such images became extremely powerful in certain development and activist circles from the 1980s. This, it seems, is because they offered a materialist discourse about women's environmental roles which suited donor and NGO pre-occupations at the time. These material dimensions were bolstered by fables about women's natural, cultural or ideological closeness to nature; varieties of 'earth mother' myths which could be, and were, used to justify women's roles, as well as to give cultural and political appeal to the notion of global environmental sisterhoods. In the first part of this article, I briefly trace the rise of this women and environment discourse, and its key assumptions. Poorly conceptualized and inherently fragile as it was, I suggest that the idea of women's inherent closeness to nature could only be sustained because of the strategic interests it served.

In the early to mid-1990s, a thoroughgoing critique of this notion of women as natural environmental carers was articulated by several vocal academics. From a range of theoretical perspectives, but with a shared emphasis on gender relations and on particular contexts, women and environment assumptions were debunked, images reinterpreted and contextualized and alternative implications

Melissa Leach, "Earth Mother Myths and Other Ecofeminist Fables: How a Strategic Notion Rose and Fell," *Development and Change* 38 (1), 67–85.

for policy and practice put forward. As one of those involved, I reflect in the second section of the article on why these critiques seemed so important at the time, whom they engaged, and the arguments presented.

The third part of the article brings us up to date. To what extent are images of nature-caring women still deployed, where and why? A brief review of some key donor and NGO documents from recent years suggests—tentatively—that such images are far less prominent than a decade ago. Statements which were then commonplace are now hard to find, and would seem slightly ridiculous. Reflecting on this observation, which initially surprised me, I conclude not that we were spectacularly successful as debunkers, but that the discourse was doomed, and only ever temporary. A flawed argument served a time-bound purpose which has diminished as broader environment and development concerns have altered. Older concerns with women and environment have now been recast in terms of property rights, resource access and control. While welcome in some respects, however, there is a danger that the baby has been thrown out with the bathwater; gender-blind environment and development work seems to be on the rise, and there is rather little evidence of a more politicized, relational perspective on gender and environment taking root.

Women and environment: natural connections?

Although building strongly on earlier concerns, and underpinned by an array of factors, the 1980s saw an unprecedented rise in global environmental concern. Amidst rising scientific and popular concern with global environmental change, many development agencies embraced concerns with environmental protection and sustainable development. In particular, this was the era of major, media-prominent droughts and famines in Africa; land and soil degradation ('desertification') and deforestation especially became key issues for policy, and on which development agencies were expected to act.

It was in this context that the notion that women have a special relationship with the environment first began to be highlighted in development circles. A series of documents and publications in the 1980s by NGOs, donor agencies and scholars writing in relatively popular presses put forward the view that women were the primary users and managers of the environment at the local level (for example, Dankelman and Davidson, 1988; Rodda, 1991). It was argued that women's work—especially in reproductive and subsistence-focused activities—involves them closely with the environment and its resources, as hewers of wood, haulers of water and cultivators of food. Women were seen to have responsibilities which make them closely dependent on and give them distinct interests in natural resources, especially as sources of food and fuel. This in turn was deemed to give women deep and extensive environmental knowledge and experiential expertise. While in the early 1980s there was much emphasis on women as victims of environmental degradation—the fuel wood gatherer walking ever-further to fulfil her roles in a deforesting landscape—by the end of the decade the positive image of women as efficient environmental managers and conservers of resources was far more prominent.

The assumptions underlying these arguments had much in common with Women in Development (WID) perspectives, which had of course been around for considerably longer. Indeed what came to be termed the Women, Environment and Development (WED) approach could, scholars later argued, be seen as a translation of WID perspectives into the environmental domain—a rather late one, given that WID was already coming under critique in other domains. Like WID, the starting point for WED was the gender division of labour, and a somewhat static conception of women's roles. As with WID, the focus was almost exclusively on women's activities, with men barely appearing in the picture. And as with WID, there was a tendency to portray women as a homogeneous group. As a result, an image of women and men operating in parallel worlds appeared, with any connections men might have with the environment invisibilized. Conceptual connections with other prevalent feminist fables can also be seen: the image of women caring for the environment as an extension of their caring roles for their families linked with ideas of maternal altruism, while (at least in Africa) arguments about women's special relationship with the environment drew heavily on stereotypes concerning female farming systems.

While those advocating WED perspectives did not necessarily believe, or state, that these women–environment connections were natural and universal, their generalized styles of writing, and the static, roles-based and women-only emphases of WED arguments often gave the impression that they were. However, it was through the alliances that developed between WED and ecofeminist arguments that the idea of natural connections became more strongly forged. Ecofeminism is based on the notion that women are especially 'close to nature' in a spiritual or conceptual sense. As a multistranded set of approaches, developed largely amongst northern academics, it is riven with theoretical differences and debates. Most stark, perhaps is the distinction between those taking an essentialist position, attributing the connection between women and nature to biological roots, and those who see it as a social or ideological construct. However, even the latter theorists have tended to portray it as a universal connection, or at least one spanning such wide cultural sweeps as to be universal. As Maria Mies put it: '[women] conceived of their own bodies as being productive and creative in the same way as they conceived of external nature as being productive and creative . . . They co-operate with their bodies and with the earth in order to "let grow and make grow"' (Mies, 1986: 56).

Ecofeminists argue that (connected) women and nature have been subjected to a shared history of oppression by patriarchal institutions and dominant western culture. Thus the scientific revolution, spanning the sixteenth and seventeenth centuries, is seen as having replaced organismic theory in which the earth (viewed as a nurturing female) lay at the centre of a cosmology in which nature and society were dynamically interconnected, with a mechanistic view of nature which upheld competition and domination as necessary to the pursuit of progress (Merchant, 1982; Plumwood, 1986; Warren, 1987). It is also argued that such western post-enlightenment images have been imposed on 'indigenous' societies in Asia and Africa through scientific and development processes. Thus Mies and Shiva (1993) reasonably portray imperialism and colonialism as bearers of a particular western, mechanistic science and rationality, but characterize this as patriarchal or 'masculinist', so 'doing violence' to women and nature. Such rationality, it is argued, undermined pre-existing conceptions which were very different, viewing people and 'nature' as interdependent and grounded in a feminine principle. In the ecofeminist view, hope for environmentally sustainable and egalitarian development lies in the recovery of this feminine principle.

While predominantly originating in the north, ecofeminism acquired a vocal international presence in the 1980s. The work of Vandana Shiva was particularly influential. Not just through widely published and accessible writings but also through presentations at international meetings, her work generalized from her particular interpretations of women's experiences and the feminine principle in Hindu cosmology to construct a notion of all third world women as still connected with the remnants of a not quite extinct feminine principle, which could be recovered.

Grounded in radical criticism of mainstream development approaches, ecofeminism has been seen as the basis for socially and politically transformative struggles and practices. Indeed, it is more strongly grounded in radical environmentalism than in mainstream sustainable development theory. In this respect, ecofeminist arguments have served to inspire a large range of social and environmental movements, from specific forms of grassroots activism around the environment, to large networks such as the Women's Environmental Network which has since the 1980s promoted green consumerism and other issues in Britain. The very vagueness and generality with which 'nature' is defined in most ecofeminist accounts, and their tendency to generalize from specific situations to posit very general connections between women and nature, seem to have facilitated the flexible deployment of ecofeminist precepts, implicitly as well as explicitly, by a vast array of movements pursuing different causes. It is perhaps this generality, too, which has allowed ecofeminist fables to draw together and unite groups and movements which might otherwise have contrasting political and material stances and aims. Thus echoes of ecofeminist discourse strongly coloured the Miami declaration adopted by a large international conference of women activists prior to the United Nations Conference on Environment and Development (UNCED) conference in Rio, 1992. It equally coloured the preamble to Women's Action Agenda 21 discussed at Rio, which linked the highly specific experiences of diverse groups of women in localized environmental protection with a broad critique of mainstream economic and military processes.

However, ecofeminist fables also came to be deployed in far less critical and politicized ways. They came to be major props to the women, environment and development approaches being developed by mainstream development agencies, largely because strong overlaps between ecofeminism and WED allowed them to be mutually supportive. One overlap lies in the lack of reference to men: Shiva's (1988) work, for example, subsumed any reference to men into 'peasants' or 'tribes'. Thus as in WED, it appears to be only women who have any environmental connection. Another is in the shared emphasis on women's environment-related 'sustenance' or 'survival' activities, and on the non-monetized reproductive sphere in general. In ecofeminism these come to be described not just as the foci of women's environmental interactions, but as repositories of spiritual and cultural value, and the central planks of a sustainable society. Perhaps one should also acknowledge that earth mother myths may have borne cultural resonance for certain (western) development experts, helping them to feel that their WED prescriptions were logical and right.

In this context, it is not surprising that echoes of ecofeminist discourse crept into the statements of donors and NGOs associated with much less radical visions. Together, ecofeminism and WED supported a view that agencies should identify women as allies—or even the prime movers—in resource conservation projects. Coming at a time when agencies were under international pressure both to address environmental concerns, and to acknowledge gender differences, this was an attractive proposition. It came to be drawn on and elaborated by a wide variety of agencies. At one extreme, the World Bank developed a synergistic or 'win–win' approach to environment and gender, arguing for a general identity of interest between women and environmental resources and thus for treating women as the best agents for ensuring resource conservation (see Jackson, 1993 for a fuller discussion). At the other, WED/ecofeminist assumptions were assimilated into the community-level 'primary environmental care' approach advocated by several NGOs (see Davidson and Myers, 1992). Women were conceptualized as the central agents of primary environmental care, which linked caring for the environment, meeting basic needs and community empowerment. Community approaches such as these were central to the broader Agenda 21 emerging out of UNCED, and 'women' and 'community' were often interchangeable terms in the documents of this period.

When translated into development practice, these women–environment links tended to come to mean one of two things: acknowledging women's environmental roles so that they could be brought into broader project activities such as tree planting, soil conservation and so on, mobilizing the extra resources of women's labour, skill and knowledge; or justifying environmental interventions which targeted women exclusively, usually through women's groups. Many, many examples of both were spawned in the late 1980s and early 1990s. To take just one example, donor agencies in The Gambia relied principally on women's labour to promote fruit tree agroforestry for environmental stabilization. In justifying this, a UNDP official commented that 'women are the sole conservators of the land . . . the willingness of women to participate in natural resource management is greater than that of men. Women are always willing to work in groups and these groups can be formed for conservation purposes'. The promotional literature of an NGO involved in tree planting echoed this perspective: 'In the Gambia, our primary focus has been on women . . . In the implementation of an environmental programme in the country, they could be deemed our most precious and vital local resource' (both cited in Schroeder, 1999: 109).

Such projects and programme approaches have had a variety of effects. As those who went on to question the assumptions have pointed out (see Green et al., 1998; Jackson, 1993; Leach, 1994), in practice, many have proved counterproductive for women or have failed to conserve the environment, or sometimes both. Project 'success' has often been secured at women's expense, by appropriating women's labour, unremunerated, in activities which prove not to meet their needs or whose benefits they do not control. New environment chores have sometimes been added to women's already long list of caring roles. At the same time, the focus on women's groups—as if all women had homogeneous interests—has often marginalized the interests and concerns of certain women not well represented in such organizations. Fundamentally, it came to be argued, the assumption of women's natural link with the environment obscured any issues concerning property and power. This meant that programmes ran the risk of giving women responsibility for 'saving the environment' without addressing whether they actually had the resources and capacity to do so.

Interventions cast in these terms have, in many cases, been actively struggled over or resisted. Thus in the Gambian garden-orchards cited above, women struggled—sometimes successfully—to regain labour control into their own horticultural activities and to sabotage the agro forestry trees whose fruit commodities their husbands would control (Schroeder, 1999). When women apparently took up such conservation tasks willingly, their motivations for doing so often proved contradictory: thus Rocheleau (1988) describes how women's groups in Kenya digging soil conservation terraces did so to secure the patron–client relations which might bring them famine relief food from the agencies concerned, not because (as those agencies stated at the time) they felt close to nature or even had much interest in conservation. However, it is worth noting that in both these cases women's participation remained open to interpretation by others as evidence of their closeness to nature, their special relationship with the environment. Re-contextualizing these actions would require a different conceptual lens, which pushes the images into new light. In a similar way, the woman head-loading fuelwood in a barren landscape, deployed as an iconic image of women's natural connections, is open to quite other interpretations through different theoretical lenses. It is to the debunking of ecofeminist/WED assumptions that I now turn.

Critiquing natural connections

From the early 1990s, a number of thoroughgoing critiques of these WED and ecofeminist perspectives began to appear. These were largely the work of academics, notably in India (Agarwal), the UK (Jackson, Leach, Joekes), the

USA (Rocheleau, Fortmann), and the Netherlands (Wieringa), although usually in the context of particular relationships with development agencies, environmentalist groups and women and men in developing countries. As one of these academics, I can recall how imperative it seemed at the time to promulgate such critiques. For while appreciating the attempt by environmental agencies to address gender, it seemed appalling and dangerous that this was occurring through an approach with glaring flaws influenced by dubious ecofeminist work. I was shocked by the jarring mismatch between WED images and my understanding of gender and environment relations from my fieldwork in West Africa's forests, and concerned that feminist insights which had proved helpful in understanding those relations—insights around intra-household dynamics, resource access, and agrarian property and power, for instance—were being so overlooked in the rush to construct women as saviours of a vaguely-defined nature.

Others may have had other motivations, but the result was a series of lectures and presentations, published articles, books and reports, which put forward a range of alternative perspectives. These carried new labels, such as feminist environmentalism (Agarwal, 1992), feminist political ecology (Rocheleau et al., 1996), and gender, environment and development, or GED (Braidotti et al., 1994; Green et al., 1998; Leach, 1994). They had particular emphases, reflecting the other literatures and preoccupations which these debunkers brought to the debate. Thus feminist environmentalism emphasized the material aspects of gender–environment relations, and their interplay with particular ideological conceptions. Drawing from the broader school of political ecology, feminist political ecology drew particular attention to the nature of gendered knowledge, questions of resource access and control, and the engagement between local struggles and more global issues. GED, on the other hand, applied the perspectives of gender analysis as developed much earlier in other domains, such as around agriculture and economy, in the environmental domain. Nevertheless they shared a number of core ideas, centring around a conceptualization of gender–environment relations as embedded in dynamic social and political relations, and an emphasis on particular contexts rather than universalisms and essentialisms. And they presented some common challenges to WED perspectives and their ecofeminist fables.

These critiques cast women's (and men's) relationships with the environment as emerging from the social context of dynamic gender relations, and thus challenged any notion that women *a priori* have a special relationship with the environment, let alone a natural and unchanging one. The critiques ran alongside (and sometimes, though not by any means always, connected with) critiques of 'nature', which emphasized the dynamism of ecologies and the social construction of resources and environmental 'problems' (for example, Leach and Mearns, 1996). Thus, if women in any particular setting appeared to be closely involved with natural resources or ecological processes, this needed to be explained. It might be explicable in terms of unequal power relations, or lack of access to alternatives: for instance if women gather wild foods, this might reflect their lack of access to income from trees on private holdings (cf. Agarwal, 1992; Rocheleau, 1988); and the fuelwood headloader might have failed to negotiate with her husband to purchase fuels as others in her village might be doing. Stemming from these critiques, then, was a challenge to the notion that women's environmental interests are synonymous, or synergistic. Women may be locked into natural resource dependence through particular relationships and feel that their interests lie in moving into other livelihood activities, as they see men do.

This body of work also unpacked 'women' as a category, pointing out the very different interactions with land, trees, water, and so on, associated with women of different ages, backgrounds, wealth and kinship positions—differences which apply to men too. Thus mothers may be able to devolve much 'environmental care' onto daughters or daughters-in-law; women with access to trading capital may reduce their dependence on natural resources, and so on. That some women become involved in environmental action does not mean that this represents all women's interest and agency (Jackson, 1993). Recognizing differences and social relations amongst women clearly undermines any notion of groups formed through homogeneity of position and interest, and forces new questions to be asked about the hierarchies and distributional issues that operate when women do form groups for environmental purposes.

In shifting the focus from roles to relationships, these critiques emphasized how relations of tenure and property, and control over labour, resources, products and decisions, shape people's environmental interests and opportunities, and how environmentally-related rights and responsibilities are almost always contingent on kin and household arrangements and the negotiations these entail. They pointed out how gender boundaries with respect to environmentally-related activities can shift with changing economic and social circumstances, whether due to changes in global market conditions or shifts in the policy context. Property relations were a particular focus of feminist environmentalism and political ecology, including an emphasis on the layers of contingent rights, and the informal practices, which underlie formal arrangements (see, for example, Agarwal, 1995; Mackenzie, 1991), and the need for policy makers to acknowledge these.

Following on from this concern with access to and control over resources, and with household arrangements, gender analyses of environmental relations pointed out the fallacy of assuming that women's participation in environmental projects is coterminous with benefit. Social institutions and negotiations can clearly deny women control over products which their own labour has produced, while diversion of women's labour without remuneration may reduce their access to own-account income. And they suggest the possibility of conflicts between environmental and women's gender interests; for example, allocating women responsibility for saving the environment could increase their workloads or reinforce regressive gender roles, rather than representing progressive change or enhanced gender equity (Jackson, 1993; Leach, 1992).

While these aspects of critique focused particularly on the material and policy aspects of WED discourse—aspects which proved remarkably easy to challenge—scholars also critiqued the ecofeminist arguments which supported them. Vandana Shiva's work came under fire in this respect. Thus, for example, Agarwal (1992) accused Shiva (1988) of unwarranted extension of principles she associates with Hinduism when she suggests that all precolonial societies 'were based on an ontology of the feminine as the living principle' (Shiva, 1988: 41). Agarwal (1992) argues that the imagery of Prakriti varies in its connotations and relevance even among Hindu groups in India, and is of comparatively little importance among non-Hindu people. Scholars have also reinterpreted the Chipko movement—which Shiva draws upon as an iconic feminine environmental movement—in other terms: not as evidence of women's closeness to nature but as a struggle for material resources in the context of gender-ascribed natural resource dependence, and women's limited opportunities to out-migrate as compared with men (Jain, 1984; Peritore, 1992). The movement can be alternatively interpreted not as feminist, but as a peasant movement which emerged at a particular historical juncture (Guha, 1989), in which men were also involved, and in which women's participation was actually conservative of their subordinate position (Jain, 1984).

Other lines of critique exposed the problems in dualisms that linked women with nature—or indeed in the assumption, in some versions of ecofeminism, that non-western societies lack such dualisms. Scholars such as Leach (1994) and Jackson (1992, 2001) referred to anthropological studies showing wide cross-cultural and historical variability in the meanings attributed to 'female' and 'male', and the ways they are linked with concepts relevant to environment (MacCormack and Strathern, 1980; Moore, 1988). At a meeting in Oxford in 1996, a group of academics assembled specifically to compare the relations between gender and spirituality in different societies, resoundingly contested the view that women are everywhere viewed as sacred custodians of the earth (Low and Tremayne, 2001). A woman's procreative roles are by no means necessarily seen to place her closer to a universally-conceived nature, and to exclude men from this relationship. As a generalized category, 'nature' certainly fails to capture complex ideas about the physical and nonphysical attributes of different micro-environments and ecological processes (cf. Croll and Parkin, 1992; Fairhead and Leach, 1996).

Related lines of critique challenged the assumption that pre-colonial, organic, sacralized views of nature went hand-in-hand with harmonious environmental practices and egalitarian gender relations. As Jackson (2001) elaborated, the relationships between religious beliefs and environmental practices are contingent and multiply-determined. At the same time, indigenous organic conceptions can evidently encompass struggle and conflict between people and certain ecological processes, as well as harmony (Croll and Parkin, 1992). That certain ecological processes are socialized in local thought, and certain resources culturally valued, does not translate into an all-encompassing respect for nature (Persoon, 1989), and often speaks to local power relations (Fairhead and Leach,

1996). Indeed, as Jackson (1995) points out, there is plenty of evidence linking organic conceptions of society and ecology with oppressive social institutions: the territorial cults which managed land and fertility concerns in late nineteenth century southern Africa have, for example, been associated with the aristocratic domination and lethal taxation of commoners, as well as the subordination of women (Fairhead,1992; Maxwell, 1994; Moore et al., 1999; Schoffeleers, 1979). Political ecology analyses of ecological knowledge and gender ideology, in contrast, locate the ways in which certain ideas are produced and debated within social and political processes, and in relation to particular groups and institutions.

Finally, critics problematized the image of western thought and colonial science as monolithically wiping out other views and knowledges (leaving perhaps a shadowy residual of the old feminized order; see Leach and Green, 1997). They pointed out how this obscures the complex content and political–economic relations of production of colonial and modern scientific discourses, and the processes through which they articulate with rural people's own. While recognizing the value of ecofeminism in drawing critical attention to the constituents of scientific epistemology and their operation through colonialism, and in raising questions about links between science and oppressive social relations, they suggested that such a critique needs to be developed through engagement with the highly diverse and contradictory theories and practices of which science is constituted (Molyneux and Steinberg, 1995: 92).

In short, these critical perspectives did not necessarily deny the events which ecofeminism interprets—the female fuel gatherer, women's involvement in some environmental movements or in conserving soil or planting trees, for instance. But they interpret these as particular to certain times, places and social relations, and interrogate the power relations which may produce them.

While it may have taken a group of vocal academics to articulate and elaborate some of these ideas, we were clearly not the only people thinking them. I vividly recall the hungry reception of my critique of WED amongst the various student, practitioner, donor and NGO audiences to which I presented these ideas in meetings in the early 1990s. It seemed that many people found the notion of women's natural environmental connections highly questionable, even ridiculous, and lapped up the critiques as speaking to their concerns. As the decade progressed, a spate of Masters and PhD studies began to appear—some but not all supervised by scholars active in the first round of debunking—which reviewed the gender and environment debate and explored field material through a gender relations, or feminist political ecology, lens (for example, Resurreccion, 1999; Schroeder, 1993).

Indeed, it seemed that by the early 1990s, even some of the donors who had promoted those images in the first place were ready to reconsider them. For example presentations by myself, Dianne Rocheleau and Louise Fortmann which eventually became the basis of published papers were first invited by Swedish SIDA, and discussed at a meeting they hosted of Scandinavian donors and conservation agencies. Many there agreed that the time was ripe for a gender perspective on environment (rather than a women-only perspective), and the workshop report reflected this emphasis (SIDA, 1992). The British government's DfID funded and attended the Oxford meeting mentioned above, which set out to challenge the image of women as sacred custodians of the earth; USAID commissioned a study from the Institute of Development Studies in Brighton to elaborate a gender perspective on environmental relations (Joekes et al., 1995); while Netherlands Development Assistance employed consultants to help their Women and Development programme, and Environment programme, to do the same (NEDA, 1997). Thus, at least to some extent, these critiques were elaborated in engagement with donor agencies, and generally found ready reception amongst them, stimulating efforts by individuals within these agencies. They were also elaborated in conjunction with certain grassroots groups. For instance, some of the contributors to Rocheleau et al. (1996) are environmental activists, or claim to represent their interests. It seems that by the mid-1990s certain activist groups, at least, were finding it more useful to present themselves in feminist political ecology terms, as engaged in struggles over rights and resources in a globalized field, than as groups of spiritually-connected earth and tree huggers.

Gender and environment in the new millennium

Having diverted my gaze somewhat from gender–environment issues during the last few years, I welcomed the opportunity to refocus it, and examine the kinds of messages present in development agency statements a decade on from the first round of WED critiques. I have not had time for a full review (although this might be worth doing), so these are only impressions, which I hope colleagues will add to or refute. My overriding feeling is that the kind of statement about women's special relationship with the environment, which was commonplace a decade ago, has all but disappeared. Earth mother myths may still be perpetuated through ecofeminist writings and certain strands of ecocentric environmental activism, mainly in the north. But they appear no longer to permeate, even implicitly, the environment and development policy and action statements of donor agencies, governments and NGOs. On the one hand, it seems, there are many fewer references to women, or gender, at all. On the other hand, when they do appear, their messages appear to be cast in more relational and rights-based terms. A brief array of examples illustrates these points.

The World Bank's World Development Report 2003—*Sustainable Development in a Dynamic World*—appeared just over a decade on from the WDR in which its famous 'win–win' approach to women and environment was publicized. Yet it barely mentions women or gender except in two paragraphs under the heading 'Nurturing women's human capital', where it states that: '[women's] largely unrecorded role in agriculture explains the survival of many traditional subsistence communities on marginal lands . . . Traditional communities depend on women and girls to fetch fuel wood and water, and to produce and prepare food' (World Bank, 2003: 71). Even these

WED-like statements are contextualized: 'in many places, traditions, limited mobility, and lack of voice or access to information make women the most marginal group. With the men seeking work elsewhere, women tend the fields and look after the children' (ibid.). While perhaps perpetuating other gender myths, such explanations begin to shift women's environmental connections from the natural and unquestionable, to see them emanating from a context of dynamic social relations.

DfID's strategy document for achieving the international development targets (DFID, 2000) contains, in its 56 pages of text, no mention at all of gender or women. Its discourse is couched entirely in terms of 'the poor' and 'communities'. In the one boxed case example where gender is mentioned—reporting on women's and men's differential involvement in a community-managed wells project in Mali—the focus is on inequalities in gendered labour allocation within the project and women's resistance to these. This is a far cry from any assumption that cleaning wells is a natural extension of women's caring roles.

A major report on biodiversity and livelihoods, prepared by IIED with DfID support (Koziell and Saunders, 2001), is couched in similar undifferentiated terms, concerned with 'the poor' and 'community knowledge and practices'. At the few points where social difference in people's relationships with biodiversity are acknowledged, the language used is of stakeholders, and their 'rights, responsibilities, rewards and relationships' (ibid.: 53).

ActionAid, once a contributor to WED statements, no longer has a policy or research programme devoted to environment. Their website's statements in the arenas which touch on environmental concerns—their programmes on food rights, and on emergencies—do not mention gender or women, and they have no publications with these in the title. What is clear is that their programme directions in these areas (for instance around the debate over genetically-modified crops, and vulnerability to drought and war) are driven by questions of rights, and of resource access and control.[1]

Water issues represent one arena where several agencies do have documents which contain a gender focus. These include reports produced amidst the follow-up processes for the 2002 World Summit on Sustainable Development, in which water was a central theme. For example, the Women's Environment and Development Organization (WEDO) convened an expert consultation resulting in the report 'Untapped Connections: Gender, Water and Poverty. Key Issues, Government Commitments and Actions for Sustainable Development' (WEDO, 2003). While starting with a WED-like statement about women's roles—'Women and girls are responsible for collecting water for cooking, cleaning, health and hygiene . . . in rural areas they walk long distances . . .' (ibid.: 3), its advocacy for 'tapping' the connections is grounded in an explicit conception of gender relations, and in concepts of gender equality in resource rights and decision-making.

A final, telling example is 'Women's Action Agenda for a Healthy and Peaceful Planet 2015', the document prepared for and discussed at the 2002

World Summit on Sustainable Development in Johannesburg (WEDO and REDEH, 2002). Formulated through regional meetings and in partnership with a diverse array of women's groups and networks worldwide, this was conceived explicitly as an updating of Women's Action Agenda 21, the 1992 Rio document that so epitomized the ecofeminist/WED discourse of that time. Its focal areas are peace and human rights; globalization; access and control of resources; environmental security and health; and governance for sustainable development. As far as I can see, there are no statements suggesting any natural connection between women and environment. While women's particular contributions to biodiversity conservation, water provisioning and so on are emphasized, this in the context of advocacy for equal property rights, inheritance rights, access to services, and for gender mainstreaming. Furthermore, and again by contrast with the 1992 document, global and international issues are much more prominent. The problems of global climate change and biodiversity loss, international militarization, and economic globalization within the neoliberal paradigm assume centre-stage, with the emphasis on how these create vulnerabilities for people, especially the poor, and the need to reinforce and reform relevant international agreements—or transform development paths more radically. While there is a notion that 'women' have specific vulnerabilities and concerns in this context, the document's language portrays women as representing, and advocating for, the more general concerns and struggles of people disenfranchised and marginalized by pernicious global processes.

What might one conclude from this brief review? If, as it suggests, images of women as natural environmental carers have receded, why might this be, and what has replaced them? First, one set of reasons may lie in broad shifts in international processes and development priorities. As concerns with war and complex emergencies, with globalization and its consequences, and with broader questions of governance have come to dominate development agendas, so environmental agendas have receded somewhat as programme priorities. At the same time, constituencies that might once have argued the 'women's case' in relation to global processes and governance, such as WEDO or DAWN (Development Alternatives with Women for a New Era), appear by necessity to be using their influence to press the more general case of the poor and disenfranchized, in a world of growing global inequality and conflict along many axes. The fuelwood head-loading woman in a barren landscape could, in these terms, become an image of the poor excluded from globalization's benefits, or of the devastation of African economies by international banks.

Second, within the environmental arena, policy priorities have, at least to some extent, moved away from the issues which first spawned the WED discourse. Fuelwood and social forestry, land degradation and soil conservation may still attract practical development attention on the ground, but have been superseded in international discussions by global climate change, biodiversity, and water—the big issues which dominated the 2002 Johannesburg meeting. Many dimensions of these involve global or transboundary processes which are seen to require international, or at least multi-levelled, approaches to governance, and new relationships with the private sector. The possible connections with women's day-to-day work and knowledge are possible to make, though far less obvious. Water is, perhaps, the exception, offering an obvious arena where gender relations shape patterns of need and provisioning at the micro-level, and it is therefore not surprising that to the extent that development discussions have pursued the gender and environment theme, it has often been in the water context.

Third, the last decade has seen a consolidation of decentralized, community-based approaches to governance and development, and a renewed focus on poverty. The environmental arena is a case in point, where a vast array of community-based and co-management approaches (to water, forests, wildlife and so on) have been launched, spawning a related social science literature reflecting on their successes and failures. In a similar vein, agencies such as DfID and UNDP (re-)discovered and pursued a debate on poverty–environment linkages. These programme approaches and the agency literature about them tend to be very gender-blind, promoting images of undifferentiated, consensual communities or 'the poor'. They have, it must be acknowledged, been influenced by discussions of resource rights, institutions and social difference emanating from other contexts, such as the sustainable livelihoods approach embraced by many donors during the last five years, and more recently, rights-based approaches to development more generally. Nevertheless, gender does not necessarily figure in these.

Fourth, just as environment and development debates have moved on, so have those about gender. The broader influence of GAD perspectives in academic, donor and practitioner circles, and the move towards gender mainstreaming, now make it difficult for statements about women's natural connections with environment (or indeed natural connections with anything else) to be made with credibility. WID was already sinking when environmental activists picked up on it to create WED in the 1980s, so perhaps it has now been fully submerged? Yet this would of course be a crass conclusion when the WID debate was, and is, multi-stranded and multi-sited. Many WID/WED assumptions, such as the idea that women's labour can be unproblematically utilized in land reclamation projects for household benefit, continue to be reproduced through practice in field-level projects. The real change may be in the rhetoric that development agencies are able to use to justify such practices.

Some conclusions

My conclusion, then—albeit tentative—is that the discourse of women as natural environmental carers had its day, but that day has passed. Emerging at a very particular moment, amidst the confluence of pressures to do something about environmental degradation amidst 1980s' droughts and in the lead-up to Rio 1992, and to address 'gender' in an era when this could more easily be taken as 'women', it offered both convenient practical prescriptions and powerful justifications for them. That the ecofeminist fables which crept into the discourse may have chimed with the beliefs and backgrounds of some of its proponents may have added to their conviction in portraying women as close to nature for otherwise quite instrumental ends. As environment and development discourses have moved on, so these ecofeminist fables seem largely to have retreated back into the world of academic writings and fringe environmental groups which originally spawned them.

While the debunking of WED's assumptions during the 1990s, through its particular engagements between academics and practitioners, may have played a role, it seems that it was swimming with a tide. However, a full assessment of both the rise and fall of WED discourses, and the influences on these, would require a proper analysis of the policy processes involved and their co-production with knowledge and research. I have not been able to undertake this here, and to do so, following leads in the large literature on policy processes in general and environment in particular (for example, Fairhead and Leach, 2003; Jasanoff and Wynne, 1998; Keeley and Scoones, 2003) would doubtless reveal forms of agency, dispute within organizations, and complexities of discourses and of actor-networks to which this article has done little justice. It would draw more detailed attention to the roles of particular political interests, funding flows, and events and meetings, and should demand attention to the interplay of research, policy and popular culture, including how mass-media influence these. Nevertheless, even the story of WED's rise and fall as I have sketched it here should alert us to the fact that what we regard as enduring myths may sometimes prove to be much more fragile, upheld less by enduring power structures than by relatively fleeting strategic interests. This is an interesting story for those who study science and policy processes, where the literature to date dwells more on the co-production of relatively enduring ideas or the gradual overturning of long-held paradigms, than on the contingent up-and-down of 'crazy' ideas. In turn, it should have implications for how feminist scholar-activists who seek to destabilize problematic gender assumptions might understand their task and go about their work.

For such a scholar-activist of a feminist environmentalist/feminist political ecology persuasion, the picture is now positive in some respects, but depressing in others. The problematic WED discourse has waned, but there is little evidence of a well-conceptualized gender relations perspective on environmental relations in policy literature. Issues of rights and resource access and control are now acknowledged, but not necessarily in relation to gender, and rarely through the relational, multi-layered lens which feminist political ecologists and gender analysts of land have seen as important. Gender-blind perspectives on community and the poor as actors in relation to ecological and global political–economic processes seem to be more prominent than ever. In academic literature, meanwhile, sophisticated studies have continued to appear which explore the intersection of gender, dynamic ecological processes and environmental politics across multiple scales (such as Li, 2002; Schroeder, 1996), showing the value of understanding people's current engagements with global processes in gendered terms. Perhaps it is time for a new round of concerted engagement with the changed world of environment and development policy which attempts to put gender back in the picture on more politicized terms.

Note

1 From their website: http://www.actionaid.org/ourpriorities/ (accessed 2 June 2003).

References

Agarwal, B. (1992) 'The Gender and Environment Debate: Lessons from India', *Feminist Studies* 18(1): 119–58.

Agarwal, B. (1995) *A Field of One's Own: Gender and Land Rights in South Asia*. Cambridge: Cambridge University Press.

Braidotti, R., E. Charkiewicz, S. Häusler and S. Wieringa (1994) *Women, the Environment and Sustainable Development: Towards a Theoretical Synthesis*. London: Zed Press/ INSTRAW.

Croll, E. and D. Parkin (eds) (1992) *Bush Base: Forest Farm: Culture, Environment and Development*. London: Routledge.

Dankelman, I. and J. Davidson (1988) *Women and the Environment in the Third World: Alliance for the Future*. London: Earthscan Publications.

Davidson, J. and D. Myers (1992) *No Time to Waste*. Oxford: Oxfam.

DfID (2000) 'Achieving Sustainability: Poverty Elimination and the Environment'. Strategies for Achieving the International Development Targets document. London: Department for International Development.

Fairhead, J. (1992) 'Indigenous Technical Knowledge and Natural Resources Management in Sub-Saharan Africa: A Critical Overview'. Paper prepared for Social Science Research Council project on African Agriculture, Dakar (January).

Fairhead, J. and M. Leach (1996) *Misreading the African Landscape: Society and Ecology in a Forest-Savanna Mosaic*. Cambridge and New York: Cambridge University Press.

Fairhead, J. and M. Leach (2003) *Science, Society and Power: Environmental Knowledge and Policy in West Africa and the Caribbean*. Cambridge: Cambridge University Press.

Green, C., S. Joekes and M. Leach (1998) 'Questionable Links: Approaches to Gender in Environmental Research and Policy', in C. Jackson and R. Pearson (eds) *Feminist Visions of Development*, pp. 259–83. London and New York: Routledge.

Guha, R. (1989) *The Unquiet Woods: Ecological Change and Peasant Resistance in the Himalaya*. Delhi: Oxford University Press.

Jackson, C. (1992) 'Gender, Women and Environment: Harmony or Discord?'. Gender Analysis in Development Discussion Paper Series 6. Norwich: University of East Anglia, School of Development Studies.

Jackson, C. (1993) 'Women/Nature or Gender/History? A Critique of Ecofeminist "Development"', *Journal of Peasant Studies* 20(3): 389–419.

Jackson, C. (1995) 'Radical Environmental Myths: A Gender Perspective', *New Left Review* 210(March/April): 124–40.

Jackson, C. (2001) 'Gender, Nature and the Trouble with Anti-Dualisms', in A. Low and S. Tremayne (eds) *Women as Sacred Custodians of the Earth?*, pp. 25–43. New York and Oxford: Berghahn.

Jain, S. (1984) 'Women and People's Ecological Movement: A Case Study of Women's Role in the Chipko Movement in Uttar Pradesh', *Economic and Political Weekly* XIX(41): 1788–94.

Jasanoff, S. and B. Wynne (1998) 'Science and Decisionmaking', in S. Rayner and E. Malone (eds) *Human Choice and Climate Change: An International Assessment. Vol 1: The Societal Framework of Climate Change,* pp. 1–88. Battelle Press.

Joekes, S., C. Green and M. Leach (1995) 'Integrating Gender into Environmental Research and Policy'. Report of commissioned study for EPAT/MUCIA, United States Agency for International Development (USAID) (September).

Keeley, J. and I. Scoones (2003) *Understanding Environmental Policy Processes: Cases from Africa.* London: Earthscan.

Koziell, I. and J. Saunders (eds) (2001) *Living off Biodiversity.* London: IIED.

Leach, M. (1992) 'Gender and the Environment: Traps and Opportunities', *Development in Practice* 2(1): 12–22.

Leach, M. (1994) *Rainforest Relations: Gender and Resource use among the Mende of Gola, Sierra Leone.* International African Library. Edinburgh: Edinburgh University Press; Washington, DC: Smithsonian Institution.

Leach, M. and C. Green (1997) 'Gender and Environmental History: From Representations of Women and Nature to Gender Analysis of Ecology and Polities', *Environment and History* 3(3): 343–70.

Leach, M. and R. Mearns (1996) *The Lie of the Land: Challenging Received Wisdom on the African Environment.* Oxford: James Currey.

Li, T. M. (2002) 'Local Histories, Global Markets: Cocoa and Class in Upland Sulawesi', *Development and Change* 33(3): 415–37.

Low, A. and S. Tremayne (eds) (2001) *Women as Sacred Custodians of the Earth?* New York and Oxford: Berghahn.

MacCormack, C. and M. Strathern (eds) (1980) *Nature, Culture and Gender.* Cambridge: Cambridge University Press.

Mackenzie, F. (1991) 'Political Economy of the Environment, Gender and Resistance under Colonialism: Murang'a District, Kenya, 1910–1950', *Canadian Journal of African Studies* 25(2): 226–56.

Maxwell, D. (1994) 'Religion and Environmental Change in North-east Zimbabwe 1880–1990'. Paper presented at the African Studies Association of the UK Biennial Conference, University of Lancaster (5–7 September).

Merchant, C. (1982) *The Death of Nature: Women, Ecology and the Scientific Revolution.* London: Wildwood House.

Mies, M. (1986) *Patriarchy and Accumulation on a World Scale.* London: Zed Books.

Mies, M. and V. Shiva (1993) *Ecofeminism.* London: Zed Books.

Molyneux, M. and D. L. Steinberg (1995) 'Mies and Shiva's *Ecofeminism*: A New Testament?', *Feminist Review* 49: 86–107.

Moore, H. (1988) *Feminism and Anthropology.* Cambridge: Polity Press.

Moore, H., T. Sanders and B. Kaare (1999) *Those Who Play with Fire: Gender, Fertility and Transformation in East & Southern Africa.* London: Allen and Unwin.

NEDA (1997) *Gender and Environment: A Delicate Balance between Profit and Loss.* The Hague: Netherlands Development Assistance.

Peritore, N. P. (1992) 'India's Environmental Crisis and Chipko Andolan', *Asian Thought and Society* VXIII(51).

Persoon, G. (1989) 'Respect for Nature among Forest People', *BOS Newsletter* 18(8): 11–27.

Plumwood, V. (1986) 'Ecofeminism: An Overview and Discussion of Positions and Arguments', *Australian Journal of Philosophy* 64: 120–38.

Resurreccion, B. P. (1999) 'Transforming Nature, Redefining Selves: Gender and Ethnic Relations, Resource Use and Environmental Change in the Philippine Uplands'. PhD Thesis, Institute of Social Studies, The Hague, The Netherlands.
Rocheleau, D. (1988) 'Women, Trees and Tenure', in L. Fortmann and J. Bruce (eds) *Whose Trees? Proprietary Dimensions of Forestry*, pp. 254–72. Boulder, CO, and London: Westview Press.
Rocheleau, D., B. Thomas-Slayter and E. Wangari (eds) (1996) *Feminist Political Ecology: Global Issues and Local Experience*. London and New York: Routledge.
Rodda, A. (ed.) (1991) *Women and the Environment*. London: Zed Books.
Schoffeleers, J. M. (1979) 'Introduction', in J. M. Schoffeleers (ed.) *Guardians of the Land: Essays on Central African Territorial Cults*, pp. 1–46. Zimbabwe: Guelo.
Schroeder, R. (1993) 'Shady Practice: Gender and the Political Ecology of Resource Stabilization in Gambian Garden/Orchards'. PhD Thesis, University of California at Berkeley.
Schroeder, R. (1996) '"Gone to their Second Husbands": Marital Metaphors and Conjugal Contracts in The Gambia's Female Garden Sector', *Canadian Journal of African Studies* 30(1): 69–87.
Schroeder, R. (1999) *Shady Practices: Agroforestry and Gender Politics in The Gambia*. Berkeley, CA: University of California Press.
Shiva, V. (1988) *Staying Alive: Women, Ecology and Development*. London: Zed Books.
Shiva, V. (1989) 'Development, Ecology and Women', in J. Plant (ed.) *Healing the Wounds: The Promise of Ecofeminism*, pp. 80–90. London: Green Print.
SIDA (1992) 'Gender and Environment: Some Interlinkages'. Report of a SIDA Seminar. Stockholm: SIDA.
Warren, K. J. (1987) 'Feminism and Ecology: Making Connections', *Environmental Ethics* 9(3): 3–20.
WEDO (2003) 'Untapped Connections: Gender, Water and Poverty. Key Issues, Government Commitments and Actions for Sustainable Development'. New York: Women's Environment and Development Organization.
WEDO and REDEH (2002) 'Women's Action Agenda for a Healthy and Peaceful Planet 2015'. New York: Women's Environment and Development Organization.
World Bank (2003) *World Development Report 2003: Sustainable Development in a Dynamic World*. Washington, DC: World Bank.

Melissa Leach is a social anthropologist and Professorial Fellow at the Institute of Development Studies, University of Sussex, Falmer, Brighton BN1 9RE (e-mail: M.leach@ids.ac.uk), where she leads the Knowledge, Technology and Society (KNOTS) team and is Director of a new ESRC Research Centre, Social, Technological and Environmental Pathways to Sustainability (STEPS). Her current research interests include science–society relations, environment and health issues.

3 In theory and in practice

Women creating better accounts of the world

Louise Fortmann

Making women visible

Throughout the nineteenth and much of the twentieth century, ideas of what was worthy of scholarship were highly gendered. Most social science research focused on men (usually white men) or male-dominated institutions. Finding "women" in an index or mentioned in the text was a considerable surprise, even an occasion for rejoicing. Then with Esther Boserup's 1970 book, *Woman's Role in Economic Development*, came a steady stream of scholarly works on women and rural development by such pioneers as Meena Acharya and Lynn Bennett (1981), Bina Agarwal (1986), Edna Bay (1982), Noel Chavangi (1984), Carol Colfer (1981), Elizabeth Croll (1985), Carmen Diana Deere (1982), Ruth Dixon (1978), Ann Fleuret (1977) Marilyn Hoskins (1980), Shimwayi Muntemba (1982), Dianne Rocheleau (1988), Gita Sen and Caren Grown (1987), Kathleen Staudt and Jane Jaquette (1982), and, of course, Irene Tinker (1976a, 1980b).[1]

Their work focused on invisible women—the invisible woman farmer, the invisible woman agroforester or forester, the invisible woman fisher—and made them visible to state agencies and donors as well as to other scholars. Much of the work focused on individual women or pried apart the notion of unitary households to identify women's roles, asking: What work do women do? What resources, in what quantities, do they have to work with? What are they responsible for? What costs do they bear? What benefits do they achieve? The picture that emerged was one of women whose work and knowledge were essential parts of rural livelihood systems and who were, in general, disadvantaged vis-à-vis men in most spheres of access to resources and power.

One could quite correctly conclude that this body of scholarship grew out of the women's movement. Under the banner of "the personal is political," analysis and critique were making gender discrimination visible in structures and processes ranging from the workplace, to the provision of health care, to access to education and credit, to prevailing discourses about intelligence and sexuality. However, this chapter looks at the body of scholarship in a different way, exploring the production of knowledge by and about women and rural development.[2] It takes as the starting point Donna Haraway's (1999, 75) assertion: "Feminists have to insist on a better account of the world." For her this is the goal of science (182), which I take to include the social sciences. How, this chapter asks, have and should we use women's theory and practice to develop better accounts of the world? It further assumes with Haraway (1999) that objectivity is not the "god-trick of seeing everything from nowhere" (176) but rather that objectivity is necessarily embodied and, therefore, all knowledges are situated and partial. Thus, the male scholars before Boserup were not able to pull off the "god-trick." Rather, their knowledge, situated in the partiality of the male gaze, led them to an account of the world that consisted of and privileged male agency, knowledge, and efficacy. Women's scholarship produced knowledge situated in the female gaze that gave a better account of the world by repicturing that world through women's agency, knowledge, and efficacy.

The chapter proceeds in three parts, each exploring how theory and practice rooted in situated knowledges can be used to develop better accounts of the world. Together they focus on what questions are asked, whose knowledge counts, and whose voices are permitted to be heard. Throughout, it is taken as given that women create knowledge through their practices including not only scholarly research and theory but also the everyday practices of producing livelihoods and raising families. The first section turns the idea of women as societal victims and perpetrators of environmental degradation inside out by asking about the adverse effects of discrimination against women on the environment. The second section explores the possibilities of analysis that combines the partial gazes of the global North and South. The third section argues for the practice of an interdependent science that eschews hierarchical notions of whose knowledge counts and what kinds of voices are permitted.

Making multi-scale effects of gender discrimination visible: an environmental example

Making multi-scale environmental effects of gender discrimination visible requires us to consider two literatures, one at the intersection of women and the environment and the other on gendered property rights. Rural women in the global South often appear in the literature on the environment in one of three roles: victims, perpetrators, or manager/stewards. The literature, often written by women, on women as environmental victims has documented the adverse effects on women of environmental degradation and environmental projects that do not take them into account. Often women must walk farther for fuelwood (Karan and Iijima 1975) and water (UNEP 2004) as a result of environmental degradation. And some environmental projects have forced women to walk farther for fuelwood (the original Campfire program in Masoka,

Zimbabwe, is a case in point) or destroyed their livelihood strategies (Schroeder 1997). The perpetrator literature portrays women's practices such as fuelwood collection as a cause (sometimes major) of environmental degradation (Van Horen and Eberhard 1995). Finally, the literature (also often written by women) documenting the knowledge and practices that enable women to be competent manager/stewards of natural resources on which they depend is extensive (Colfer 1981; M. Leach 1994; Rocheleau and Edmunds 1997).

Much of the literature on property and environment has focused on the debate about environmental outcomes of different property regimes: common, public, and private property (Ostrom 1991; McCay and Acheson 1987). Attention to gender in this literature is rare (Meinzen-Dick and Zwarteveen 2001). The literature on gendered property regimes demonstrates that generally women are less likely than men to have access to land in their own right or to own or control land or the crops they produce and are more likely than men to be landless or to lose access to land with a change in their marital status; when they are landowners, women have less land than men on average (Bruce 1990; Maboreke 1990; Ngqaleni and Makhura 1996; Verdery 1996; Agarwal 1994; Simbolon 1998; Dore 2000; Deere and Leon 2001; Whitehead and Tsikata 2003; Casolo 2004).[3]

These literatures pose situated knowledge at different scales. The literature on women as manager/stewards takes a micro-level view, focusing on the knowledge, practices, and livelihood strategies of individual women and their households.[4] The macro level appears in this literature in the form of institutions and structures that have adverse micro-level effects on women. The property regimes literature generally proceeds as if macro-level property institutions were gender neutral and focuses on their environmental effects. When the environmental degradation literature takes a gendered view, it focuses on macro-level environmental effects and looks to women's micro-level practices (real or imaginary) as a cause of environmental degradation. The following case study connects these partial and situated knowledges in asking the question: Do the adverse micro-level effects of macro-level institutions also have adverse effects at the macro level? Do the focus on and implicit willingness to accept gendered adverse outcomes at the micro level mask adverse societal level effects?[5]

An example of adverse environmental effects of a gendered property regime

In a 1991–92 study of 27 percent of the households in two villages[6] in central Zimbabwe, 56 percent of the respondents had planted at least one tree in the homestead. But only 44 percent of the women planted trees in their homestead, in contrast to 83 percent of the men. To analyze homestead tree planting, logit models were used.[7]

The analysis showed that women, regardless of class, were significantly less likely to plant trees in the homestead than men. Taking the average value of each variable over the entire sample, the predicted probability of planting a tree is 58 percent. With all other variables held at their mean (average), men had an 83 percent probability of planting a tree in the homestead while the probability for women was only 43 percent. Wealth was not statistically significant at the .05 level. Poor and mid-level farmers were as likely as the wealthy to have planted at least one tree, although they were not necessarily planting for the same reasons. Poor men have a positive (and highly significant) probability of planting a tree. This analysis suggests that gender plays a more important role than wealth in the decision to plant trees on homestead property.

Why might gender adversely affect tree planting? It is not physical strength, as anyone knows who has done women's work: planting groundnuts, hauling water, collecting firewood, making groundnut butter (smooth, not chunky!) with a grinding stone. It is not women's lack of knowledge about or need for trees and tree products: in the study area for all but two categories of use, women knew far more tree species than men did. Neither age nor education had any statistical significance. Tree planting is neither culturally proscribed for women nor prescribed for men in the study area. Insecurity of land and tree tenure resulting from a gendered land tenure system is by far the most persuasive explanation.

This interpretation is strengthened by two additional pieces of data. First, women who are divorced in the village (all of whom had lost all rights to the trees they had planted and tended during their marriage even when they stayed in the village) were emphatic that they would not plant trees in a new marital compound lest they once again be discarded and once again lose everything. Second, gender did not affect tree planting in the community woodlot where women retained their rights after a divorce as long as they continued to reside in the village. The latter finding must be approached with caution since women's tree planting in community woodlots may be done not on their own behalf but as an emissary of the household. Nonetheless, it is instructive that gender has adverse effects on tree planting when women's tenure is insecure and apparently has no such effects where their tenure is secure.

Although we must be cautious in drawing conclusions from a single study, these data are certainly suggestive. To the extent that the ecological stability and health of a society and its production systems depends on women's willingness to invest their labor in long-term investments (what Blaikie and Brookfield [1987] call landesque capital) such as tree planting, terracing, irrigation infrastructure, and fences, property and tenure systems that discriminate against women will have negative societal consequences. Since in many parts of Africa the productive rural population is still disproportionately made up of women, this finding should give pause to those concerned with maintaining or improving ecological conditions that can continue to sustain agricultural livelihoods. Clearly, property rights must enable and encourage women to be ecological stewards.[8]

The study also shows the importance of recognizing the partiality of one's knowledge. Better accounts of the world require not only making women visible but also making visible the multi-scale adverse effects that result from micro-scale adverse effects on women.

Making visible the commonalities of women in the global North and global South

This section focuses on the need for and the possibilities of combining the partial knowledges of both theory and practice in the global North and global South. The following vignette exemplifies the problem at hand.

At a 2002 international meeting on community forestry networks,[9] the suggestion that participants break out by regional groups was met with a pained query from a prominent scholar and practitioner of development: "But what would the people from the North talk about?" The reply, "We would talk about community forestry networks in the North," left many dumbstruck. Practitioners and development scholars from the South as well as those from the North whose scholarship and practice were limited to the South could not imagine that rural people in the South had anything in common with rural people in the North.[10] The mirror image is the common belief of those in the North that the South has no relevance to them, despite the fact that the northern community forestry movement is firmly rooted in the community forestry experience of South and Southeast Asia.

These reactions are not surprising when one acknowledges that, with some exceptions,[11] scholarship on women (and other issues) in the global South and in the global North have been more or less distinct. Of course, some edited volumes (Rocheleau et al. 1996; Perry and Schenck 2001) as well as conferences include both. But more often than not, discussion of women in one geographical area takes place as if the other areas do not exist in any relevant way. This may reflect the time and money required to do comparative research effectively or the narrow focus of many scholarly studies. It may well mean that the everyday lives of women in one region are not imaginable to women in another. Or it may be because the altogether appropriate insistence that we pay attention to social embeddedness has led to a failure to ask if the macro-level structures and processes within which local communities operate are similar across regions. For example, all too often the working assumption seems to be that communities in countries where the Bretton Woods institutions influence (or dictate) policy or initiate projects are qualitatively different from communities in countries where they do not. Perhaps the position of a community is conflated with the power of the country where it is located. Whatever the reason, these partial gazes result in a failure to look at the similarity of the effects of institutions such as transnational corporations and NGOS on communities in the North *and* South.

Practice may provide the most effective pathway to better accounts of the world though combining the partial gazes and situated knowledges of the North and South. There are numerous examples of bridging practice involving institutional practitioners. An Indian forester with experience with Joint Forest Management in India directs a U. S. national community forestry organization. Community forestry practitioners from the North and South met at the Johannesburg summit and made common cause, forming the Global Caucus on Community Based Forest Management. The successful women's saving circles of the Grameen Bank came to the United States in the form of the ShoreBank in Chicago.

Commonalities are made visible at the personal level. When villagers from resource-dependent rural communities in the South visit rural resource-dependent communities in northern California, both groups almost inevitably make the same comment: "We never realized that they have the same problems we do. Now we know we are not alone." Thus members of the Hupa Nation in northern California and the villagers of Bawa, Mozambique, recognized that they both struggle to maintain culturally based natural resource use and management against government interference. On one occasion an African NGO staffer quipped that he never would have guessed that his country would have so much in common with a small California forest community; namely, that in both places major decisions are made in Washington, D.C., without consulting local people. The instant the words were out of his mouth, the political relationships that Cameroonians and the residents of Hayfork had in common became visible to everyone.

To return to scholarly practice, imagine what we might learn if we broke out of our geographic ghettos and did, for example, a study of the economic and social strategies of poor female household heads in the North and the South, or a study of the reasons for and effects of excluding wives from ownership of lineage land in parts of rural Africa and family farm corporation in the U. S. Midwest, or a study of the cultures and social consequences of physical and sexual abuse of women wherever they are.

Much mutual learning between North and South is clearly to be done. It is to further possibilities of mutual learning between civil and conventional scientists that the next section turns.

Making interdependent science visible

The twenty-first century's most privileged form of knowledge production is conventional science. Conventional scientists are formally educated and use prescribed experimental and observational techniques. Their findings, often validated by statistical tests, networks of other scientists, and journals, are intended to be generalizable and may not translate easily into useful solutions to local problems.[12] Conventional science has been criticized for frequently being hostile to women and for privileging a narrow spectrum of possible ways of producing knowledge including both the questions that may be asked and the methods that may be used (Merchant 1980; Haraway 1999; Maddox 2002; Bug 2003; Conkey 2003; Gowaty 2003).

Civil science is a different way of producing knowledge. Civil scientists by and large work informally using experimental and observational techniques they and their predecessors have developed themselves. Their science depends on their knowledge of a particular set of social-ecological relationships. Their findings, validated by utility, are well suited to providing useful solutions to local problems but may not be generalizable. Civil science encompasses indigenous knowledge, traditional knowledge, local technical knowledge, and the ethnosciences.

Given the situated and partial nature of knowledge, one way to develop better accounts of the world would be through the collaboration of civil and conventional sciences. For the most part, there is relatively little interaction between the two modes of knowledge production but there are exceptions. For example, in the field of medicine, conventional scientists have used civil science as a source of information. A familiar example is the common or Madagascar periwinkle, *Catharanthus roseus*, which was traditionally used in different parts of the world to stop bleeding, as an astringent and a diuretic, and to treat diabetes, wasp stings, coughs, lung congestion and inflammation, sore throats, and eye irritation and infections. Two alkaloids, vincristine and vinblastine, found in the Madagascar periwinkle are used in drugs made by Eli Lilly to treat leukemia and a variety of cancers. While the Madagascar periwinkle story, without question, has had good outcomes for cancer patients, civil-conventional science interactions of this unidirectional sort are likely to have undesirable features. The relationship is generally hierarchical, privileging conventional science. It may also be extractive as demonstrated all too well by the problems of bioprospecting and biopiracy (Chapela 1994; Fairhead and Leach 2003; see also note 9 below).

A different mode of interaction is possible through interdependent science, a set of knowledge-producing practices intended to provide better accounts of the world through collaboration between conventional and civil scientists. Interdependent science does not privilege conventional science. Rather, it recognizes that good science integrates and acknowledges different actors and actions, including different ways of knowing, such as civil science. It recognizes practice as a mode of knowledge production (Chaiklin and Lave 1993; Holland and Lave 2000). It recognizes, as this essay has argued from the outset, that all science is embodied and that objectivity comes not from the "god-trick" but from the recognition of partiality. Its metric of objectivity is knowledge and explanatory power. It recognizes the multiple female and male voices in which better accounts of the world emerge, voices ranging from sparse passive-voice prose to chatty accounts and explanations, from the chants of a ritual healer to the thick description of an ethnographer and many more. Since its goal is the co-production of knowledge, it is likely to utilize participatory methods.[13]

Examples of interdependent science involving women and men demonstrate its potential. The combination of farmers' knowledge with the knowledge and technologies of conventional scientists in participatory plant breeding has shortened the time required to develop a new variety and increased the adoption of new varieties. For example, in Rwanda, "Farmer bush bean selections outperformed their own mixtures with average production increases of up to 38 percent; breeder selections in the same region on average showed negative or insignificant production increases" (Sperling 1996:45). Conventional scientists who have been involved in participatory plant breeding say they would never go back to doing plant breeding on their own because conventional plant breeding is more time consuming and less effective (Robin Buchara personal communication, 2000). Sally Humphries (personal communication, 2004) reports the commercial release of an improved bean variety by a research team of women and men Nicaraguan farmers working in collaboration with a conventional science agronomist. In the face of a dearth of knowledge about the ecology of commercially important floral greens, a team of women and men floral greens harvesters worked with a forest ecologist to develop experimental trials on the effects of harvesting on floral green production (Ballard et al. 2002; Ballard and Fortmann 2004). This research has provided information on which to base management policies. A team of women and men Zimbabwean villagers working with a rural sociologist and a botanist documented the use and management of 122 varieties of indigenous trees (Chidari et al. 1992). This was an important complement to the approach of the Forestry Commission, which concentrated on growing a few exotic species. In addition, the team also documented the extensive knowledge that women had of trees and their uses, knowledge that generally was greater than men's.

In addition to producing knowledge, the practice of interdependent science can empower the civil scientists including women. After the women and men of the Zimbabwe village research team presented their research findings at a well-attended village meeting, the Chairman of the Grazing Scheme (the *de facto* village head) rose to his feet and said, "I never thought we could learn anything from a woman, but we have." While this did not overturn patriarchal social relations in the village, it was an important step, a moment that no one could take away from the women research team members whose work and knowledge had been acknowledged.

Conclusions

This chapter has shown that women's research and practice has led to better accounts of the world through making neglected social actors and social relations visible, by bringing actors and social relations thought to be distinct into conversation with each other, and by embracing and utilizing difference and collaboration rather than hierarchy and extraction. To be sure, this is an ongoing project requiring the iteration of making things visible in each of the ways discussed here. For example, if women are to be participants in interdependent science, their knowledge of a particular subject in a particular place may first have to be made visible, the first step discussed above. Clearly, much remains to be done. Equally clearly, much has been accomplished. In theory and in practice, in the academy and in the household, farm field and forest, women have produced knowledge that we need for the lives of our children and grandchildren. One of our jobs as scholars is to collaborate in keeping that knowledge visible and validated.

Notes

I would like to acknowledge the research assistance of Katariina Tuovinen and the critical comments of Emery Roe.

1 Despite its length, this list only scratches the surface of the numbers of brave and energetic women who did research on and published about women even when it put their own careers in jeopardy.

2 The focus of this chapter is limited to rural development both for reasons of manageability and because the literature on women and rural development has often been more applied than other feminist scholarship and research in women's studies. For genealogies of scholarly work on women in development see M. Leach (1994) and Rocheleau et al. (1996).

3 Mexico provides a case to the contrary. Hamilton (2002) argues that although land titling eliminated women's inheritance rights, because of ejidatarias' social status as mothers deserving their children's cooperation and the respect and aid of the community following their husband's death, they have generally been well provided for.

4 Meinzen-Dick and Zwarteveen (2001) note the general focus at the household or micro level and call for attention to the community or meso level. This chapter calls for further scaling up to the macro level.

5 Put in a more pithy fashion: is oppressing mother also bad for Mother Nature (and everybody else)?

6 The research project documented the management of trees and woodlands and their commercial and domestic uses as well as the factors affecting tree planting. There were 48 men and 106 women in the final sample. There are more women because men in many households worked in town most of the year and came home only occasionally.

7 Logit models are appropriate for situations in which individuals must make a choice between two options, in this case: to plant or not plant, and can be used to estimate probabilities—in this case, the probability that a person will plant a tree. This study is described in detail in Fortmann et al. 1997.

8 Although it is beyond the scope of this chapter, it is worth noting that land races developed, often by women, over a number of years using deliberate agronomic practices and seed selection are not recognized as deserving of legal protection (Kameri-Mbote and Cullet 1999). Thus, it can be argued that the right of women to intellectual products of their agricultural labor is as insecure as their rights to the physical means of production and their products. Kameri-Mbote and Cullet maintain that the current lack of protection of intellectual property rights at the level of the individual farmer "has contributed to the erosion of the genetic base necessary for the further development of agrobiodiversity" (24).

9 Community forestry is a set of institutional arrangements in which communities are involved wholly or in part in decision making about and benefits from forest management as well as contributing knowledge and labor to achieve healthy forests and social well-being (Cecilia Danks personal communication, 2003).

10. Similarly, many participants from the global South who attended a 2004 traveling workshop on community forestry in Alabama and California were quite surprised to discover that there were both forests and poor people in the United States.

11. Refugee studies are an exception in that they address questions of women from the South who find themselves in the North. These studies tend now, however, to address women from the North.

12. For an in-depth discussion of conventional, civil, and interdependent sciences see Ballard and Fortmann (2004).

13. There is informal evidence (but no systematically collected data) that suggests women are more likely to use participatory methods. Whether this is true is a question worth researching.

References

Agarwal, Bina. 1994. *A Field of One's Own: Gender and Land Rights in South Asia.* Cambridge: Cambridge University Press.

——. 1986. *Cold Hearths and Barren Slopes: The Woodfuel Crisis in the Third World.* London: Zed Books.

Ballard, H., D. Collins, A. Lopez, and J. Freed. 2002. *Harvesting Floral Greens in Western Wasbington as Value-Addition: Labor Issues and Globalization.* Meeting of the International Association for the Study of Common Property, Victoria Falls, Zimbabwe.

Ballard, Heidi, and Louise Fortmann. 2004. "Collaborating Experts: Integrating Civil and Conventional Science to Inform Management of Salal *(Gaultheria shallon).*" In *Fostering integration: Concepts and Practice in Resource and Environmental Management,* edited by Kevin Hanna and D. Scott Slocombe. Oxford: Oxford University Press.

Bay, E. 1982. *Women and Work in Africa.* Boulder: Westview Press.

Bruce, John. 1990. "Legal Issues in Land Use and Resettlement." World Bank Background Paper for Zimbabwe Agricultural Sector Memorandum. Harare: Zimbabwe.

Bug, Amy. 2003. "Has Feminism Changed Physics?" *Signs* 28 (3): 881–99.

Casolo, Jennifer. 2004. "*Voz y Voto* In Deed? Reworking of Land Rights, Gender and Power in Post-Hurricane Mitch Honduras." Unpublished master's thesis. Department of Geography, University of California at Berkeley.

Chaiklin, Seth, and Jean Lave, eds. 1993· *Understanding Practice: Perspectives on Activity and Context.* Cambridge: Cambridge University Press.

Chapela, I. H. 1994. "Bioprospecting in the Information Age: A Critical Analysis of Pharmaceutical Searches through Biodiversity." In *Emerging Connections: Biodiversity, Biotechnology and Sustainable Development,* edited by J. Feinsilver, 27–46. Washington, D. C.: Panamerican Health Organization/Interamerican Institute for Cooperation in Agriculture.

Chavangi, N. A. 1984. *Cultural Aspects of Fuelwood Procurement in Kakamega District.* Working Paper no. 4. Nairobi: Kenya Woodfuel Development Programme.

Chidari, Gift, Francisca Chirambaguwa, Patricia Matsvimbo, Anna Mhiripiri, Hilda Chanakira, James Chanakira, Xavier Mutsvangzwa, Angeline Mvumbe, Louise Fortmann, Robert Drummond, and Nontokozo Nabane. 1992. "The Use of Indigenous Trees in Mhondoro District." Centre for Applied Social Sciences, University of Zimbabwe NRM Occasional Paper 5.

Colfer, C. 1981. "Women and Men and Time in the Forests of East Kalimantan." *Barneo Research Bulletin,* 75–85.

Conkey, Margaret W. 2003. "Has Feminism Changed Archeology?" *Signs* 28 (3): 867–80.

Connelly, M. Patricia, Tania Murray Li, Martha MacDonald, and Jane L. Parpart. 2000. "Feminism and Development: Theoretical Perspectives." In *Theoretical Perspectives on Gender and Development,* edited by Jane L. Parpart, M. Patricia Connelly, and V. Eudine Barriteau, 51–160. Ottawa, Canada: International Development Research Center.

Cooper, William. 1991. "Overview." In Joint Economic Committee, *China's Economic Dilemmas in the 1990s: The Problems of Reforms, Modernization, and Interdependence.* Washington, D.C.: USGPO: 335–39.

Cornwall, Andrea. 2002. "Making a Difference? Gender and Participatory Development." In *Shifting Burdens: Gender and Agrarian Change under Neoliberalism,* edited by Shahra Razavi, 197–232. Bloomfield, Conn.: Kumarian Press.

Council of Europe. 1998. *Gender Mainstreaming: Conceptual Framework, Methodology and Presentation of Good Practice.* Strasbourg: Council of Europe Publishing.

Croll, Elisabeth. 1985. *Women and Rural Development in China: Production and Reproduction.* Geneva: International Labor Office.

Deere, Carmen Diana. 1982. "The Division of Labor by Sex in Agriculture: A Peruvian Case Study." *Economic Development and Cultural Change* 30 (4): 795–811.

Deere, Carmen Diana, and Magdalena Leon. 2001. *Empowering Women: Land and Property Rights in Latin America.* Pittsburgh: University of Pittsburgh Press.

Dixon, R. 1978. *Rural Women at Work: Strategies for Development in South Asia.* Baltimore: Johns Hopkins University Press.

Dore, E. 2000. "Property, Households and Public Regulation of Domestic Life: Dinomo, Nicaraugua, 1840–1900.' In *Hidden Histories of Gender and the State in Latin America,* edited by E. Dore and M. Molyneux, 147–71. Durham: Duke University Press.

Fairhead, James, and Melissa Leach. 2003. "Practicing 'Biodiversity' in Guinea: Nature, Nation and International Convention." *Oxford Development Studies* 31 (4): 427–40.

Fleuret, A. 1977. "The Role of Women in Rural Markets: Lushoto, Tanzania." Paper presented at the First Women and Anthropology Symposium, Sacramento, California.

Fortmann, L., C. Antinori, and N. Nabane. 1997. "Fruits of Their Labors: Property Rights and Tree Planting in Two Zimbabwe Villages." *Rural Sociology* 62 (3): 295–314.

Gowaty, Patricia Adair. 2003 "How Feminism Changed Evolutionary Biology." *Signs* 28 (3): 901–21.

Hamilton, S. 2002. "Neoliberalism, Gender and Property Rights in Rural Mexico." *Latin America Research Review* 37 (1): 119–43.

Haraway, Donna. 1999. "Situated Knowledges: The Science Question in Feminism and the Privilege of Partial Perspective." In *The Science Studies Reader,* edited by Mario Biagioli, 172–88. New York: Routledge.

Kameri-Mbote, P., and P. Cullet. 1999. *Agrobiodiversity and International Law.* Biopolicy International Series no. 22. Nairobi: African Centre for Technology Studies.

Karan, P. P., and Shigeru Iijima. 1975. "Environmental Stress in the Himalaya." *Geographical Review* 75 (1): 71–92.

Leach, M. 1994. *Rainforest Relations: Gender and Resource Use among the Mende of Gola, Sierra Leone.* Washington, D. C.: Smithsonian Institution Press.

Maboreke, M. 1990. "The Gender Dimensions of the Land Question in Zimbabwe." Paper presented at the Conference on Land Policy in Zimbabwe after Lancaster, Harare, Zimbabwe.

Maddox, Brenda. 2002. *Rosalind Franklin: The Dark Lady of DNA.* New York: Harper Collins.

McCay, B., and J. Acheson. 1987. *The Question of the Commons.* Tucson: University of Arizona Press.

Meinzen-Dick, Ruth, and Margreet Zwarteveen. 2001. "Gender Dimensions of Community Resource Management: The Case of Water Users' Associations in South Asia." In *Communities and the Environment: Ethnicity, Gender, and the State in Community-Based Conservation,* edited by Arun Agrawal and Clark C. Gibson, 63–88. New Brunswick: Rutgers University Press.

Merchant, Carolyn. 1980. *The Death of Nature: Women, Ecology and the Scientific Revolution.* San Francisco: Harper and Rowe.

Muntemba, S. 1982. "Women as Food Producers and Suppliers in the Twentieth Century: The Case of Zambia." *Development Dialogue* (1–2): 29–50.

Ngqaleni, M., and M. T. Makhura. 1996. "An Analysis of Women's Status in Agricultural Development in Northern Province." In *Land, Labour, and Livelihoods in Rural South Africa.*

Ostrom, E. 1991. *Governing the Commons: The Evolution of Institutions for Collective Action.* Cambridge: Cambridge University Press.

Perry, Susan, and Celeste Schenck. 2001. *Eye to Eye: Women Practising Development Across Cultures.* London: Zed Books.

Rocheleau, Dianne. 1988. "Women, Trees and Tenure: Implications for Agroforestry." In *Whose Trees: Proprietary Dimensions of Forestry,* edited by L. Fortmann and J. W. Bruce, 254–72. Boulder: Westview Press.

Rocheleau, Dianne, and David Edmunds. 1997. "Women, Men and Trees: Gender, Power and Property in Forest and Agrarian Landscapes." *World Development* 25 (8): 1351–71.

Rocheleau, Dianne, Barbara Thomas-Slayter, and Ester Wangari. 1996. "Gender and Development: A Feminist Political Ecology Perspective." In *Feminist Political Ecology: Global Issues and Local Experiences,* edited by Dianne Rocheleau, Barbara Thomas-Slayter, and Ester Wangari, 3–23. London: Routledge.

Schroeder, R. A. 1997. "'Re-claiming' Land in the Gambia: Gendered Property Rights and Environmental Intervention." *Annals of the Association of American Geographers* 87 (3): 487–508.

Sen, Gita, and Caren Grown. 1987. *Development Crises and Alternative Visions: Third World Women's Perspectives.* New Delhi: DAWN (Development Alternatives for Women for a New Era).

Simbolon, I. J. 1998. "Peasant Women and Access to Land: Customary Law, State Law, and Gender-Based Ideology, The Case of Toba-Batak (North Sumatra)." Unpublished Ph.D. dissertation: Landbau Universiteit Wageningen.

Tinker, Irene. 1980. "New Technologies for Food-Related Activities: An Equity Strategy." In *Women and Technological Change in Developing Countries,* edited by Roslyn Dauber and Melinda Cain, 22–34. Boulder: Westview Press.

——1976. "The Adverse Impact of Development on Women." In *Women and World Development,* edited by Irene Tinker and Michelle Bo Bramsen, 22–34. New York: Praeger. Reprinted in *Peace Corps Program and Training Journal* IV (6) 1977. Translated as "Le developpement contre les femmes" in *Question feministes,* no. 6, September 1979.

UNEP (United Nations Environment Program). 2004. "Women and The Environment." Nairobi: UNEP.

Van Horen, C., and A. Eberhard. 1995. "Energy, Environment and the Rural Poor in South Africa." *Development Southern Africa* 12 (2): 197–211.

Verdery, K. 1996. *What Was Socialism and What Comes Next?* Princeton: Princeton University Press.

Whitehead, Ann, and Dzodzi Tsikata. 2003. "Policy Discourses on Women's Land Rights in Sub-Saharan Africa: The Implications of the Return to the Customary." *Journal of Agrarian Change* 3 (1 and 2): 67–112.

4 On communication among "unequals"[1]

Carol J. Pierce Colfer

Introduction

Professionals in development fields are beginning to design more people-oriented, grassroots kinds of programs. We have been trying to foster the participation of local people in their own development, fashion effective links between their own and national goals, and encourage increased local responsibility for the improvement of local conditions.

An important (and somewhat neglected) component of this attempt has been the necessity to communicate effectively with the members of such populations. If we are to collaborate with them, we must interact with them and we must gain access to the way they perceive the world. Much evidence has been accumulated documenting the difficulties of such communication (Rogers 1973). And by now we should recognize the importance of designing programs that are responsive to and congruent with local behavior and beliefs. An excellent, and unfortunately still timely, source on this topic is Paul's (1955) compendium of case studies, *Health, Culture, and Community*. See also Chambers (1978) or Colfer (1977a; 1979); Vayda, Colfer, & Brotokusumo (1980).

This paper provides a theoretical perspective for understanding some of the dynamics of interaction between "unequals," e.g., rural peoples and planners, including some important mechanisms by which the elite are ordinarily denied access to the world views of "the poor." I present three ethnographic examples of this process in operation, for illustration; and conclude with four policy recommendations that follow from this analysis.

As world views—and status—differ

Because of the ubiquity of what has been characterized as the hypodermic model of social change, I would like to devote a few pages to the explication

Carol J. Pierce Colfer, "On Communication among 'Unequals,'" *International Journal of Intercultural Relations* 7 (3), 263–283.

of some differences between that approach and the one exemplified by this analysis. Bateson's (1972) words introduce the crucial difference:

> It all starts, I suppose, with the Pythagoreans versus their predecessors, and the argument took the shape of "Do you ask what it's made of—earth, fire, water, etc?" Or do you ask, "What is its *pattern*?" Pythagoras stood for inquiry into pattern rather than inquiry into substance. The controversy has gone through the ages, and the Pythagorean half of it has, until recently, been on the whole the submerged half.
>
> (p. 443)

Bateson is discussing an early example of the kinds of disagreements that separate the "hard" sciences from the "soft" sciences. In terms that are more relevant to development, this ancient controversy is manifest in opposing approaches to research, to policy, and to program goals.

The dominant social scientific position is represented by those who maintain that human behavior is best understood using the conceptual tools available to us from the physical sciences. From this perspective, human behavior is subject to the same kinds of laws of cause and effect as are falling rocks and ripples in water; and the task of scientists is to isolate causal factors in order to control them and their effects.

In this paper, I am presenting an alternative, a more Pythagorean approach. I will be looking at patterns of behavior, specifically patterns of behavior in contexts where people have *different* statuses, and relating the findings to the possibility of change. Again in Bateson's (1972) words:

> (Difference which occurs across time is what we call "change.") A difference, then, is an abstract matter. In the hard sciences, effects are, in general, caused by rather concrete conditions or events—impacts, forces, and so forth. But when you enter the world of communication, organization, etc., you leave behind the whole world in which effects are brought about by forces and impacts and energy exchange. You enter a world in which "effects"—and I am not sure one should still use the same word—are brought about by *differences* . . . In the world of mind, nothing—that which is *not*—can be a cause. In the hard sciences, we ask for causes and we expect them to exist and be "real." But remember that zero is different from one, and because zero is different from one, zero can be a cause in the psychological world, the world of communication. The letter which you do not write can get an angry reply.
>
> (p. 452)

In the framework of this paper, the status and power that you do *not* have can influence your behavior in interaction with those who *do* have such status and power, and can influence their behavior in turn.

It might also be worth pointing out that I will be looking at people's models of reality from a relativistic point of view. Saral's (1976) words nicely summarize the perspective of this analysis:

> It is . . . apparent that there is no absolute reality, nor is there a universally valid way of perceiving, cognizing and/or thinking. Each world view has different underlying assumptions. Our normal state of consciousness is not something natural or given, nor is it universal across cultures. It is simply a specialized tool, a complex structure for coping with our environment.
>
> (p. 6)

Wolfe (1979) elaborates a similar perspective, relating to health.

By looking at several specific cases of interaction between "unequals," some commonalities among such cases emerge. I offer some theoretical interpretations of these data that bear on planned change efforts and on any interaction between "unequals" (whether citizens of superpowers with Third World nationals, men and women, urban elites with rural folk, or proponents of "hard" vs. "soft" scientific paradigms).

This paper was originally motivated by my observation that certain patterned qualities seemed to characterize interaction between people of unequal status, in a wide variety of settings. The crucial element in these patterns was the fact that normally articulate lower status people, when in the company of their "superiors," were less able to communicate their views, less free to express themselves, perhaps less lucid. This pattern struck me as very important in a variety of development settings—particularly now that the importance of local participation in development is being increasingly recognized. A contribution to our understanding of the interactional dynamics that result in the "silent masses," and "the inarticulateness of women" (E. Ardener, 1975), would be significant. This paper is intended as a tentative step in that direction.

I will be discussing situations of culture (or subculture) contact, specifically situations where one culture is viewed as the dominant culture. I am using "culture" as a convenient cover term to include both cognitive and behavioral systems which are subject to change. See Moore (1975) on the necessity to attend both to the regularities of culture and to the processes of situational adjustment; similarly Turner (1977). For instance, in contact between middle class urbanites and an enclave of Hutterites in the U.S., the former group and its belief system would be considered dominant. Similarly, in contact between males and females in a given culture or group, the male system would be dominant. I agree with Ortner (1974) that "The secondary status of woman in society is one of the true universals, a pan-cultural fact" (p. 67). Within the scientific community, a unicausal "hard" scientific paradigm would be dominant over those of the "softer" social sciences. If this paper—as a "soft" contribution—should be judged "inarticulate" it will perhaps supply further evidence to support the view herein expressed! (See example III below for a discussion of interaction between proponents of these two approaches to human inquiry.)

The concept of "inarticulateness" was developed by E. Ardener (1975, orig. 1968) to explain the paucity of anthropological data on women. According to E. Ardener,

> . . . if ethnographers (male and female) want only what the men can give, I suggest it is because the men consistently tend, when pressed, to give a bounded model of society such as ethnographers are attracted to.
>
> (p. 2)

He goes on to hypothesize that

> . . . to the degree that communication between ethnographer and people is imperfect, that imperfection drives the ethnographer in greater measure toward the men.
>
> (p. 2)

A group of British female anthropologists have elaborated on E. Ardener's work (S. Ardener, 1975). Hardman (1975), a member of this group, proposed the terms *muted group* and *counterpart model* to refer to lower status people and their cognitive systems, respectively; I have adopted her usage.

S. Ardener (1975) contends, with regard to E. Ardener's concept of inarticulateness, that:

> The implications are that a society may be dominated or overdetermined by the model (or models) generated by one dominant group within the system. This dominant model may impede the free expression of alternative models of their world which subdominant groups may possess, and perhaps may even inhibit the very generation of such models.
>
> (p. xii)

One of the points I hope to support in this paper is the extent to which dominant models interfere with, and preclude, the *expression* of alternative models. I question, however, the degree to which model *generation* can be inhibited. MacCormack (1977) rejects E. Ardener's theoretical framework, arguing that building on his interpretation, "there logically can be no place for women, except at the most menial levels of production and reproduction, in the development plans of nations" (p. 100). If one accepts the notion that muted groups do not or cannot generate models, her conclusion is indeed logical, and I too would reject Ardener's work. Building on part of his analysis, however, I am suggesting that muted groups (whose models are routinely unavailable to dominant groups) possess valuable information and perspectives that can (a) aid planners and practitioners in building effective programs, and (b) help researchers in developing a more coherent and inclusive understanding of human behavior and beliefs—if only such perspectives can be accessed by members of dominant groups.

That models (including assumptions, structures, relationships, content) are necessary for human thought is generally accepted in the literature on cognition. We also must contemplate the implications of the fact that muted groups have to integrate at least two models of reality at any one time, when dominant groups need integrate only one. In discussing these "models of reality" I want to emphasize that I see them as mutable and dynamic, consistent with Kelly's (1963) view of human psychology (see also Colfer 1974, for a discussion of cognitive process). As Eastman (1979) points out, of course, group status is subject to change, just as are models.

There is ample evidence in the ethnographic literature that peripheral peoples participate in alternate and separate subsystems. Indeed, the discovery of such systems was a major impetus in the development of the field of anthropology. In recent years the particular subsystems in which women participate have also been described (see e.g., Colfer, 1977b; Murphy & Murphy, 1974; and the following collections: S. Ardener, 1975; Reiter, 1975; Rosaldo & Lamphere, 1974; Tinker & Bramsen, 1976; and the special issue of *Signs* on "Women and National Development," 1977). The degree to which such alternate systems are integrated with dominant systems varies depending on frequency of interaction and specifics of the cultural systems in question.

In the literature there is also scattered evidence in support of the idea that muted groups know more about the dominant groups than the reverse. Such evidence does not form a coherent body of data, but exists in the form of a telling remark here, a related aside there, some of which are presented below.

In my own field experience, in Bushler Bay, Washington (see pp. 70–72), there was no question that the subdominant group (Locals) knew more about the dominant (Public Employee) lifestyle and orientation than the reverse. Locals were involved in economic activities that required some integration into, and in many cases, frequent contact with, Public Employees and their belief systems. Locals had gone and their children were now going to a public school dominated by Public Employee ideology and behavior patterns. A close approximation to the Public Employee lifestyle was performed continually on the television set. Public Employees had no comparable access to knowledge of Local beliefs and behaviors; person-to-person communication of a substantive nature was rare between Public Employees and Locals.

Likewise, women's awareness of and integration into male belief and behavior systems is far greater than the reverse. Leavitt, Sykes and Weatherford (1975) discuss this understanding of dominant systems in evaluating Kaberry and Goodale's (1939) Australian work.

> . . . in the male ethnographies of the Australian aborigines the "anthropologist's categories" predominate; the societies are represented as male-dominated, with women in a subordinate, degraded status. However, in *Aboriginal Woman* and *Tiwi Wives*, Phyllis Kaberry and Jane Goodale succeed in combining the "anthropologist's categories" with those of the native. Their theory and methodology seem to stem from "double-consciousness,"

> a concept that W.E.B. DuBois evolved to define the special awareness of black people in a racist society. Themselves women in a society that is also sexist, Kaberry and Goodale have the special sensitivity that members of subordinated groups must, if they are to survive, develop to those who control them, at the same time as they are fully aware of the everyday reality of their oppression, a quality the superordinate groups lack.
>
> (p. 112)

Okeley's (1975) work on gypsies documents the same phenomenon. She quotes a gypsy, referring to the gypsy concept of pollution: "People say we're dirty . . . they don't see that we think they're dirty" (p. 61). At another point, Okeley says: "Perhaps the gypsies may be more aware of the stereotypes which Gorgios [dominant group] attach to them than is the case in reverse, and exploit these." *(Ibid,* p. 82)

The participation of women and Third World peoples—particularly in interaction with more dominant, higher status groups—in two systems has a number of results, one of which (I would maintain) is the comparative "inarticulateness" discussed by E. Ardener. Two important processes are at work: cognitive and political.

First, participation in, and awareness of, two (or more) often conflicting cultural systems mitigates against acceptance of the straightforward and "articulate" models of society that have dominated in social science.

An illustration used to explain parts of Einstein's theory of relativity to a lay public will help to clarify the relevant distinction: If one is bouncing a ball and watching it, the ball appears to be going straight up and down. On the other hand, if one is observing someone on a moving platform bouncing a ball, the ball appears to arc. To carry this analogy back to the question at hand, one can say that the higher status person, as ball bouncer, need not be aware of the alternate appearance, or interpretation, that is available to the lower status person. The lower status person, as participant in both systems, is aware of both interpretations or views; I am suggesting, in this paper, that the lower status person, if he/she is to construct an "articulate" model, must be

Figure 4.1a Person bouncing a ball while standing still.

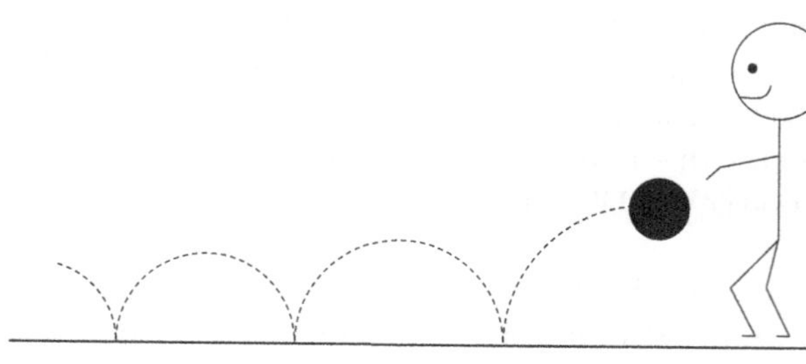

Figure 4.1b Person bouncing a ball on a moving train as seen by a person standing still.

much more creative and comprehensive than must the higher status person who has only one cultural system to account for (Leavitt, Skyes & Weatherford, 1975).

Lakoff (1975) has a similar observation in her discussion of the predicament of professional women who must switch back and forth in daily life between a system of largely male professional sociolinguistic rules and a system of domestic and female sociolinguistic rules.

Related to this issue is the fact that women often marry into an alien cultural system. In Iran (where residence is patrilocal) for instance, or in the rural U.S., women take on the subsystem of their husbands. In these circumstances, females must have a sufficiently permeable (in Kelly's [1963] sense) world view to accommodate such a change. This kind of permeability would seem to be related to E. Ardener's (1975) "inarticulateness," mitigating against the construction of "bounded models of society." In a similar vein, Lakoff (1975) says:

> It may be that the extra energy that must be (subconsciously or otherwise) expended in this game [switching between sociolinguistic systems] is energy sapped from more creative work, and hinders women from expressing themselves as well, as fully, or as freely as they might otherwise. (p. 7)

S. Ardener (1975) points out that muted groups may carry a greater commitment to their own system, which adds another dimension to the issue:

> . . . the principal measure for social success or for other satisfactions in the counterpart model may differ from that of the model of the dominant group, and therefore their acquiescence at being placed low down on the latter's scale for success may occur because the placing seems unimportant or irrelevant to them, since they may not necessarily be "unsuccessful" or "unsatisfied" according to the logic of their own muted model. (p. xvii)

Moving from the cognitive to the political, we see another dimension to this question. Lower status people are not as free to *express* their perspectives, even if they are *able* to construct coherent, bounded models/perspectives. The consequences of displeasing higher status people can be severe (see Krivonos [1976], for instance, on this phenomenon in hierarchic settings); and provision of information that does not fit neatly into the dominant model, or contradiction of a dominant person's views is often dangerous. Additionally, there may be tactical advantages to keeping lower status people's strategies from higher status people (Okeley [1975], on gypsy survival strategies).

At this point some ethnographic examples can illustrate the processes of interaction between muted and dominant groups, which reflect and reinforce these differing models of reality. I will draw from a variety of settings.

Villagers and urbanites in Iran

Since we are discussing power differentials, Iran is a good starting point. Power in human relations is straightforward there, and easy to see.

In 1972 I spent 5 months in Iran supported by Fulbright-Hayes, beginning research on informal avenues to power available to Muslim women. In this case, as well as the ones to follow, pseudonyms are used to protect people's privacy. I settled in a Qashqa'i village (alias Dolatabad) in southern Iran. After some stiff negotiations, I was accepted into the home of the village headman (Mukhtar), where I resided for a month (July 1972). The friendly demeanor of my host was tempered slightly by the fact that he was used to obedience and respect. Both he and his wife were proud tribal people who walked with their heads high, and good-naturedly but persistently corrected me for my cultural gaffes.

The fact that I was a foreigner and that I paid a higher rent than would usually be the case, as well as my curiosity value in the village, gave *me* power in our relationship; and their knowledge of the local context gave *them* power in our relationship. (Because this paper deals with situations marked by power differentials, I am focusing on that aspect of the relationship, to the exclusion of the affective components.) The fact that I "allowed" them (indeed encouraged them) to correct me also enhanced their prestige with other villagers, as a continual demonstration of their power over a foreigner/city dweller. All in all the relationship was nicely balanced, and each party was apparently pleased with the arrangement.

In daily life among the villagers, there was a clear disdain for Persians and city people, who were considered effete, weak, and incompetent. The opportunity to observe Qashqa'i-Persian interaction in the village arose one night when a group of four or five Persians and Americans was expected for supper. When 13 people arrived (the Persians had independently invited other relatives along), Mukhtar (the headman) was put in a very difficult situation, as the family had not obtained or prepared sufficient food. They had to borrow food from their neighbors, on the sly, and prepare more, while the guests waited. Although innocent in this whole fiasco, I received a mild rebuke from Mukhtar later (indicating his self-confidence in interaction with me).

During the interaction, however, Mukhtar was an attentive, hospitable host, making certain everyone was comfortable and happy. His Qashqa'i pride and his disdain for Persians were masked. He was operating with the strong hospitality code of the Middle East in mind; and he was making certain that he did not offend these Persians and Americans who surely had more power, wealth, and prestige in the wider sphere than did he.

In August I became ill, and it soon became apparent that I would have to leave the country. My partner (also ill) distributed most of our possessions to Mukhtar's family, and invited him to come to Shiraz for our remaining possessions.

When Mukhtar came, his behavior was almost unrecognizable. He bowed continually to me; he was obsequious, overly polite. He took on the behavior of urban dwellers of the lowest position. Gone were his proud, erect carriage, his confident air, and his manifestations of disdain for things urban and Persian. I attribute this dramatic change to two things: First, by giving him valuables, I (inadvertently) displayed a new and greater power over him; and second, by my leaving the village, he lost (or felt he lost) the power he had over me. Now in the social context of the dominant (Persian, urban) society, he was separated from his traditional sources of power, and he felt he had to conform to the dictates of urban culture which place a tribal village headman far down in the status hierarchy.

We have now seen Mukhtar in three contexts. In daily village life, he is self-confident, proud, dignified. He assiduously corrects my behavior, even treating me like a child at times. When a group of foreigners and urbanites come, Mukhtar is more careful not to offend, bowing in welcome and seeing to the needs of his guests. Surrounded by his familiar human and natural context, however, he retains much of his overt dignity. In the city, Mukhtar is a different man—obsequious and embarrassingly respectful toward me.

Compare this example to Okely's findings in the Gypsy community:

> Often thought of as 'underprivileged' and to be ripe for 'retraining' and 'resocialization' by some well-meaning Britons (an idea sometimes appearing to gain support from the Gypsies when they adopt 'a *subservient* and *humble posture*' in compliance with the expectations of the dominant model) the Gypsies are seen [by Okely] to possess a private view of the world, a counterpart model, in which members of the dominant group are not only NOT ideals to be respected or emulated, but, on the contrary, are seen as polluting.
>
> (S. Ardener 1975, p. xii, emphasis added)

Although I doubt that ideas of pollution are important in Mukhtar's case, surely the "subservient and humble posture" and the lack of sincere respect for members of the dominant group are exactly comparable.

Mukhtar understands the subordinate role of a rural Qashqa'i headman in the Persian urban model. He also understands the relative power of urban Persians as compared to rural Qashqa'i. The powerful can (and do) preempt the right to define situations according to their (dominant) model.

The asymmetry of the interaction is clearest when we compare Mukhtar's subservient city behavior with the visitors' relaxed, confident behavior in the village. The urban Persians do not grant Mukhtar the "proper" respect. Some are ignorant of the village reality and have no motivation to learn it (indeed such ignorance is a source of pride, indicating urbanity). Because of their power, their gaffes are not gaffes at all, but statements which, in the culture contact situation, undermine the existence, coherence, and logic of the village subsystem.

In many Middle Eastern contexts urbanites with village roots display their adherence to the dominant model with a vengeance. Goodenough (1963) in discussing identity change, particularly in developing countries, notes the same phenomenon. American folk knowledge likewise recognizes this pattern: "There's no one as Catholic as a convert."

Local and public employee women and men in rural America

Between December 1972 and September 1975, I did ethnographic research in a rural American village (alias Bushler Bay). In Bushler Bay, two important realities co-exist. The relations between adherents of the respective realities are not so dramatically related to personal power as in the Persian case, but important parallels exist.

The dominant reality divides the community into social classes based primarily on social, economic, and educational achievement. In this view, the wealthy, well-educated folk then comprise a top category (upper class). The poor, uneducated folk comprise the lower class; and everyone else is lumped together in the middle class. The people of Bushler Bay are all perceived as participants in *one* (and only one) mainstream American way of life. The differences between people derive from their differential accomplishment of (predominantly urban) universalistically determined goals.

This reality is supported by the dominant universalistic American ideology and by a long tradition of social scientific endeavor that is built on the assumptions inherent in this reality (see Colfer with Colfer 1978 for a discussion of this; also see Cohen 1971, for a good discussion of universalism and particularism). The widespread acceptance of, and the power behind, this perception of reality, grants it a stature approaching truth.

However, ethnographic research in the community reveals another reality. A.M. Colfer and I have termed this other reality *Local* in contrast to the *Public Employee* reality described above. Our recognition of these two systems is, in effect, support for the Local viewpoint. Such support is reasonable in light of the evidence that groups of people of lower status (from the dominant perspective) recognize the existence of various systems, whereas higher status people often do not have access to knowledge about muted groups. The specific behavioral and cognitive (in the sense of world view) differences between Locals and Public Employees are discussed at length in Colfer with Colfer (1978) and Colfer and Colfer (1979).

An interesting point here is the response of community members to an earlier draft of the above-mentioned 1979 report, regarding their community. Public Employees, without exception, felt the division within the community had been exaggerated ("There is diversity, but not division"). Locals were particularly supportive of the chapters which dealt with the "fact" of two systems at work in Bushler Bay. The existence of coherent (and legitimate) subsystems is logically incompatible with the universalistic world view of Public Employees. Locals must be able to operate in both systems.

To this point, we have neglected the question of women and women's models of reality (or lack thereof). Indeed, the Local–Public Employee split in Bushler Bay derives primarily from male occupation. Women grow up with the awareness that the course of their lives will depend in large part on the man they choose to marry. The teenage daughter of a Local family not infrequently marries the teenage son of a Public Employee family; and vice versa. A woman's cognitive system, therefore, must be sufficiently mutable or permeable that she can incorporate the new system and learn to function as part of it.

Maria, a local woman, was the mother of three preschool children. She was industrious, intelligent, attractive, and determined to make this (her second) marriage work. She was also cohesively integrated in the group of Local female age-mates who dominated Bushler Bay's Preschool Co-op and Preschool PTA. At the meetings of these all-female organizations, Maria was vocal, inventive, industrious and cooperative. Female leadership in Bushler Bay is diffuse and somewhat difficult to identify, but Maria's ideas and contributions were generally received with respect, and she was well-liked.

At a meeting or a child's birthday party (at which women traditionally gather to feast and frolic together), the inadvertent or unavoidable arrival of a man on the scene would immediately quell the conversation. The laughing and animated faces would be composed into shy smiles and downturned eyes. Some brave soul might venture a conversational inanity directed at the intruder, but generally all breathed a sigh of relief when he left.

Some of the talk among women revolved around their relations with their husbands. The dominant Bushler Bay model of male–female interaction includes an acceptance of male dominance. Among Locals, particularly, male dominance is overt and much displayed. When women are alone together they often share the techniques they have used to avoid doing what their husbands told them to do, or how they have managed to accomplish something their husbands have prohibited. Like Arab women (Fernea, 1965) and Japanese women (Salamon, 1975), Bushler Bay Local women and men limit interaction; they lead largely sex-segregated lives. And the behavior of women in the presence of men is markedly different from the behavior of women in the presence of other women—as different, I would venture, as Mukhtar's behavior in the city and in the village, respectively.

When Maria is with her husband, she is, for the most part, submissive, respectful, ladylike. She defers to his wishes and accepts his interpretations, as well as fetching cigarettes and coffee whenever she is so instructed. Because of her particular wish to make this marriage work, she conforms a bit more closely to his dictates than do most of her friends. Her freely expressed desire for the financial security he can offer is such that she is willing to participate more fully in the dominant male model (which gives her lower status and less autonomy) than are her friends. It is also worth noting that she has three children, and her friends all have one or two. One could postulate that the difficulty (and poverty) of single motherhood in the rural U.S. increases exponentially with the number of children one has.

It should be becoming clear that these women must invariably learn to operate in two systems, and often in more than two systems. The lower status person (as defined by the dominant reality—a Local woman in Bushler Bay—learns the system of behavior for Local women, and in order to function in daily contact with a more powerful person (her Local husband), she learns the system of behavior and belief for Local males. She then is thrown into contact with Public Employee women who manifest different behaviors and beliefs and who are granted greater status in the dominant American system. The Local Bushler Bay woman can (and does) limit her contact with Public Employee women, but she cannot avoid contact altogether. She has to develop some strategy for dealing with this—in some sense—higher status person. Maria, like many Local women, tends to adopt a shy, closed, blank expression, averting her eyes if possible, discouraging further interaction, waiting for the Public Employee woman to go away. In my early days in Bushler Bay (as an educated newcomer of ambiguous status), I was frequently treated to this form of interaction with Local women. Although contact with Public employee males is less frequent, it occasionally occurs, and this shy behavior is intensified.

In the most general terms it seems that insofar as a low status person interacts regularly with people of higher status, low status people must understand and be able to operate in the system accepted by the high status people. If regular and integrated interaction occurs, the cognitive situation of the lower status woman could be represented thusly:

$$♀a_2 \geq (♀a_1)\,(♂a_{1,2})\,(♀b_{1,2})\,(♂b_{1,2}) + \ldots (♀n_{1,2}) + (♂n_{1,2})$$

On the other hand, if she can segregate her interactions, the relationship could be expressed thusly:

$$♀a_2 \geq (♀a_1) + (♂a_{1,2}) + (♀b_{1,2}) + (♂b_{1,2}) + \ldots (♀n_{1,2}) + (♂n_{1,2})$$

In the above formulae, *a, b,* and *n* refer to variant models; the subscripts *1* and *2* refer to the behavioral and cognitive aspects of the models respectively; and ♀ and ♂ refer to the sex of the actor. Each successive model in the formula (in this case, *a, b,* through *n*) is the next superordinate social group's model, such that $n_{1,2}$ is the highest, or most dominant, model extant.

If this relationship holds, an obvious conclusion of this is that, *the lower the status of a person, the greater the cognitive complexity required of that person to function adequately in situations of culture (or system) contact.* It is important to recognize that this is not the same as saying that lower status people cannot function adequately because they cannot handle the cognitive complexity. (I am grateful to Eastman [1979] for reminding me of this possible interpretation.) This latter interpretation implies genuine inferiority among members of the muted groups; whereas my conclusion is that members of the muted groups have to cope with a more complex set of data (several separate models of reality) than do members of dominant groups.

In Colfer (1974), I summarized the literature on limits to cognitive complexity, and analyzed ethnographic data on a Bureau of Indian Affairs residential school in terms of the cognitive complexity required of Native American teenage girls as they strove to function in the white school system. That analysis likewise lends support to the more general formulation proposed here.

Another dimension of this is the tendency (perhaps necessity) for lower status people (as defined by the dominant view of reality), whenever they interact with people of higher status, always to *react*; to be alert to and knowledgeable of the reality of the situation as perceived by the higher status person, and to act accordingly *in response*.

On scientists interacting

To this point, I have been discussing interaction between rural people and some person or group with a higher status and more power in a wider sphere. I would now like to extend this discussion to the sorts of cross-cultural contact that occur between the elites of various nations. To do this, I will draw on observations made at a research center (alias R.C.) in the U.S. Since these elites also happen to be scientists, perhaps we can gain some insight into the notion of scientific imperialism. I hope the fact that processes similar to those described below occur regularly in negotiations about the uses to which foreign aid can be put, will not escape the reader's notice.

My participation in the seminar in the following discussion was made possible by an award from the research center in question. Seminar X was composed of 33 participants (including R.C. staff members), 10 of whom were non-American. Racially 6 of the 33 could be considered non-Caucasian. Seven of the participants were female. The seminar lasted for a week, and involved the formal presentation of academic papers with brief discussion periods after most papers.

The discussion periods were dominated by a group of seven American men of middle age. These men had all been participants in an experimental, computerized communications system, and had thereby had the opportunity to develop a set of shared assumptions, approaches, and common vocabulary which set them apart from the other participants. My estimate, as a trained and experienced observer, is that their internal, though public communication accounted for three-fourths to four-fifths of the speaking turns and speaking time, during discussion periods.

Now I would maintain, congruent with the material presented in the Iranian and American examples, that these men were able to dominate the proceedings so effectively because of (1) the higher status of the scientific paradigm they were representing (manipulations of sophisticated, mathematical models), (2) their higher status as representatives of "America the Powerful," and (3) their higher status as middle aged, white males. The facts that they had established a shared perspective and that they all spoke English as a mother tongue were not irrelevant to this dominance. Superlan (1979), another seminar participant, notes,

> these men . . . have had previous contacts—and thus they have more or less formed an organized group through their using of similar models of thought and symbols during the discussions in the conference. In their relationships to other participants, they are not acting as individuals but as a group. While on the contrary, other participants who are new ones in the conference—and thus . . . are dispersed and [acting] as individuals in the discussions. As individuals, they are over-powered by the group in their play in the arena of social interactions.

I would, however, also venture at this point, that their overall high status, as elite, middle aged, white, male citizens of a very powerful nation allows them the comparative luxury to operate predominantly *within one cognitive system*. As powerful, high status beings, they have no need to recognize their ignorance of alternative realities (and in any event the data they would need for such awareness is not available to them, under normal circumstances). Their models of reality (e.g., ♂x_2) need only encompass the complexity necessary for ♂x_1, which in this case is equal, or very close to ♂n_1. Or using Einstein's analogy, they need never leave their position as ball bouncer, to see the arcs made by someone else bouncing the ball.

Returning to Seminar X, another interesting phenomenon is worthy of note. Despite widespread recognition of certain conceptual problems with the models offered by these men (particularly as regarded links to data), the only challenges issued were by other members of the dominant group. These challenges were not epistemological in nature, but rather questions of techniques and other details well within an established, shared, dominant scientific paradigm. Several factors seemed important to me in this regard. First, those whose views differed from the dominant one (ascertained principally in informal, more private discussions outside the formal seminar), were either female, Third World, or scientifically committed to more Pythagorean sorts of analysis. The dangers of challenging a dominant model derive, as noted earlier, from differential access to power by members of the dominant and muted groups. Specific risks in this case perceived by some participants included the possibility of not being invited to R.C. again, and the possibility that one's scholarly reputation might be tarnished by being branded part of the lunatic fringe.

Second, my analysis suggests that a dominant viewpoint, particularly one as close to ♂n_2 as the one presented at the seminar, is integrated, cohesive and easy to defend. In the Ardeners' terms, it is bounded. The more complex models of lower status persons, which must incorporate divergent realities and behaviors, are less easily enunciated and defended, particularly off the cuff, and in a public forum. The cohesiveness of the dominant view seemed effectively unassailable, though wrong, completely partial. From the perspective of a member of a muted group, the paradigm that dominated ignored important empirical factors; yet the confidence of the dominant group in the logic and consistency of their model of reality combined with their aforementioned sources of personal power, made effective communication or real dialog impossible.

The final factor is related to the particular, culture-specific interaction style of many eastern Americans. The brash, aggressive, confrontive style that forms many people's stereotypes of Americans would seem, conceivably, to be reinforced by these very factors. Surely a cohesive and internally consistent set of assumptions about the world, combined with a reticence on the part of lower status persons to challenge such assumptions, would tend to reinforce a sense of superiority and self-confidence.

In interaction with people who behave in this brash, aggressive manner, one can be silent (as most seminar participants were). Or, one can enter into the fray, adopting those very characteristics oneself. Silence reinforces the self-confidence of the dominant group and encourages the maintenance of the dominant, incomplete yet cohesive world view. (A major part of this incompleteness derives from the dominant group's lack of access to counterpart models, reflective of the life circumstances of muted groups.) Yet adoption of such an interaction style is, to many, abhorrent. This dilemma effectively precluded meaningful participation by members of muted groups in the discussion periods, and denied the members of the dominant group access to valuable critiques and substantive input.

As with the rest of the seminar, the concluding remarks were generally made by these seven gentlemen. One eulogized the seminar, judging it the most valuable he had ever attended. He said he particularly appreciated the chance to hear from people from other countries—he seemed genuinely unaware of the irony of his remarks. I see this as an indication of the near-impenetrability of the dominant reality (perhaps therein lies its strength).

Another, in apparent goodwill, eagerly picked up on someone's suggestion that the ties established at the seminar should be maintained. He turned to an Asian scholar and said, "Yes, by all means, send me your proposals. I'll be glad to critique them for you." He failed to recognize the asymmetry implicit in his offer.

Conclusion

In this paper I have begun exploring some of the implications and ramifications of cross-cultural, cross-disciplinary, and cross-paradigmatic interaction (cf. Maruyama, 1973, for more on this topic), in situations where clear status differences exist. I have tried to demonstrate some parallels in communication between rural peoples and urban elites, between females and males, and between adherents of "soft" and "hard(er)" scientific disciplines. These generalizations are offered here, because of the ubiquity of attempts to communicate across status lines in planned change efforts.

The conclusions drawn in this paper are essentially as follows:

1 That differential power and status tend to interfere with the free expression of ideas from lower status people;
2 That the inhibition of such expression reinforces the integrated but incomplete world views held by the powerful, by denying them access to alternative perspectives and to information that does not fit neatly into their cognitive models; and
3 That one implication of the necessity of lower status peoples to adapt to the dominant models at the same time as they retain their counterpart models, is that greater cognitive complexity may be required of them for adequate life functioning than is required of elites.

Since lower status people are dependent on the good will of the powerful, in many areas of life, members of the subdominant groups do not, realistically, have the option of taking the risk of alienating the powerful on any large scale. An important responsibility—if improved communication is to result—then, lies squarely with the elites, to create a supportive environment which encourages the expression of counterpart models. I see such improved communications as critical to the success of development programs (see also Vajrathon [1976]; and Clark [1979]. McCaghy, Skipper and Lefton (1968) make a similar point:

> It is our contention that there is a need to listen more carefully to those who speak from the margin and try to understand their own perspectives and explanations of their behavior, unhindered (as much as possible) by preconceived albeit sophisticated models or frames of reference.
>
> (p. vi)

Although clearly many dominant people have no motivation to create such a supportive environment, or to encourage or learn other perspectives, many others do have such a motivation and are unaware of their complicity in muting the expression of counterpart models.

Policy recommendations

I have developed the following list of policy recommendations, congruent with the above analysis and designed to alleviate some of the communicational difficulties which limit the effectiveness of development efforts, and which contribute to Third World perceptions of imperialism in most foreign aid attempts. (These were developed at the suggestion of Dr. A. Manoharan.)

1. In training programs for extension agents, researchers, planners, and other development personnel, we need to incorporate components addressing these barriers to accurate appraisal of the perspectives of such rural people and other would-be participants in development projects. Such components could very

nicely complement the current interest in "action research" and other participatory approaches to research and development (e.g., Bajracharya 1982, 1983; Morse et al. 1983; Shaner, Philipp and Schmehl 1982; and others).

2. We need to develop and utilize a cadre of liaison personnel, skilled or talented in communicating across these existing status lines. Some people have a natural sensitivity to such considerations; anthropologists are trained to overcome such barriers and gain experience in the course of ethnographic fieldwork; people who have straddled two cultures often possess skills of this kind (the East-West Center Culture Learning Institute has a series of analyses related to this latter topic; see Bajracharya [1982] for a recent discussion of such a function, based on personal experience).

3. We should decentralize problem identification and planning responsibilities to minimize the *necessity* to communicate complex systems with largely local relevance to central office personnel who have much higher *status than members* of the populations to be served, and who in most cases will have inadequate understanding of local lifeways.

4. An effort should be made to minimize status differences between members of local populations and development personnel. Two tacks can be taken in this attempt. First, we can try to minimize interaction between groups of people who are defined as unequal (for instance, by hiring "change agents" who are as like the population to be served as possible, or by encouraging communication and technical assistance *among* Third World nations). This is consistent with the practical implications of Rogers' (1963) work. Or, second, we can try to eradicate or minimize people's acceptance of the legitimacy of status differences. Although this may appear to be utopian, I personally see this as a more viable solution, in the long run. Many development personnel are beginning to become convinced—some grudgingly—of the importance of local knowledge of local conditions (environmental, social structural, agricultural, etc.). Elite awareness of the utility of such knowledge can form a base for the respect for "disadvantaged" people that I would maintain is a critical cornerstone for development "with a human face."

Note

1 This research was funded by the National Institute of Education via a contract with Abt Associates. Inc., who employed the author. Requests for reprints should be sent to Dr. Carol J. Pierce Colfer, People who have significantly influenced my thinking, as reflected in this paper, include Dr. A.M. Golfer, Dr. C. Eastman, Dr. D.L. Kincaid, Dr. Sarah King, Dr. A. Manoharan, Dr. Parsudi Suparlan. Dr. A.P. Vayda, and Dr. Robert Wolfe. The innocence of all of the above, as regards any remaining errors in this analysis, of course is complete. *By "unequals" I refer to categories of people with differential access to resources and power; *not* to any fundamental or inherent patterned inequality. In other words, I do not imply that women, rural peoples, or Pythagorean analysts *are* inferior.

References

Ardener, E. Belief and the problem of women. In S. Ardener. Ed., *Perceiving women*. London: Malaby Press, 1975.

Ardener, S. Introduction. In S. Ardener. Ed., *Perceiving women*. London: Malaby Press, 1975.

Bajracharya, D. Energy for rural enhancement. *The Rising Nepal*. November 21. (Also available as a longer, working paper: Experiences as a *Lami:* Energy for rural enhancement in Gorkha, East-West Resource Systems Institute, Honolulu, Hawaii, 1982.)

Bajracharya, D. Rural organization for energy assessment and development. In Morse, Islam and Soesastro, *Rural energy to meet development needs: Issues and methods*. Denver: Westview Press (in press), 1983.

Bateson, G. *Steps to an ecology of mind*. New York: Ballantine Books, 1972.

Chambers, R. Simple is sophisticated. *Development Forum* VII(6): 1–2 July, 1978.

Clark, N. *Education for development and the rural woman*. New York: World Education, 1979.

Cohen, J. The shaping of men's minds: Adaptations to the imperatives of culture. In Wax, Diamond, and Gearing, Eds., *Anthropological Perspectives on Education*. New York: Basic Books, pp. 19–50, 1971.

Colfer, A.M., & Colfer C.J.P. Becoming American: Life and learning in a rural school. Final report to the National Institute of Education (June 15). Cambridge: Abt Associates (Contract No. OEC-0-72-5245), 1979.

Colfer, C.J.P. High roads bypass the poor. *Development Forum* VI (December) 1977a.

Colfer, C.J.P. Women's communication and family planning in rural America: The case of Bushler Bay. Honolulu: East-West Communication Institute (No. 4), reprinted 1978, 1977.

Colfer, C.J.P. Integrating anthropology into public health. *Pacific Health* 12: 16–18, 1979.

Colfer, C.J.P., & Colfer, A.M. Inside Bushler Bay: Lifeways in Counterpoint. *Rural Sociology* 43(2): 204–220, 1978.

Eastman, C. Critique of earlier draft of this paper (6/79), 1979.

Fernea. E. *Guests of the sheik. An ethnography of an Iraqui village*. Garden City, NY: Anchor Books, 1969.

Goodenough, W. *Cooperation in change*. New York: Russell Sage Foundation, 1963.

Kaberry, P. & Goodale, J. *Aboriginal women. Sacred and profane,* and *Tiwi Wives*. London: G. Routledge and sons, 1939.

Hardman. In S. Ardener, Ed., *Perceiving women*. London: Malaby Press, 1975.

Kelly, G. *A theory of personality: The psychology of personal constructs*. New York: Norton and Co., 1963.

Krivonos, P. *Distortion of subordinate to superior communication in organizational settings*. Paper delivered at the International Communication Association meetings. Portland, Oregon, April 1976.

Lakoff, R. *Language and woman's place*. New York: Harper Colophon Books, 1975.

Leavitt, R., Sykes, B., & Weatherford, E. Aboriginal woman: Male and female in anthropological perspective. In Rayna Reiter (Ed.), *Toward an anthropology of women*. New York: Monthly Review Press, 1975.

Maruyama, M. Paradigmatology and its application to cross-disciplinary, cross-professional, and cross-cultural communication. (Proofs from a paper delivered at the International Congress of Ethnological Sciences, Chicago, 1973.)

MacCormack, C. Biological events and cultural control. *Signs* (3(1): 93–100, 1977.
McCaghy, C.H., Skipper, J.K., & Lefton, M. *In their own behalf: Voices from the margins*. New York: Appleton-Century-Croft, 1968.
Moore, S. Epilogue: Uncertainties in situations, indeterminacies in culture. In S.F. Moore & B.G. Meyerhoff (Eds.), *Symbol and politics in communal ideology: Cases and questions*. Ithaca: Cornell University Press, 1975, pp. 210–239.
Morse, R., Islam, N., & Soeastro, H. *Rural energy to meet development needs: Issues and methods*. Denver: Westview Press (in press), 1983.
Murphy, Y., & Murphy, R. *Women of the forest*. New York: Columbia University Press, 1974.
Okely, J. Gypsy women: Models in conflict. In S. Ardener (Ed.), *Perceiving women*. London: Malaby Press, 1975.
Ortner, S. Is female to male as nature is to culture? In M. Rosaldo & L. Lamphere (Eds.), *Woman, culture, and society*. Stanford: Stanford University Press, 1974.
Paul, B. *Health, culture and community*. New York: Russell Sage, 1955.
Reiter, R. *Toward an anthropology of women*. New York: Monthly Review Press, 1975.
Rogers, E. *Communications strategies for family planning*. New York: The Free Press, 1973.
Rosaldo, M., & Lamphere, L. (Eds.), *Woman, culture and society*. Stanford: Stanford University Press, 1974.
Salmon, S. The varied groups of Japanese and German housewives. *Japan Interpreter* 10(2): 151–170, 1975.
Saral, T. *Consciousness theory and intercultural communication*. Paper presented at the International Communication Association meetings. Portland, Oregon, April 1976.
Shaner, W., Philipp. P., & Schmehl. W. *Farming systems research and development: Guidelines for developing countries*. Denver: Westview Press, 1982.
Suparlan, P. Critique of earlier draft of this paper (7/79), 1979.
Tinker, I., & Bramsen. M. (Eds.), *Women and world development*. Washington, DC: Overseas Development Council, 1976.
Turner, V. Process, system and symbol: A new anthropological synthesis. *Daedalus* (summer), pp. 61–80, 1977.
Varjrathon, M. Toward liberating women: A communications perspective. In I. Tinker & M. Bramsen (Eds.), *Women and world development*. Washington, DC: Overseas Development Council, 1976.
Vayda, A.P., Colfer, C.J.P, & Brotokusumo, M. Interactions between people and forests in East Kalimantan, *The Impact of Science on Society* 30(3): 179–190 (reprinted by EAPI, EWC, Honolulu. No. 13, 1982), 1980.
Wolfe, R. *Realities: Health and culture*. Presented at the Thursday Symposium. School of Public Health, University of Hawaii, April 26, 1979.
Women and National Development. [Special issue] *Signs*. 3(1), 1977.

5 Discordant connections

Discourses on gender and grassroots activism in two forest communities in India and Sweden

Seema Arora-Jonsson[1]

Winner of the 2009 Catharine Stimpson Prize

In 2005 Wangari Maathai was awarded the Nobel Peace Prize, and the issue of women's rights in environmental matters appeared briefly on the international agenda. In a joint article, Maathai and Lena Sommestad, both active in the network of female environmental ministers, emphasized that women's interests must be taken into account when environmental policies are formulated (Sommestad and Maathai 2005). They position equality between men and women as a crucial development question—necessary for effective and sustainable development, especially in poorer countries.

The importance of gender equality and of the relationship between third-world women and the environment is evident. Development and a certain standard of welfare make these issues appear to be less urgent in a wealthier country such as Sweden. However, my research with women's groups in forest communities in India and Sweden showed otherwise. First, questions of gender and power in environmental management are relevant not only in a poorer country such as India but also in a richer country such as Sweden. In the latter they can take forms that make gender discrimination more difficult to contest. Second, development discourses about equality and empowerment of oppressed third-world women not only affect how gender equality is conceptualized and practiced in the global South but also shape the possibilities for gender equality in the North. Understanding how this takes place opens an opportunity for interruption in an order and in a space (a global/social order and a developed and gender-equal space) that appears to have become narrower under the umbrella of development, welfare, and growth. It brings into question the category of development not only in a Southern but also in a Northern context, where the North, especially Sweden, is taken as a referent for development and gender equality.

In this article, I trace some of the contradictions and connections in the ways in which gender equality and women's empowerment are conceptualized in women's struggles in villages in the Nayagarh district in Orissa, India, and in the village of Drevdagen in western Sweden. Examples of women's grassroots activism from the two case studies give material form to abstract discussions about the possibilities for women's agency in different cultural settings. In both study sites, women in villages in rural and peripheral areas (in relation to policy- and decision-making centers) formed organizations parallel to the male-dominated organizations in their villages. In absolute terms, the women in Sweden were far better off: in health care, wealth, availability of food, choice to work and marry, and geographical mobility. Yet their meetings as women were hedged with ambiguity and sparked resistance in the village. Power and discrimination were veiled and subtle in Sweden. Paradoxically, the rhetoric of gender equality that is pervasive in Sweden serves to mask forms of subordination and makes it difficult to question the purported neutrality of given structures. The Indian women were more vocal about discrimination against them. Outside intervention in gender issues was acceptable in the Indian development context but was regarded as interference in what were considered to be personal relations in the case in Sweden. Gender and environmental relations were conceptualized in ways that were specific to each context but also carried the impress of outside forces. Strong normative assumptions about development, gender equality, and empowerment resonated with one another in both places. A complex movement of ideas at the global level—about what it means to be developed, rural or urban, empowered, and independent—found expression in the everyday practice of development in both places.

A relational analysis

Third-world feminists and others have challenged universal conceptions of gender. They emphasize that understanding the different ways in which meanings are produced and challenged reveals the complex and specific political choices that caution us against ahistorical and universalizing categories (see Mohanty 2003). Gender is a historically and culturally variable category. Yet, as my studies show, overlapping discursive contexts and recurring practices in the politics of everyday life also reveal links between distant places, links that in turn pose new questions for the study of gender and power in an interconnected world. Ideas about gender and gender equality are formed by various local material practices, but as the two cases indicate, they are also a hybrid of different influences. Despite shifting meanings and contextual relations, the practices of gender equality in particular places echo one another, are marked by the histories and relationships of power that structure the world, and are linked to ideas about modernity and development. According to Cindi Katz, situated knowledge assumes knowledge at a single point, the knowing subject. This, she argues, has tended to facilitate a collapse of dimensionality. She calls instead for a topography to elucidate the intersection of processes with others elsewhere and thereby inspire a different kind of politics, one in which crossing space and "jumping scale" are obligatory rather than overlooked (2001, 1230–31).

Crossing space by reading stories from India and Sweden made it possible for me to develop a relational analysis of the particular places. The methodology I employed to study different analytic frameworks and different geographies included three elements: freezing time, reversing the gaze from North to South, and embracing critical subjectivity (Arora-Jonsson 2005, 60–61). I scrutinized language and actions as indications of underlying structures of meaning, indications that tell us not just about the person speaking and acting but about a wider discursive context that goes beyond the micropolitics of the villages. I studied not only collective action but also how individual women define their subjective positions within a collective. I used the case study from India as a frame of reference rather than accepting conventional assumptions of gender equality where Northern principles are the reference points. As a researcher working in both the global North and the global South, I became aware of a global discourse on gender and the need for reflexivity in situations where I am both an academic and an active participant.

When one is discussing gender, equality, or empowerment, the challenge is to look beyond macrogeneralizations while also avoiding a fixation on difference between the places. By arguing that relationships of power can determine the flows of ideas on equality, I do not imply that the North or the South is symbolically and ideologically fixed or that there is a homogeneous body of ideas. However, there are dominant ways of thinking about development in which the West is a referent (cf. Mohanty 2003). Nor do the connections between the two places imply a one-way flow. Conversations between them have unexpected effects. The relational analysis calls into question prevailing metaphors and categorical divides while keeping in sight the relationships of power that organize the world. It has been possible to locate each case in its context yet carry out a dialogue between the two across what might be regarded as the development divide.

People in the two locations in India and Sweden were linked by the fact that both were affected by similar discussions on gender and development, although their experiences were not the same. This can be seen in Western influence in policy making on gender in India. But a colonial discourse on oppressed women is also palpable in how Sweden positions itself in terms of gender equality, as unique in its relations to others. Though Sweden was never a colonial power itself, the contemporary effects of colonization and of immigration to Sweden clearly reflect thinking about the other (de los Reyes, Molina, and Mulinari 2002). Women in Nayagarh, India, gave substance to wider social debates about women and development in unanticipated ways as they took up issues of discrimination and argued for their cause. In Drevdagen, Sweden, the categories of development and gender equality were used to stall change and disregard inequalities.

In the following sections, I move between different geographical levels as I analyze the two places. The first section is a discussion of public policy making and dominant discourses on gender equality and empowerment that framed the women's activism. I study how national discourses and policies on gender, equality, and empowerment played an important part in forming women's subjectivities and modes of identification. In the second section I provide background on the two places and the women's organizing. In the third, I go on to examine how beliefs about gender, equality, and empowerment are given meaning in the material and discursive practices of the men and women in both study sites. In the fourth and last sections, I analyze how these beliefs influenced dimensions of difference in conceptualizing collective and individual agency and in what counted as gender equality or empowerment in the two contexts, and I trace the connections between the two places.

Jämställdhet in Sweden and empowerment in India

Striking in discussions about gender in Sweden is how male dominance and the different treatment of women and men have become illegitimate as a basic social principle, both in state policies and in wider discourses in society. Discourses of the new fatherhood and gender equality are culturally dominant today, regardless of actual practice (Plantin, Månsson, and Kearney 2000). "Contemporary mainstream researchers, decision makers and journalists often describe the transformation of Sweden following the introduction of general suffrage as having created a society which is both more egalitarian and more women-friendly than most others. Public policies aim at making it possible to achieve gender neutrality, defined as equal opportunities for women and men in the labour market, the family and political life" (Gustafsson, Eduards, and Rönnblom 1997, 42). Sweden's path to gender equality has been through the labor market, and women's presence in the labor force has been an accomplishment. Gender-neutral policies were so effectively promoted as the way to achieve gender equality that gender equality has become conflated with neutrality. The state is seen as the main source of economic and moral support for gender equality through its welfare policies and through the public-sector labor market, which employs large numbers of women.

A sense of uniqueness, of having come far in questions of gender equality, permeates Sweden. The subtle yet significant presence of the not-quite-developed South is evident in how gender equality is defined in policy but also in the subjectivities of women in the Swedish village I worked with, as I demonstrate below. The following lines taken from the Swedish government memorandum on gender equality epitomize this thinking:

> We in Sweden have come a long way compared with other nations, yes, in fact, we are far ahead of the rest of the world. We are glad to share our experiences; we are glad to export our Swedish model for gender equality. But this, our first place, should not lead us to think that we are done. As yet there is a lot to be done in many areas. . . .

> We all have a responsibility to ensure gender equality. All the ministers, all the parliamentarians, and Swedish citizens in general must feel like Sweden's gender equality ministers.
>
> (Integrations och Jämställdhets Departmentet 1999, 6)[1]

The nationalistic tone in dominant policy and research discourses on Sweden's exceptional record on gender equality has led to a line of feminist argument that claims that the emphasis on gender equality has shifted attention from the real problem, that is, discrimination against women. The word for gender equality, *jämställdhet*, that gained currency in the 1990s is considered problematic by several feminists. The basic reason, argues Malin Rönnblom, is that a word symbolizing a vision is being used to name a problem (2002, 213). The term is infused with positive connotations and with a focus on goals and aspirations without naming the group that is disadvantaged. This consensus term (Tollin 2000) is used to discuss and explain power relations between women and men and is linked to democracy and justice. Since it refers to a vision, there is a great deal of variance in how the term is interpreted in practice. To speak of injustice in society by linking power and gender becomes difficult when the official word available is a description of a political ideal and one that emphasizes harmonious interdependence. Yet sexual difference is implicit in political discourse, not least in the framing of *jämställdhet*, which, in contrast to *jämlikhet* (equality), is a term used specifically for equality between men and women. Thus, a discourse is created where the rhetoric of equality and assumptions about political decision making as a gender-neutral activity conceal the existence of a gendered political order and the continued subordination of women (see Gustafsson, Eduards, and Rönnblom 1997).

Problematizing this further, postcolonial feminist writers in Sweden maintain that *jämställdhet* has been used to distinguish Swedes from immigrant populations and has been established as a basic part of the Swedish self-image, in relation to the rest of the world as well as to the immigrant populations (de los Reyes, Molina, and Mulinari 2002, 306).[2] According to Diana Mulinari and Anders Neergard, Sweden has long been characterized by a form of welfare state nationalism that is based on a "we-pride" (2004, 210) in contrast to a world that is created as irrational, chaotic, and filled with conflict. They write that the image of the generous and tolerant Swedish identity has been weakened in the past twenty years in the context of the shrinking welfare state and through events such as the Palme murder, the Göteborg fire, and Nazi murders that have acquired symbolic significance.[3] Mulinari and Neergard (2004) write that against this background, *jämställdhet*, or gender equality between men and women, is the only successful cultural product (de los Reyes, Molina, and Mulinari 2002) that can be used as an ethnic marker against those who are constructed as "the other."

A diffracted image of gender equality in Sweden appears in the Indian national context. Here gender is "an 'issue' in another sense of the word: a crisis, a problem, a scandal" (Sunder Rajan 1999, 2). Rajeswari Sunder Rajan traces this conceptualization to the milestone report on gender equality published in 1974 (*Towards Equality*, produced by the government of India) that pointed to the skewed demographics of the female/male population ratio, the decline in women's workforce participation levels and in levels of literacy and health, all of which showed how poorly women were faring in independent India. Gender gaps in development, employment, nutrition, land distribution, and inheritance became impossible to overlook. Thus, across the board in government concerns, social movements, discourses, disciplines, and sites of action, gender began to figure as an issue as well as a category of analysis (Sunder Rajan 1999). Debates on gender have meshed directly with and have actively constituted prevailing conceptions of India's national identity, conceptions that include the reconfigured primacy accorded to development (John 1999, 110). The fate of women became linked to mainstream development agendas: "Development experts cite 'gender bias' as the cause of poverty in the 'Third World'; population planners declare their commitment to the empowerment of Indian women; economists speak of the feminisation of the Indian labour force. . . . There is a sense, therefore, in which the new visibility is an index of the success of the women's movement. But clearly this success is also problematic" (Tharu and Niranjana 1999, 494).

Increasingly, women are represented as efficient workers and economic subjects, reflecting a national discourse and international influences. Nevertheless, policies and programs still treat women as needy, and images of victimhood and incapacity persist. The help provided to women, such as family planning programs or sanitation, is in keeping with stereotypical ideas about women's roles. However, there have been unexpected exceptions. Awareness-raising programs have led women to organize against oppressive state organs, resulting in "a peculiar situation of the state sponsoring women's struggles against itself" (Lingam 2002, 317).

Outside influences on how gender is construed can be felt more tangibly in India than in Sweden, reflecting an overt power difference between the North and South. International donor agencies and nongovernmental organizations are important in shaping policies and development activities. Women's marginalization from centers of power is widely discussed, not least because the large amount of development literature has helped to create a space for so-called women's issues. International feminist networks have been able to influence where and how various forms of development aid are channeled. Northern aid agencies often demand a gender perspective or gender component to their programs in the South. Although the validity and the effects vary, they often succeed in treating women as a special category. This has led to the co-option of such agendas by bureaucracies. Gender becomes a technocratic measure, resulting in its depoliticization as it is turned into a matter of monitoring and planning rather than struggle (Baden and Goetz 1998). Empowerment through self-help groups or support for women's groups has often been bureaucratized and interpreted by government and nongovernmental organizations or aid agencies in a simplistic manner. Underscoring these measures is an assumption of inequality and of an obvious male dominance that characterizes much of the discussion on gender. Formal structures such as administrative arms of the government, forest committees, or village organizations evidently lack women and are not necessarily seen to represent women as a group.

Images of the gender-neutral and gender-equal Sweden, as opposed to the patriarchal nature of the state and society in India, have played out in various ways in women's organizing in both Nayagarh, India, and Drevdagen, Sweden, in their respective modes of self-identification as well as in the actions taken by women in both sites. When seen in relation to each other, the two cases make clear women's possibilities for action in different national and discursive settings.

A house of dreams and many small threads

Drevdagen is situated in the sparsely populated region of western Sweden, with modest service provisions and few avenues for employment. The village is ethnically homogeneous with no major differences in wealth and education among the inhabitants. Since the 1950s, the migration of people to urban centers has been a cause for concern. To counter these trends, men and women in Drevdagen formed a village association in 1995. The women took up what they called the social issues: the village shop, the school, tourist initiatives. The men spearheaded the struggle to get rights from the state to manage surrounding forests to generate employment. This struggle for the forests brought them in touch with some male colleagues at my university. My colleagues' interest in the forests led the men heading the association to focus on the forests to the neglect of the issues that the women had taken up. The women also found it difficult to make themselves heard on questions concerning the forests. This led most women to lose interest in the association. When I first visited the village in 1998, my desire to conduct a participatory inquiry with the women on issues of development and local forest management coincided with their need to meet and discuss village development and the forests on their own terms. They transformed the space of the inquiry to create what they came to call the *kvinnoforum,* a women's forum. The *kvinnoforum* became a space where women from their village and beyond met, socialized, supported one another in their projects, and worked for what they called a "living countryside." We met in an abandoned building where the village shop used to be and that the women had renamed the "house of dreams," a space from which they would make their dreams come true. The forum and my research on the process as a participant member led to resistance from the men in the village association and from my colleagues at the university.[4] They felt that the women's organizing disrupted gender harmony in the village. My colleagues pointed out to me that this was not the global South and that gender was not a problem here in the same way that it is there. In their opinion, interference in their forest project and my writing about what they considered to be criticism by a few disgruntled women could jeopardize their chances to effect larger forest policy changes.[5]

Such tensions shadowed the women's organizing and become more obvious when seen in relation to a similar process in Nayagarh, India. The women in Nayagarh were organizing themselves in microcredit groups as part of a women's development program when I came to the villages in 1998.[6] Since the 1980s, a community forestry movement had spread across the entire district. The villages had been protecting forests, and this had resulted in considerable forest cover. Women had been active participants in the movement, especially since the male leaders advocated what they called "total development," which included social reform, and greater involvement of women.[7] As the movement took the shape of a formalized network of forest organizations, it began to get a small amount of funding from Oxfam. The number of forest protection networks expanded, and they together formed the Nayagarh forest federation. Women, who had never formally been part of the decision-making groups, became conspicuously absent. The forestry networks came to be dominated by elderly men from the general castes.[8] Oxfam began to provide funds for a women's development program to bring women into the forest organizations. The training for this took the shape of classes on tailoring, family planning, and sanitation and eventually, in 1994, in assistance in setting up women's groups, *mahila samitis*, to carry out a microcredit program.

While some *mahila samitis* were groups on paper only, others became informal credit centers. Some went beyond the agenda of the program and transformed the collectives into a space from which they could take action on behalf of the village and on behalf of themselves as women. They challenged how funds were spent and how decisions were made by the forest organizations. Like the Swedish women, who also needed a physical place to call their own, one *mahila samiti* set up an herb garden that was their own space materially as well as symbolically. It was the only place where the women could gather and discuss issues outside and not in someone's home. In Nayagarh, the women's groups had to navigate a more heterogeneous terrain. Differences in age were complicated by caste and class differences. The groups were normally organized along lines of caste and also according to age. Despite these differences, the women were conscious of the strength they derived from the existence of other such groups. As one woman put it, "small, small threads make a big piece of cloth . . . small tools make a big factory."

Women from a number of *mahila samitis* in neighboring villages related how they organized themselves and forced several landowners to part with land needed to build a road to the village. The women lay down in a public space on a hot summer morning in May and refused to move until the landowners agreed to talk. This resolved a fifteen-year-old problem that, as the women mentioned frequently, the men had been unable to resolve in their committees. Stories from this incident and others, such as that of women protecting forests from loggers and miners and forcing district officials to provide services that were rightfully theirs, contributed to building up their collectives as political and not merely social forces. As one of the coordinators of the women's program said, "What is the point of saving money and making mixtures [snacks] when everything else stays the same?"

The women's actions evoked mixed reactions in the villages. The forest federation aimed to integrate the *mahila samitis* into the federation (not without opposition from many men). Even among those who supported the women, several believed that there was little that "a group of illiterate women could do without guidance from the men."[9] But the women thought differently: "We can't keep waiting for them to decide when they feel that our issues are important to take on." Tensions rose when they began to take up questions of violence and dowry and to speak of the need for a women's federation parallel to the men's committees. The *mahila samitis* and especially the staff of the women's program active in coordinating the groups began to meet with resistance from the men in the forest organizations.

Modes of self-identification

Meeting in the *kvinnoforum* in Sweden or in the *mahila samitis* in India was neither natural nor self-evident. Both were spaces that the women had consciously constructed. The making of these spaces was not only the result of the women's social location. Social boundaries between them—such as class, age, and their place in the labor force or caste—were not always erased as the women sought to organize together, but these boundaries were often acknowledged as the women found ways to relate to one another over them. The experience of collectivity proved to be their strength rather than a sign of their weakness as individuals who were unable to act individually and independently. However, a fear of being seen as weak and incapable as a result of choosing to organize in their own group troubled the women in different ways in both study sites.

Ideals about equality and about the gender neutrality of common spaces were present in Drevdagen in several ways in the women's own doubts about organizing separately and in the opposition to their organizing. It became apparent that the women saw themselves as autonomous, but they also talked about the male dominance ingrained in much of the *föreningsliv*, the associational life of Swedish villages. However, none of the women thought of themselves as being personally disadvantaged simply by being a woman. On personal and individual levels, they saw themselves as equal with men. In spite of this (or perhaps because of this), when it came to forming the women's group, there was a certain amount of tension. Taking on an identity as a member of a women's group was uncomfortable for some since officially there were already neutral committees in the village in which both men and women could be involved.

To complain in what was regarded as a gender-equal system could be seen as being hysterical. For the women in Drevdagen, the ideas on equality thus functioned as a form of self-restraint, preventing the women from openly voicing their discontent if they were excluded or discriminated against. The women themselves spoke of being strong women. For them, acknowledging discrimination and thereby assuming the role of a victim would be antithetical to that identity. In justifying their group, the women chose consciously to build on their strengths rather than emphasize disadvantage. They considered themselves equal with the men. This did not mean that they did not see disadvantage, but that it expressed itself differently for different women. Age and their life experiences (the kind of employment that they had or their level of education) made a difference in how they regarded and dealt with feelings of injustice. Some women who had lived in the city before felt that it was easier to deal with municipal authorities than others who felt that the authorities regarded them as backward, rural women. The women's relations to the men in the village were also varied. They felt it important to emphasize their differences. There was thus a tension in acknowledging unequal power relations. It meant acknowledging discrimination, which could imply loss of self-worth and power in a system in which everyone was supposed to be equal. Tensions such as these sometimes made it difficult for the women to articulate a common identity (or rather the identity needed to be reaffirmed each time) or to make their activities and discussions more stable and continuous. Their need for informality stemmed partly from this. It put together women who were different and wanted to do things differently. For example, some felt that it was important to involve men in their forum whereas others did not.

In comparison, the women's identity as members of women's groups in Nayagarh was not in question. The women claimed that it was through their group that they saw themselves as making a difference. They were working purposively for activities that needed organizing. They were doing more than just going about their daily chores, which they do at the individual or family level and which are often governed by tradition and custom (regulated by the men and, for the younger women, also by older women). In the groups they communicated and acted together for something that concerned most of them or when individual women needed the support of the others. They seemed to have more of a sense of identity as a coherent group in comparison to the women in Drevdagen.

The Swedish women drew on a repository of meanings that existed outside them as they described their experiences. Initially, in get-togethers they sometimes explained women's lack of participation in village and forest committees as a result of low self-esteem. However, nobody spoke of herself as having low self-esteem. This image was always used to describe someone else: the old-fashioned, shy, and traditional woman, an image that the women saw themselves working against. They did not see their own roles as mothers or wives as conservative, a label they applied when discussing the previous generation.

I described my research in Nayagarh to members of the *kvinnoforum*, prompting a discussion on what it meant to be modern in India with respect to gender relations and equality. One woman remarked, "I have heard that in India women have become stronger . . . have developed quite a lot, ever since India has become modern. Perhaps that is why they have been able to do these things together." Another woman was offended at being associated with the women in Nayagarh. "But we are different," she told me. "We are working women."[10] The Swedish women's subjective positions as modern women were built very much in contrast to the earlier generation as well as to the not-so-modern women of the third world who were not working women in the same sense. Not surprisingly, this attitude was absent in the stories of the older women (in their sixties, seventies, and eighties) who spoke of the strong women of the past, women in their mothers' generation who had moved to these mountains and built up farming communities.

The importance of being working women recurred often in our conversations. The women's involvement in the workforce, mostly in service positions, was an important part of their identity as independent women, reflecting Sweden's efforts over the years to bring about gender equality through the labor market.[11] This was an important component of their status: being modern working women provided a sense of liberation from the drudgery of the past.

Yet, in the stories of the younger women (ages 30–60) in the forum, contradictory images abound. There were tensions in how the women regarded their working lives. They were proud of working outside the home, but many felt confined by not being able to do community work for the village and by having to travel long distances to work. Although many worked outside the home, they also spoke of not being able to do the work they wanted on their own terms and of being trapped in the jobs they had. Their involvement in the workforce—which in some respects provided them with economic freedom and an identity—also made them prisoners of a system over which they had little or no control. Several worked in the health sector and spoke of their difficult positions and the fact that they "were made to do all kinds of strange things" in the bouts of rationalization of health care services undertaken by the state during the 1990s. Many women insisted that motherhood, in comparison to work, was a relationship that was rewarding, in which they were appreciated. This was not a glorification of motherhood but reflected the practicalities of their lives and was "central to many women's ideas about themselves" for which many felt there was "little or no validation in dominant discourses" (O'Connor 1998, 131). A strong notion of mothers as primarily responsible for children in Sweden is juxtaposed with ideas about gender equality, employment, and dual responsibility (Elvin-Nowak and Thomsson 2001). Dual responsibility is a strong discourse in Sweden regardless of actual practice (Magnusson 2001). For some women, living and working in the village meant that negotiations over the use of their time and juggling responsibilities and work were for their own benefit rather than that of their employers, suggesting that to work in paid jobs was not always rewarding.

For the women in Nayagarh, discrimination was self-evident. They linked their disadvantage in the villages to wider orders beyond their immediate or local context. Mamta Tiwari, one of the field organizers for the women's program, expressed this in a report: "The government has reserved 33% of seats for women. It may rise to 50% in future. But this reservation will not solve our problem. Unless we represent our problems effectively. Women should not only conduct environmental work, they should protect themselves, build up awareness in society. The purpose of the mahila programme should not be only confined in the discussions, meetings, rather it should be translated into action" (Tiwari n.d.).

Dimensions of difference

The women's modes of self-identification were reinforced by several dimensions that bring out the contextual differences between the two places: the formality or informality of the groups, the rhetoric they used, and the ways in which they took action. Being publicly women, that is, presenting themselves formally as women's groups, was possible for the women in Nayagarh in a way that was not for the Drevdagen women. Since the *mahila samitis* were part of the women's development program, they followed certain rules and regulations: they had officeholders, documented their activities, and carried out economic activity together. But what really made them formal was that they presented themselves as a women's group and were accepted as such by the rest of the village. The *mahila samitis* also provided a separate space sanctioned by the forest organization and the village. The women's program accorded them a legitimacy that the women would not have had otherwise. It drew them together outside their households into a public space sanctioned by the community and the state. In acquiring even limited visibility as a formal group, the women had nondomestic reasons to meet, to establish linkages, and perhaps to build nascent ideas of solidarity (see Krishna 2004, 33).

The formality of the program gave the groups the opportunity to become stable entities within the village (for a while) and to link up with other groups. The *mahila samitis* differed from one another to the varying extent to which they were active and in terms of the questions with which they chose to work. Their formality sometimes resulted in greater bargaining power for women vis-à-vis other groups such as landlords, forest committees, violent husbands, or in-laws.

In Drevdagen the women found it necessary to work informally. They had few prospects for influencing discussions in the formal structures on questions important to them. The issues they took up were often considered subordinate to what were regarded as more important questions. The men in the village associations worked hard to maintain existing power relations by asserting the gender neutrality of common spaces. "Nobody has stopped the women from attending the village associations," retorted one of the men in the association when he was told that many women felt apprehensive about attending meetings.[12] For the women, it was difficult to challenge a system that was suffused with notions of equality and welfare for everyone, that existed under constant threat from the urban center, and where there was a culture of self-imposed guilt at going against the order. Speaking of discrimination or gender made you someone who was looking for problems. It was against village harmony. This was reflected in the women's ambivalence about speaking (or limitations in being able to speak) from within or as an all-women group.

It was not as if disadvantages for women were not recognized at all. But as the quotation above makes evident, the problem was individualized. Women who did not attend meetings or speak out were considered lacking in self-confidence, disinterested in village affairs, old-fashioned, or passive. It was made into the problem of individual women and not one of an order of gender and power. In this conception, there were other "normal" ways of dealing with problems that women in the village may have with certain men, through proper channels. Formal equality, however, precluded an analysis of substantive inequality. This was one reason for the need for informality. While in Nayagarh it was through the formal nature of the group that the women were able to wield greater bargaining power vis-à-vis other groups, in Sweden it was the group's informal nature—its not really being a group at all, nor being a women's project—that was important. The *kvinnoforum* came into being whenever the women met rather than being a stable entity. This strategic informality made the *kvinnoforum* more difficult to pin down and oppose. The informality enabled the group to be more inclusive of different women and to take up a range of issues.

The women in Nayagarh often got their demands met by confronting formal authorities such as male landlords or district officials, as in the cases described above. By taking action publicly, the women were demolishing the myth of being victims and beneficiaries as they demanded what they considered rightfully theirs. They communicated their message in a bodily way, exhibiting bodies that are otherwise meant to be confined to the home. By taking part in the action, they affirmed their own identity. Although some of these demonstrations initially were started by one *mahila samiti*, women from other castes and villages also joined in, making their identity as women (rather than as caste members or village residents) important for the moment. As Zygmunt Bauman writes, "In action one can be a victim without shame" (1995, 73), and the women of Nayagarh were using and demolishing their victimhood in these terms.

Politics in Drevdagen did not have space for such action. Similar actions by women would be seen as embarrassing because women in Sweden were understood to be equal, definitely not victims, even though they were regarded as passive by both men and women. The importance of the women's bodies in Nayagarh and the way in which the women made use of them are significant. In Sweden, in an assumed neutrality of male and female, the implications of the body are to be denied. Here it was something far less dramatic, merely meeting as women, that became controversial. The collective brought to light unequal gender relations in Sweden in a way that was already obvious in India. Perhaps that also explains why it appeared more threatening. Difference was given a political meaning.

In keeping with this was a noticeable difference in the rhetoric used by the *mahila samitis* in Nayagarh and by the *kvinnoforum* in Drevdagen. There was a certain self-restraint in the language in the *kvinnoforum*. Although their actions suggested that the reason for organizing was the need for this space as women, the women were quick to emphasize that they were working for the benefit of the village and its development. Except for one occasion, their reticence at the prospect of publicly challenging male domination and exclusion from decision making in the village association was the result of an effort to not draw undue attention to themselves as women—a position that would make them vulnerable in a society built on notions of neutrality.[13] Nor did they speak out publicly as a women's collective, unlike the women in Nayagarh.

The women's modes of action in both places were conditioned by the role of external intervention and by the village men's varying responses to their organizing. Ideals and discourses about gender relations were reflected in the images the women had about the men. When I asked the women in a Nayagarh village what made them successful as a group, one answer was, "Our men are good. The men in other villages get jealous when their women get more advanced and try and stop them from acting together." This was contradictory to the views of women in the *kvinnoforum*. They often spoke of how old habits die hard, implying a backward-looking view on gender relations among many men in the rural areas. They compared the village men to my colleagues working with the forest project. The women believed that the city men did not know about the discrimination in the village, would not behave in this manner themselves, and would do something about it if they did know.[14] There was a belief that times had changed and that everyone was in fact equal. The unequal relations that kept recurring were regarded not as a reflection of present values but as remains of the past.

Although not quite indulging in gender wars, the married women in India were much more outspoken about expressing conflictual relations with the men in their households and villages. In Sweden, personal relationships with male partners were rarely the subject matter of group discussions. When they were, they were characterized as a love contract (Magnusson 2001), where relations are negotiated through love, making the relationship personal and unique. This is not to say that there are no wars in Swedish households or no love in Indian ones. Rather, these images are descriptive of normative ways of talking about male-female relations. Marriage in India is more a social arrangement than only a private matter between the couple, as it is regarded in Sweden.

It is necessary to theorize not only the separate social interests that women's groups may have but also the deep interdependencies between men and women, which are vital for understanding gender relations. Differences in the men's responses were as important as the differences among the women and had a role in shaping women's activism. In Sweden, many projects spear-headed by the women could be carried out because of the support of men and women in the village. Both in Sweden and in India there were men who believed that the women needed their own groups. In the view of the Swedish women, there was a generational difference in this respect among the men in the village. It was considered easier to speak to the younger men. It was through younger men, often relatives, that the women sometimes tried to influence association meetings. The opposition expressed by some of these younger and "good" men (to use the terminology of the women in Orissa), however, contributed to the uncertainty among the women. These men's hostility may have sharpened the sense of discrimination. It contradicted the commonplace assumption in Sweden that gender equality is being brought about via a generational shift.

Different men responded in different ways, although they cannot necessarily be placed in different categories (such as age) in their relation to the women's organizing. The male leadership of the forest organizations in Orissa, mainly the older generation, spoke of the need to make the women aware, to impart training. The younger generation (though far from all) spoke about needing to involve women in the forest federation for reasons of both efficiency and equity. The Oxfam officer responsible for the funding of the women's development program also held this view. Men's response to the need for the women to have their own group depended on the particular situation and the men's involvement in that situation, although a normative order was also being negotiated. For example, in Drevdagen, the younger men in the association spoke about the importance of women's networks and considered it important to work toward gender equality. Yet the presence of a *kvinnoforum* in their own village was deemed dangerous, especially if it interfered in forest matters, traditionally a male domain. Women's activism in Nayagarh was sometimes aided by the men ("our men are good"), and at other times it was a response to violence against women. Forest committee members at the forestry offices often invited me to come and see how well the *mahila samitis* in their villages were working. This did not necessarily mean that they believed in them for the same reasons as the women, but it does indicate support for the women's organizing. As the coordinator for the women's program said to me, "As it is now, some men are supportive in some places while in others they do not want the women to get together or go for training camps or get-togethers. They feel that the women just go there to eat and get smart and then destroy the household. They are afraid that the women won't listen to the men after having organized. *Mahila samitis* can be strong if they are supported by the men."[15] The development agents who worked with the men in the communities in Drevdagen and Nayagarh also happened to be men. In Drevdagen, the development practitioners insisted that women were included in the village associations. However, they regarded the women's critique as being a result of personal politics that the women needed to be able to solve with the men in their village. In Nayagarh, although the development practitioners wanted to support the women in the *mahila samitis*, they felt unable to relate to them directly instead of through village organizations. But for a brief period the women had support in their organizing from Oxfam-funded programs implemented through the forest organizations.

The women's relations with the men in their villages once again reflected larger discourses in their societies. Nayagarh was considered male dominated, and individual men who were different were seen as progressive. In Sweden, regarded as a leader in gender equality internationally but also in its own image, men in rural areas who exercised power over others were seen as relics from the past. This power, however, is not a phenomenon of the past but is rooted very much in the present. References to the past to explain unequal relations make acceptable inequalities that are embedded in and have their own history in existing relations of power.

Women are like boats: the conundrums of personal and collective agency

Studying the trajectories of the two groups in relation to each other directs attention to how the personal and the collective take shape in different ways. One of the men in the forest federation in Nayagarh said with a sigh that women refused to come to meetings and training workshops unless several were invited together. "Women are like boats," he concluded, referring to the convoys of catamarans that went out to fish in the sea. "They always want to go out together. When they will be able to go out alone, become educated and do things on their own, they will be independent . . . empowered."[16] But was it this independence that constituted empowerment for the women in the two places?

Naila Kabeer (2001) writes that one facet of empowerment is the ability to choose. In this sense empowerment may be seen as a somewhat normative idea. The women in Sweden were able to choose (more or less) where they lived, whom to marry, and how many children to have, choices that were not available for many women in India. In that respect, the women of Sweden were already empowered.[17] But their space for collective agency was circumscribed by the dominant discourse on equality and modernization. The ideal and discourse of gender equality, by making discrimination illegitimate, had the negative effect of creating barriers for what women may or may not do or question in their everyday lives.

In Nayagarh, women's agency was not always recognized. The notion of individual empowerment through modernization was not consistent with the idea of the women organizing in groups. True enough, the women's activism in Nayagarh did not always lead to a better life for them. On a personal level, once home from meetings, they were still expected to cover their heads and not to speak in the presence of older males. These were the same women who had fought for their cause with male moneylenders and forest officials. Their confidence in themselves and their success, in certain instances, did lead to changes in household gender relations, but there was no direct cause and effect. In a draft evaluation report on the community forestry groups commissioned by Oxfam, the authors narrate the story of the president of one of the women's groups whose husband beat her for taking up the cause of a village woman who had been cast out by her own husband (Mitra and Patnaik 1997). The authors use this single case to dismiss what the women were saying or were doing in everyday village life: solving disputes, dealing with violent husbands and nasty mothers-in-law, and working with problems related to dowry. Espousing the idea of the individual empowered woman, they claimed that the women's groups were ineffective in changing gender domination in the home.

Frances Cleaver puts it succinctly: "As 'empowerment' has become a buzz word in development, an essential objective of projects, its radical, challenging and transformatory edge has been lost. The concept of action has become individualized, empowerment depoliticized" (2001, 37). Third-world feminists have criticized the focus on "a singular women's consciousness" (Mohanty 2003, 81). Chandra Talpade Mohanty (2003) writes that the strategy is to speak from within a collective. Doris Sommer "identifies the 'plural' or 'collective' self of Latin American women's testimonials as 'the possibility to get beyond the gap between public and private spheres and beyond the often helpless solitude that has plagued Western women even more than men since the rise of capitalism'" (Mohanty 2003, 82, quoting Sommer). Collective organizing may seem to jeopardize gains made by individual women who are able to find a place for themselves in the system. But, both in India and in Sweden, acting together on their own terms constituted empowerment for the women at the collective and village levels, highlighting that there is not always a correlate between individual and collective empowerment or between the strategic and the practical.

On the one hand, by regarding women as incapable, the men in Nayagarh held attitudes that limited the women. On the other hand, the experience of discrimination due to poverty gave both men and women a lens through which to develop a structural analysis of certain inequalities and the need for collective action. The men and women here, by and large, were spared the kind of individualism that attributes every inequality to personal failures on the part of the less rewarded. Understanding caste and class discrimination perhaps helps one understand the fact of sex discrimination, and understanding the need for an independent community forestry movement might help one to understand the need for women's groups. This may be one reason (apart from the fact that the women's group was a program that was part of the community forestry movement) why several men active in the movement tended to support the women's groups although they had their own views on how the groups were to function in relation to themselves (cf. Jane Mansbridge's [1999, 300] argument distinguishing support for the women's movement among white and black men and women in the United States). The women's groups tended not to position themselves in opposition to individual men but in opposition to what they regarded as male behavior, especially in cases of violence and dowry, because their men or some men could be good. In their view, development was indeed incomplete without gender. But for them, development was a transformation not only in which women became independent or empowered but in which jealous men became good.

In an ideal world there would be more direct correlation between personal and collective agency that leads to empowerment. However, discrimination takes many different forms, and though it may be systematic, there is not necessarily a cause-and-effect relationship between the personal and the collective or the strategic and the practical. We need to be open in imagining what women's liberation will be like in the multiple spaces in which women may choose to act individually but also collectively.

Clearly, women in both places worked not only for themselves but also for an equitable society for women and men. But issues of power were highlighted more often in India. As one of the field organizers wrote in a report, "there is no point talking about the forests when women do not have power themselves" (Tiwari n.d.). The Indian women described themselves as a women's group and took up questions that may be seen to be more challenging of gender relations and were more specifically related to discrimination against women in comparison with the women's group in Sweden. This does not imply that power and discrimination were not issues in the Swedish village. Rather, the space for taking up questions of power and discrimination was smaller. But the women were resisted all the same, not because of what they said but because of what their collectiveness suggested.

To organize as women in the village in Sweden felt illegitimate. A rhetoric of collaboration and gender harmony hid unequal power relations. This was reinforced by wider orders of meanings in policy and institutions. Women's organizing became a "forbidden action" (Eduards 2002). An important contextual distinction between the two places is that in India there was a marked perception and acknowledgement of difference whereas in Sweden that difference was expected to be minimized for reasons of equality. It is a contrast not so much in perception of difference but in the response to it.

Although their actions were by no means seamless or noncontradictory, by stressing the need to have their own space the women of the *kvinnoforum* produced what Patti Lather calls "new configurations of social relations rather than reproduc[ing] . . . the behaviors that instil dominant values in us" (1991, 96). The very act of forming the *kvinnoforum* was an act of interrupting dominant practice, although nothing was said overtly about challenging the status quo. The women's plans and everyday activities such as organizing village festivities or keeping the village clean as part of the *kvinnoforum* or the *mahila samitis* were not very different from the work they may have done in the absence of these groups. It was, instead, the structure of the alternative spaces they created for themselves that made their organizing political and caused antagonism. By organizing as women, the women made men visible as men and not as neutral individuals.

The "surprise factor" (Smith and Smith 1983, 114) of experiencing their subordination as a result of a structure that went beyond their immediate relationships was not relevant in the case of the women in Nayagarh, as it was in Drevdagen. The Nayagarh women did not expect to have much direct influence over the committees as individual women or over state agencies. They had less to lose and much to gain by speaking from within a women's group.

For several activists and scholars in the Swedish context, violence against women is the perpetuation of male power over women. Maud Eduards writes that in women's collective action related to the body, violence and power are particularly challenging for democracy because it is men who are singled out as responsible, and this in turn because they are men: "The gender power order is revealed in its nakedness" (1997, 21). In India, however, this connection is far from simple. In the perception of the women in Nayagarh, mothers-in-law (and other female relatives) were often instrumental in instigating, abetting, and sometimes taking part in violence against a daughter-in-law. The women's organizing in Nayagarh was directed as much against inhumane mothers-in-law as, for instance, against husbands and fathers-in-law.

Men from outside the villages played an important role in both places. Development projects, especially those with participatory approaches, have been criticized for reinforcing unequal relationships by prioritizing those who are most vocal at meetings or by constructing certain male members as "the community" (see Guijt and Shah 1998). As compared to the case in Sweden, however, in India it was permissible for development practitioners to challenge unequal relationships. The reason is that these inequalities are more obvious in India but also, importantly, that it is more permissible to challenge inequalities in a rural, third-world society that is saturated with development discourses and in which inequality is a premise to begin with. Outside intervention did not cause the change, but, based on an understanding of gender inequalities, it provided a little extra space for the women, which some of them used to negotiate power relations. Intervention in gender relations from outside had more legitimacy.

The relational analysis in this article has uncovered new ground by dwelling on the connections between the Swedish and Indian contexts. The connections played out in ways that were contextual but also discordant. Development discourses about oppressed women and the need for women's empowerment in the global South shaped the spaces for conceptualizing gender equality in both sites: in India, a country where assertions for the need to empower its rural women are frequent, but also in developed Sweden, with its self-image of being far ahead in gender equality in relation to other countries. Statements about victimized and incapable women who need help circumscribed the actions of women in India. At the same time, the discussion of male dominance and victimized women gave the Indian women space to bring up discrimination and to press for change. In Sweden, the debates centered on equality and participation. This vision of equality encouraged the idea that the system was good, that women just needed to be added to it to make it perfect. The appreciation of being developed was echoed in how women in Drevdagen formed their subject positions as modern working women in contrast to traditional women in the past and in third-world countries as well as in their ambivalence about whether and how much to challenge the system. It was also reflected in the ease with which my university colleagues regarded gendered inequality in environmental management as a problem in the global South but not relevant in Sweden. Crossing space and jumping scale reveal how discourses are reproduced at different levels, from the international and national policy levels to village politics in Sweden and India. The use of the term "development" showed connections in dominating assumptions both in India and in Sweden about what development is and how gender relations ought to be organized in a developed society. My comparative analysis of the gender politics in both places challenges the linearity inherent in discussions of development and modernization, the notion that women's empowerment follows from economic development or that gender equality comes from a linear progression into modernity. The focus on individual equality in Sweden put the onus on women because it made exclusion and discrimination their responsibility. The notion of the individual, empowered modern woman was also held up as an example to follow in India, thus negating the gains made in collective action. The two instances present examples of the travel of ideas and their specific material forms. They reflect the "scattered hegemonies" (Grewal and Kaplan 1994) of ideas about gender equality and development, not only in the South but also in the North.

Department of Urban and Rural Development
Swedish University of Agricultural Sciences

Notes

1 I would like to thank Alison Wylie for helping me structure my arguments for this article and Louise Fortmann for all her support and important insights. For reading the draft at various stages and giving invaluable feedback, I would like to thank Stefan Jonsson, Mary John, Nivedita Menon, and the seminar group at the Centre for Gender Research at Uppsala University, especially Katarina Petersson for long-distance help with references. Thanks also to the *Signs* reviewers and the journal editors, who have helped to improve the manuscript with their questions and advice.

2 All translations from the Swedish are my own.

3 Some researchers in Sweden believe that a postcolonial perspective (where race and ethnicity play an important part in the conceptualization) is important to understanding Swedish society. For example, in their book *Sverige och de andra* (Sweden and the others), Mc Eachrane and Faye (2001) write that although Sweden has never been a colonial power, the Swedish self-image has been formed by being inscribed into the cultural history of the West. This self-image has inevitably been colored by colonialism in Sweden's relations to other non-Western countries and in its relation to its nonwhite others (such as immigrants to the country).

4 The killing of Prime Minister Olof Palme in 1986 gave rise to a number of conspiracy theories and the brief arrest of a right-wing extremist and a Kurdish dissident living in Sweden. In 1999, the picture of Swedish neo-Nazis as working-class youths upset over immigration and often provoked by violent immigrant youths was shattered as a result of neo-Nazi attacks against journalists, trade union activists, and homosexuals, most of them native Swedes. The Göteborg fire in October 1998 took place in a disco; several young people died, most of whom were immigrants. There was tension because many believed that Swedish youths had set the fire. Eventually four Iranian men were found guilty.

5 Between 1998 and 2008, the research process went through a number of cycles. In 1998 I interviewed twenty-three women and six men in Drevdagen. This was followed by a collaborative inquiry with the women (1999–2000). My role in the inquiry was to share facilitation, document the process, interview individual women, and contribute with examples and theories to discuss together. We met every six weeks over a period of two years. I kept a record of the stories, minutes, and discussions. After 2000 we met less frequently in the group and mainly to discuss what I was writing. The most recent phase of our process (2007–8) was the analyses of our different experiences of the research in

two articles, one by the *kvinnoforum* (Bergelin et al. 2008) and one by me (Arora-Jonsson 2008). In my PhD dissertation (Arora-Jonsson 2005) I used material generated from my field journals, interviews with other women and men in Drevdagen, anecdotes of actions taken, and reports of our get-togethers and discussions in the group, as well as my material from Nayagarh. This participatory approach provided me with unusual insights into how wider politics of gender equality were shaped at the village level. The politics in Drevdagen resonated with the analyses of gender equality, a politically correct gender discourse, and a sense of uniqueness about Sweden described by other researchers as well as with accounts of resistance encountered by women-only groups in other parts of Sweden. Such long-term participatory research and the analytical practice of reversing the gaze of collective and individual subjectivities in the process of organizing (described in detail in Arora-Jonsson [2005]) also provided unusual insights into the gendered relations of everyday life in the Swedish countryside. All quotations from Drevdagen in this article are from the group discussions and other conversations during our get-togethers as part of the collaborative inquiry between 1999 and 2000. The transcripts are with the author. All other quotes from individual interviews and conversations are specified in separate footnotes.

6 Telephone conversation with one of my colleagues at the university, Uppsala, October 1999.

7 In Nayagarh (first in 1993–94 and then in 1998–99), I carried out group discussions and individual interviews (which I filmed) with women and men in the villages, attended meetings, and interacted informally during my stay there. All quotations from my research in Nayagarh are taken from group discussion with *mahila samiti* members between 1998 and 1999 unless specified otherwise. Transcripts are with the author.

8 Interviews with two of the founding members of the movement: Kesharpur, Orissa, March 1993, and Bhubaneswar, Orissa, July 1993.

9 The general castes consist of the Khandait, Chasa, Kumithi, and the Telegus among some others; they are low in the caste hierarchy.

10 Interview with one of the officeholders of the forest federation, February 1999. However, not all the women were illiterate, just as not all the men were literate.

11 Interview with a woman in Drevdagen at her home in April 1999.

12 Almost ninety percent of Swedish women work in the labor force. Despite this involvement in the labor force, Sweden has one of the most gender-segregated labor forces in Europe, with women in lower-paid and part-time professions.

13 Discussion with two men and one woman about the women's dissatisfaction with the village association at the woman's home in Drevdagen, June 1999.

14 This was a strategic move that may be seen as the exact opposite of strategic essentialism.

15 Unfortunately, not only did the men from the university know about these discriminatory practices, but they participated in making women's efforts to resist them irrelevant by referring to them as personal and as village politics.

16 Interview with the women's groups' coordinator at the Office of the Forest Network in Kesharpur village, Orissa, February 1999.

17 Interview with an officeholder of the Nayagarh Forest Federation at the Forest Federation office, Nayagarh town, Nayagarh district, Orissa, February 1999.

18 Interestingly, in the context of Swedish academic or policy debate, empowerment is not a concept used in terms of gender relations or in discussions of equality for women (with some exceptions). A search in the research literature (2004–5) shows that the concept has been used (with some exceptions) primarily in research with disabled people; in the context of immigrant women; in some cases in *byggforskning* (building research), mainly in the case of immigrant settlements; and in business management literature.

References

Arora-Jonsson, Seema. 2005. "Unsettling the Order: Gendered Subjects and Grassroots Activism in Two Forest Communities." PhD dissertation, Swedish University of Agricultural Sciences.

———. 2008. "'Research sounds so big . . .': Collaborative Inquiry with Women in Drevdagen, Sweden." In *Participatory Research in Conservation and Rural Livelihoods: Doing Science Together*, ed. Louise Fortmann, 130–45. Oxford: Wiley-Blackwell.

Baden, Sally, and Anne Marie Goetz. 1998. "Who Needs [Sex] When You Can Have [Gender]? Conflicting Discourses on Gender at Beijing." In *Feminist Visions of Development: Gender Analysis and Policy*, ed. Cecile Jackson and Ruth Pearson, 19–38. London: Routledge.

Bauman, Zygmunt. 1995. *Life in Fragments: Essays in Postmodern Morality*. Oxford: Blackwell.

Bergelin, Åsa, Margareta Emretsson, Anna Lundgren Halvarsson, Ewa Halvarsson, and Anna Ryen. 2008. "För oss är naturen en lisa för själen (Where Peace Comes Dropping Slow): The Forests and Nature for Us." In *Participatory Research in Conservation and Rural Livelihoods: Doing Science Together*, ed. Louise Fortmann, 146–61. Oxford: Wiley-Blackwell.

Cleaver, Frances. 2001. "Institutions, Agency and the Limitations of Participatory Approaches to Development." In *Participation: The New Tyranny?* ed. Bill Cooke and Uma Kothari, 36–55. London: Zed.

de los Reyes, Paulina, Irene Molina, and Diana Mulinari, eds. 2002. *Maktens (o)lika förklädnader: Kön, klass och etnicitet i det postkoloniala Sverige* [Power's various disguises: Gender, class, and ethnicity in postcolonial Sweden]. Stockholm: Bokförlaget Atlas.

Eduards, Maud. 1997. "Interpreting Women's Organizing." In Gustafsson, Eduards, and Rönnblom 1997, 11–22.

———. 2002. *Förbjuden handling: Om kvinnors organisering och feministisk teori* [Forbidden action: On women's organizing and feminist theory]. Malmö: Liber.

Elvin-Nowak, Ylva, and Heléne Thomsson. 2001. "Motherhood as Idea and Practice: A Discursive Understanding of Employed Mothers in Sweden." *Gender and Society* 15(3):407–28.

Grewal, Inderpal, and Caren Kaplan. 1994. *Scattered Hegemonies: Postmodernity and Transnational Feminist Practices*. Minneapolis: University of Minnesota Press.

Guijt, Irene, and Meera Kaul Shah, eds. 1998. *The Myth of Community: Gender Issues in Participatory Development*. London: Intermediate Technology Publications.

Gustafsson, Gunnel, Maud Eduards, and Malin Rönnblom. 1997. *Towards a New Democratic Order? Women's Organizing in Sweden in the 1990s*. Ed. Gunnel Gustafsson. Stockholm: Publica.

Integrations och Jämställdhets Departmentet (Ministry of Integration and Gender Equality). 1999. "Jämställdhetspolitiken inför 2000 talet" [Gender equality policy for the decade of 2000]. Swedish Government Memorandum 24. Stockholm: Swedish Government.

John, Mary E. 1999. "Gender, Development and the Women's Movement: Problems for a History of the Present." In *Signposts: Gender Issues in Post-independence India*, ed. Rajeswari Sunder Rajan, 100–124. New Delhi: Kali for Women.

Kabeer, Naila. 2001. "Resources, Agency, Achievements: Reflections on the Measurement of Women's Empowerment." In *Discussing Women's Empowerment: Theory and Practice*, 17–57. Sida Studies no. 3. Series ed. Anne Sisask. Stockholm: Swedish International Development Cooperation Agency.

Katz, Cindi. 2001. "On the Grounds of Globalization: A Topography for Feminist Political Engagement." *Signs: Journal of Women in Culture and Society* 26(4): 1213–34.

Krishna, Sumi. 2004. "A Genderscape of Community Rights in Natural Resource Management." In *Livelihood and Gender: Equity in Community Resource Management*, 17–63. New Delhi: Sage.

Lather, Patti. 1991. *Getting Smart: Feminist Research and Pedagogy with/in the Postmodern*. New York: Routledge.

Lingam, Lakshmi. 2002. "Taking Stock: Women's Movement and the State." In *Social Movements and the State*, ed. Ghanshyam Shah, 310–34. New Delhi: Sage.

Magnusson, Eva. 2001. "Politics, Psychology, and Gender in Heterosexual Couples: Welfare States as Laboratories for the Future?" Paper presented at the 109th convention of the American Psychological Association, San Francisco, August 25.

Mansbridge, Jane J. 1999. "'You're Too Independent!': How Gender, Race, and Class Make Many Plural Feminisms." In *The Cultural Territories of Race: Black and White Boundaries*, ed. Michele Lamont, 291–317. Chicago: University of Chicago Press.

McEachrane, Michael, and Louis Faye. 2001. *Sverige och de andra: Postkoloniala perspektiv* [Sweden and the others: Postcolonial perspectives]. Falun: Natur och Kultur.

Mitra, Amit, and Sanjoy Patnaik. 1997. "Community Forest Management in Orissa and the Role of Oxfam: A Review." Unpublished draft of report. Oxfam, Bhubaneswar.

Mohanty, Chandra Talpade. 2003. *Feminism without Borders: Decolonizing Theory, Practicing Solidarity*. New Delhi: Zubaan.

Mulinari, Diana, and Anders Neergard. 2004. "Syster, du är ju annorlunda: Migration, institutionell förändring och kön" [Sister, you are different: Migration, institutional change, and gender]. In *Framtiden i samtiden: Könsrelationer i förändring i Sverige och omvärlden* [The future in the present: Gender relations in flux in Sweden and the world around it], ed. Christina Florin and Christina Bergqvist, 204–31. Stockholm: Institutet för Framtidsstudier.

O'Connor, Pat. 1998. "Women's Friendships in a Post-modernWorld." In *Placing Friendship in Context*, ed. Rebecca G. Adams and Graham Allan, 117–35. Cambridge: Cambridge University Press.

Plantin, Lars, Sven-Axel Månsson, and Jeremy Kearney. 2000."Mäns föräldrarskap: Om faderskap och manlighet i Sverige och England" [Men's parenthood: On fatherhood and masculinity in Sweden and England]. *Socialvetenskaplig Tidskrift*, nos. 1–2, 24–42.

Rönnblom, Malin. 2002. "Ett eget rum? Kvinnors organisering möter etablerad politik" [A room of one's own? Women's organizing meets established politics]. PhD dissertation, Umeå University.

Smith, Barbara, and Beverley Smith. 1983. "Across the Kitchen Table: A Sister-to-Sister Dialogue." In *This Bridge Called My Back: Writings by Radical-Women of Color*, ed. Cherríe Moraga and Gloria Anzaldúa, 113–27. New York: Kitchen Table/ Women of Color Press.

Sommestad, Lena, and Wangari Maathai. 2005. "Fredspristagare: Genombrott för kvinnors betydelse i miljöarbete" [Peace prize winner: Breakthrough for women's importance in environmental work]. *Göteborgs-Posten*, February 26, debate page.

Sunder Rajan, Rajeswari. 1999. "Introduction." In her *Signposts: Gender Issues in Post-independence India*, 1–16. New Delhi: Kali for Women.

Tharu, Susie, and Tejaswani Niranjana. 1999. "Problems for a Contemporary Theory of Gender." In *Gender and Politics in India*, ed. Nivedita Menon, 494–525. New Delhi: Oxford University Press.

Tiwari, Mamata. n.d. "Progress Report of Women's Group Formation: Initiative of BOJBP" (in English, translated from the original in Oriya). Unpublished report, Friends of Trees and Living Beings, Nayagarh.

Tollin, Katharina. 2000. "Det måste finnas män med för att det skall bli jämställt: En utvärdering av mainstreamingprojektet inom Arbetsmarknadsverket i Västerbotten" [There have to be men for it to be gender-equal: An evaluation of the mainstreaming project within the Labour Market office in Västerbotten]. Report. Umeå University.

Part II

Knowledge, community forestry, and grouping in South Asia

South Asia has produced a large number of gender experts and some of the most insightful gender analyses. The two papers we selected for Part II, by Nightingale (2003) and Agarwal (2002), highlight material of particular relevance for gender theory and forests. Here, we first highlight the contents of each paper. We then conclude with a brief analysis of their differences and why we chose them.

Nightingale's methodological piece goes to the heart of forestry research. It echoes and extends feminist critique of the dominance or hegemony of positivist, statistical, and 'scientific' knowledge (like much forestry research) through methodological reflections from her own research on "how cultural understandings of forestry and the social-political contestations embedded within forest use shape the implementation of a resource-based development program" (Nightingale 2003, p. 80 in the original). Her work reinforces the view that knowledge is inevitably partial and a product of the context in which it is situated.

Using aerial photographs and oral histories in relation to each other to explore forest change, she illustrates the value of mixing methods in uncovering the inevitable partiality and situatedness of methodologies. "It was only by analyzing the incompatibilities between the photos (that showed minor change) and the histories (that insisted resources are much more accessible) that I was able to appreciate the inherent silences, and hence, the partiality, of both methodologies" (p. 85 in original).

Going further, she suggests feminists have had an uneasy relationship with 'quantitative' methodology – condemning it for failing to unravel the complexities of social relations and/or reclaiming it for the purposes of 'triangulation' and shedding light on the plurality of 'truths'. Instead, she argues that only "when different kinds of knowledges are taken seriously and all are critically interrogated [are] richer results . . . generated, [do] new interpretations emerge and [is] the supremacy of any one kind of knowledge . . . challenged" (p. 87, original) – a point made differently by Norgaard (this volume) in the USA, as well.

Agarwal's (2002) chapter focuses on the growing wave of community-based forestry, including the decentralization of forests to local communities, which professes to address forest degradation and enhance rural livelihoods. She assesses how these waves fare from a gender perspective through a detailed and layered analysis of data from community forestry in Nepal and Joint Forest Management in India. She points out that, like Reed's findings in Canada (this volume), women's voice and influence in decision-making processes have been marginal at best. Where women are present, they are 'nominated' and remain as largely silent spectators. She demonstrates how women's low participation can undermine the efficiency, effectiveness, and equity of these policies and institutions (see Fortmann, this volume, on similar gender-related shortcomings in Zimbabwe). Women's lack of participation is a product of restrictive rules, social norms, social perceptions, personal and household endowments, and attributes. She calls for a 'bargaining framework' to identify how women's position can be improved and more women-inclusive outcomes realized.

There are considerable theoretical and practical implications of Agarwal's work. She builds on and extends Amartya Sen's 'patriarchal bargain' to explain why women are excluded from seemingly inclusive regimes and how their position can be improved through bargaining (explicitly and implicitly) with the state, community, and household. "Women's ability to change rules, norms, perceptions and endowments in a gender-progressive direction would depend on their bargaining power – with the state, the community and the family" (p. 202, original; see also in this volume: Li on Indonesia and Singapore; Schroeder on the Gambia). Agarwal also questions Elinor Ostrom's widely acclaimed work on collective action by pointing to the instrumental ways in which social relations are treated and social inequalities ignored. For Ostrom, social relations matter insofar as they support or undermine collective action and institutional efficiency. But as Agarwal points out, "there has been relative neglect of whether or not the outcomes of collective action (in terms of, say, cost and benefit-sharing) are equitable and how their outcomes impinge on sustainability of collective action" (p. 195, original).

Agarwal's work has considerable practical implications for gender-inclusive design, delivery, and advocacy. She calls attention to the importance of building a "critical mass of . . . women" (p. 205, original) to demand greater recognition of women's rights and claims and the importance of focusing reforms ultimately at the mixed-gender group level rather than creating parallel women's groups (though see Schroeder for examples of women acting collectively and very effectively), and points out how seemingly benign rules defining membership, distribution, and protection, amongst others, can serve to systematically exclude marginalized women.

The rationale for including the work of Agarwal and Nightingale in this volume is to illustrate both the continuities and the contestations in feminist engagement with forestry. Agarwal's work is indicative of how and why forestry is intricately linked to the social and political relations in which it is inevitably situated. Nightingale reminds us that feminist concerns over power, autonomy, and authority extend not only to studying forest-dependent women 'out there', but also to interrogating critically the research process itself. The two researchers differ in that Agarwal is a firm structuralist committed to making visible both women and the ways women are discriminated against in forest governance. Nightingale, in comparison, draws on postmodern and post-structural influences within feminist and gender studies to unravel the complex and contradictory ways in which social and cultural processes produce and reproduce inequalities. Agarwal has inspired a generation of researchers and practitioners to consider gender equity more seriously in forestry research and action, whereas Nightingale is credited with pushing the boundaries of feminist theoretical engagement in forestry and natural resource management more broadly.

References

Agarwal, B. 2002. The Hidden Side of Group Behaviour: A Gender Analysis of Community Forestry in South Asia. In J. Heyer, F. Stewart, and R. Thorp, eds, *Group Behaviour and Development: Is the Market Destroying Cooperation?* Oxford, UK: Oxford University Press, 185–208.

Nightingale, A. 2003. A Feminist in the Forest: Situated Knowledges and Mixing Methods in Natural Resource Management. *ACME: An International E-Journal for Critical Geographies* 2 (1): 77–90.

6 A feminist in the forest

Situated knowledges and mixing methods in natural resource management[1]

Andrea Nightingale

Introduction

Donna Haraway's (1991) notion of partial and situated knowledges has been a major influence on feminist methodological debates within geography. Her concept has been applied to feminist work to emphasize that an omniscient, detached observer stance is not possible within any kind of scholarly research (Gibson-Graham, 1994; McDowell, 1992; Moss, 1995, 2002; Rose, 1997). Central to the concept of situated knowledges is the idea that there is no one truth out there to be uncovered and, as a result, all knowledge is partial and linked to the contexts in which it is created. In the following, I expand upon the idea of situated knowledges by arguing that geographers can interrogate the partiality of knowledge through mixed method research design.

While most feminist and critical geographers embrace the notion of situated knowledges, few have attempted to examine situatedness to the extent that Haraway (1991) did in her early work on primate biology. Haraway demonstrated how interpretations of primate group behaviour were highly gendered and reflected the positionality of the researchers as much as they reflected the dynamics of primate social groups. She used this work to argue for a new understanding of objectivity that takes seriously different kinds of knowledges and explicitly recognizes that academic work is situated, political and partial. These insights are now taken for granted among feminist geographers, and indeed in any critical work attention to how positionality and power influence the knowledge production process is expected. As a consequence, feminist methodological debates have engaged with issues of power in the research process (for example, Dyck, 1993; Elwood and Martin, 2000; Gibson-Graham, 1994; Katz, 1996; McDowell, 1992; Moss, 2002; *Professional Geographer*, 1994; Staeheli and Lawson, 1995; Wolf, 1996) and have provided insights into the day to day interrelationships that both 'blur' and 'specify' the distinction between researcher and researched, as well as the broader epistemological and institutional contexts within which researchers operate. The importance of these insights not

Andrea Nightingale, "A Feminist in the Forest: Situated Knowledges and Mixing Methods in Natural Resource Management," *ACME: An International E-Journal for Critical Geographies* 2 (1). Copyright 2003 Andrea Nightingale. Reproduced with permission.

withstanding, here I focus explicitly on partiality itself as an object of interest: my objective is to argue that many feminist geographers have squandered opportunities to challenge 'scientific knowledge' by eschewing quantitative and other 'hard science' methods. In keeping with this objective, I shift the debate away from power and positionality by discussing some of the epistemological and methodological implications of mixing methods.

The next section briefly reviews the quantitative/qualitative methods debate within feminist geography to highlight the methodological issues at stake in mixing methods. I then briefly describe the research design I used to investigate community forestry in Nepal to ground the succeeding discussion on epistemology in relation to the methods I used. I draw from my case study in Nepal to illustrate my argument and conclude with the importance of engaging directly with 'scientific knowledge' to challenge its hegemony.

Quantitative and qualitative methods

In 1995, several feminist geographers explicitly engaged with issues of positionality, power and quantitative methods in a special issue of *Professional Geographer* entitled 'Should Women Count?' As Mattingly and Falconer Al-Hindi (1995) made clear, it is the positivist epistemology of quantitative methods and their operationalization within research designs that are of primary concern. In particular, the claims of objectivity and neutrality made by the vast majority of researchers working with quantitative methods are considered to be problematic.

To my mind, this debate brought to the surface a tension within feminist geography between scholars who were trained before postmodernism and cultural studies came to dominate much of feminist geography and those trained more recently. Prior to (roughly) the 1990s, most feminist geographers were engaged in a project to make women and women's contributions to geographical processes visible. In this largely structuralist work, counting was critical as geographers were able to systematically show the ways in which women are discriminated against in a range of social contexts and the failure of many centrally collected statistics to capture these issues (see critiques by Hanson, 1992; McDowell, 1991; Radcliffe, 1991; Safa, 1981; Sassen-Koob, 1984; and Staeheli and Lawson, 1995).

More recent work in feminist geography has emphasized the construction of space and difference such as gender, class, race and identity, noting the various ways in which power and knowledge are embodied and situated (Gibson-Graham, 1994; Kobayashi, 1994; Longhurst, 2001; McDowell, 1999; Moss, 2002). This research has derived primarily from post-modernist theoretical traditions and rather than explicitly seeking to reveal that men and women are positioned differently, geographers seek to understand the social and cultural processes that (re)produce such inequalities (Longhurst, 2001). Qualitative methods are well suited to this kind of work as they are able to capture both issues of power and oppression and to interrogate how gender is embedded in the construction of knowledge itself (Lawson, 1995; Mattingly and Falconer Al-Hindi, 1995).

As a result of these theoretical and methodological trends, the merits of quantitative methods were questioned. Yet most geographers acknowledge that what is most important is not which methods are used, but *how* they are used to ask *which* kinds of questions and how the results are interpreted (Rocheleau, 1995; Sheppard 2001). Others have pointed out that quantitative methods are not by necessity positivist (Barnes and Hannah, 2001; Kwan, 2002). Many feminist geographers, however, prefer to use exclusively in-depth interviewing and other qualitative techniques as they provide rich, detailed and contextually grounded data consistent with research questions that seek to understand embodied and situated issues (Moss, 2002).

In moving the earlier debate on quantitative and qualitative methods in a new direction, I want to explore how different methods can be used to illustrate the partiality of knowledge. While feminist geographers do mix epistemologically diverse methods, they rarely use them to analyze a set of research questions at the same scale or give the methods equal importance. Rather, quantitative data are most often used to provide the context and qualitative data to explore more nuanced research questions. It is this insistence that qualitative methods are more suited for in-depth, feminist work that I want to challenge. I argue two main points; first, that mixing methods in the way I propose can yield rich insights by analyzing the discrepancies between the results. Second, linking methods provides opportunities to examine the partiality of knowledge produced in different theoretical and methodological contexts.

Mixing methods

Dianne Rocheleau (1995) has discussed the merits of mixing methods from different epistemological traditions to address a set of research questions, emphasizing the value of triangulating the results to produce a narrative that is able to capture gendered differences in access to, control over and knowledge of resources. Her work demonstrates some of the creative ways in which methods such as interviews, maps and surveys can be mixed to produce narratives that are sensitive to gender, power and context and incorporate alternative knowledges. And yet, understood in these terms, triangulation is a technique whereby the results from one method are compared in relation to another method to ensure the results are *consistent* or *corroborate* each other, thereby validating the data (Denzin and Lincoln, 1998; Silverman, 2000; Yin, 1994). For example, interview and observation results can be triangulated to ensure that they produce the same results; if not, the results of one method or both would be drawn into question. In consequence, an overarching framework of intelligibility is produced.

While I support this form of triangulation in some contexts (and have used it in my work), here I am interested in the silences and incompatibilities that become evident when data sets produced by diverse methodologies are brought together. This form of triangulation, or "mixing methods," allows for the notion that such knowledges are partial and that different vantage points – for example interview participants' perspectives versus researchers' results from observation – will produce different views of particular processes and events, such as those constituting community forestry.

In my own work on community forestry in Nepal, I used qualitative, ethnographic techniques, such as oral histories, participant-observation and in-depth interviewing, as well as aerial photo interpretation and quantitative vegetation inventory. In addition to highlighting the situatedness and partiality of knowledge, the Nepali case study also helps to show the importance of challenging 'dominant' representations of forest change – in this case aerial photo interpretation – not by rejecting them outright, but by demonstrating explicitly how they provide only one part of the story of forest change. This is a particularly important project in Nepal where increasingly remote sensed data are used to determine changes in forest cover, land use and environmental degradation (Soussan, Shrestha and Uprety, 1995; Pradhan and Shrestha, 1997; Bitter and Shrestha, 2000).

Silences and incompatibilities in community forestry

Community forestry in Nepal is a government-sponsored development program that turns management of forests over to village user-groups. It is designed to promote sustainable use of forest resources while providing for villagers' basic needs. In the past fifteen years, the program has also been seen as a vehicle for promoting economic development and democratic institutions throughout Nepal (Arnold, 1998; Gilmour and Fisher, 1991; Graner 1997). Between 1993 and 1999, I worked with a user-group in northwestern Nepal, consisting of three villages, Chain, Hernikanth and Sangkhola, which together manage Pipledi community forest.

My research sought to interrogate nature-society issues by focusing on how cultural understandings of forestry and the social-political contestations embedded within forest use shape the implementation of a resource-based development program. I examined how resource use and management is a site for the contestation and reproduction of social difference such as gender, class and caste and the implications of this for development policy and practice (Nightingale 2001, 2002). These processes are also mutually constitutive of ecological conditions and one of my challenges was to explore both theoretically and empirically these inter-relationships (Nightingale, 2003).

This project rested on two key theoretical commitments. First, I conceptualised constant change in cultures and ecosystems, recognizing that there is a dynamic, complex relationship between communities and their land (Botkin, 1990; Scoones, 1997; Turner et. al., 1990; Zimmerer, 1994). Second, as I have argued above, *all* knowledge is partial and situated (Haraway, 1991; Rose, 1997). Methodologically, these theoretical commitments required me to think through the research design consequences of analysing complex change and incomplete knowledge of those processes. For me, mixing methods was appropriate, in that no one method could hope to capture both ecological change – influenced by and 'independent' of human actions[2] – and the social-political complexities that are co-productive of ecological conditions. And, the resulting data sets produce insights into the social construction of knowledge within this case study context. Providing different methods are considered robust on their own terms (i.e. within their own paradigms), the silences and discrepancies between the results can be usefully assessed.

In keeping with my theoretical commitments, I decided to mix ecological oral histories with aerial photo interpretation. Ecological oral history is a qualitative method that allows an analysis of landscape change from the perspective of people who have used that land over time (Rocheleau, 1995) and are intimately embedded within people's life experiences. When I asked people to talk about the way the forest used to be, they never told me only about the trees, understory plants and other ecological conditions. Instead, their assessments of ecological conditions were woven together with accounts of personal, political and community change. The narratives present a view from a very particular place (see also Cope, 2002). Therefore, the variations in the histories different people told to some extent reflected how they were positioned differently in relation to the land and control over it, and to me. Although I found I was told much richer stories by people who knew me well (Nightingale, 2001), I still could not account for all the differences in the narratives. Epistemologically, oral histories have been used by feminists to tell alternative histories and to present multiple perspectives by interpreting the values, symbols and contradictory histories contained in individual accounts (Behar, 1993; Nagar, 1997). While these narratives are snap shots in a sense, they also continuously link the past with the present through the words and experiences of the individuals telling them.

Aerial photo analysis, while also dependent on interpretation, is embedded within a different epistemology.[3] The research task is to 'correctly' interpret the land cover and land forms in the images, aided by checking areas on the photos with areas on the ground, known as "ground truthing," systematic categorisation of different textures, colours and shades within the photos, and experience. Aerial photos are in some sense the quintessential Cartesian view from "no-where" (Haraway, 1991; Sheppard, 2001). The technique produces an image of land cover change that is flat, remote and static. Aerial photos are quite literally snap shots, yet they are often assumed to represent the land over a longer period of time because many vegetation types change relatively slowly. In my own work, I analyzed aerial photos from 1978 and 1996 and mapped the changing forest cover of the area (Figure 6.1).

Importantly, aerial photos require specialized knowledge to interpret (see Demeritt, 2001). The photos themselves take quite a bit of time to read effectively, and the maps generated often require relatively strong map reading skills to interpret. Figure 6.1 shows an image I produced from the aerial photo interpretation as an example of this. The areas of different land cover have been mapped onto the only topographic map available for the valley,[4] and through this mapping some of the place names have been obscured. Despite producing the original hand-drawn version of this map, I nevertheless have a hard time rectifying this image with the forest and valley in which I worked.

I began with the premise that these two methodologies were equally valid –that they both 'correctly' mapped the phenomenon in question (Denzin and Lincoln, 1998; Silverman 2000 p. 91). Given this validation, what if the histories and the photos did not tell the same story? What would that tell me about both oral histories and aerial photo interpretation? Placed in conjunction with each other, the resulting data sets revealed far more about the political and social struggles around claiming forest land and ecological improvement than either one did in isolation. Specifically, I was able to understand how small areas of forest improvement symbolise the success of the user-group and are crucial to the user-groups' control over that land.

The oral histories talked of wild and thick forests that dominated many parts of the valley when my respondents were children: roughly 35–60 years ago. At this time, the forest was managed by the village headman and many people ascribed the good forest conditions to their cultural traditions that ensured everyone followed the rules. About twenty-five years ago, however, the district forest office (DFO) took over management of the forest and ecological conditions rapidly declined. As one woman said:

> Before the community forest it was very difficult. But in 1976 we had a good forest . . . After 1976 the forest became really bad . . . People would cut a small sapling for firewood, and all they would get would be three pieces of firewood. For three pieces of firewood, why cut a whole tree? Leaf litter – or anything else – was not available. After that we all got together and made the community forest. After the forest came into our own hands it is much better.
>
> (Interview with a Brahmin woman, February 20, 1999)

Other people spoke in great detail about how much more readily available firewood, timber, leaf litter and other resources had become since the forest was returned to village control in 1991 through the community forestry program.

The oral histories were not wholly consistent. Some people indicated that recent improvements were few, but insisted that community forestry had made their lives easier as the following exchange indicates.

> *Andrea*: What do you like best about the CF [Community Forestry]?
>
> *Chetri woman*: It has become very easy (*sukka*) for us now that we have the committee. It is not difficult/troublesome [literally, 'there is no *dukka*']
>
> *Another Chhetri woman jumped into insist*: There isn't that much *dukka*, there also isn't a lot of *sukka*, it's ok [only].
>
> (Interview with a group of Chhetri women June 14, 1999)

In short, then, the histories indicate that the forest was over-harvested after the village lost control of it in (roughly) 1976, but conditions have improved in recent times. The timeline suggested by different participants varied, but all were generally consistent with decline under the DFO and improvement after the establishment of the community forest.

The aerial photos also show improvement between 1978 and 1996 (see Figure 6.1), but this is balanced out by areas that have been cleared. By 1978, the district forest office had control over the forests, and by 1996 the community forest had been operating for five years. Thus, the early photos capture the forest as it was under DFO management and the later photos show five years after village management began. Because of the scale (1:50,000) and the topographical distortion of the photos, only limited interpretations can be made, but it is possible to show changes from one land use type to another. I have flattened out some of the variation in this image (other, more complex images were produced from the data, but they are harder to read) by including both regeneration of forest from cleared land and an obvious increase in forest density in the 'regrowth' category. Similarly, I have included areas that have been dramatically thinned as well as those that have been completely cleared in the 'clearing' category. 'Forests' and 'open' are those areas that have not changed. The three villages that manage the forest in question are named in the north and eastern side of the map (Chain, Hernikanth and Sangkhola). Areas that have begun to regenerate are a relatively large section near Sangkhola and Hernikanth, and just above Chain. While harvesting is currently forbidden in

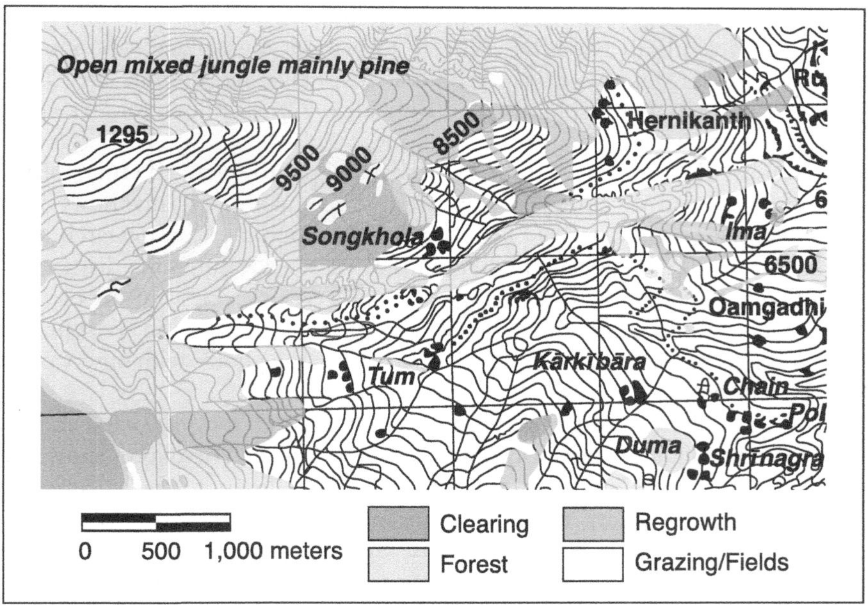

Figure 6.1 Forest cover change in Pipledi forest and surroundings 1978–1996 (colour version available at https://ojs.unbc.ca/index.php/acme/article/view/709).

the latter locality, people look forward to harvesting firewood in another 5–10 years. Near Sangkhola, land has been cleared for agriculture but other parts of the forest that were cleared or partially cleared in 1978 have begun to regenerate. The total change is an increase of 4.7% in relatively dense forest area, with a 1.8% decline in open or cleared land, and 2.9% decrease in patchy or thin forest (Nightingale, 2001). Thus the total change is quite small but shows an overall increase in forest cover and small areas of significant regeneration since 1978.

To ensure that the results from these two methods were robust *on their own terms*, I used several different tests of reliability. For the histories, I used theoretical saturation—asking the same questions to different people until I was getting no new information (Krueger, 1998)—particularly for the information about the forest conditions. I also interpreted the histories without using this criterion in order to analyze other aspects of them, especially important differences in the narratives that ensued from people's positionality in relation to the rest of the user-group, myself and my research assistants.

I also triangulated (after Denzin and Lincoln, 1998; Yin 1994) the interview results with participant observation to the extent that I could. Many of my research participants discussed very important information about forest change when we harvested firewood or examined the community forest boundaries. In this way, the oral histories were validated by triangulation with other methods of similar epistemological origins.

The aerial photo interpretation was done by mapping the boundaries of the different land cover types for each set of photos on top of each other. This allowed for a calculation of percent change in area of each cover type without needing to calculate absolute area. I did not try to correct quantitatively for the distortion as these calculations are extremely complex and time consuming and do not sufficiently increase the accuracy to warrant their usage. The kind of analysis I undertook avoids the problems of calculating absolute area from photos with high topographical relief (Bolstad, 1992), yet allows for an analysis of percentage change.

To ensure the robustness of the data the aerial photos were also triangulated with a vegetation inventory of the forest. This quantitative inventory used plots, sampling a total of 5% of the area; within each plot, the research team counted total trees by species in different size classes, measured the diameter at breast height of each tree and catalogued ground cover species and any obvious disturbance. While doing this survey I also "ground-truthed" the aerial photos by checking the classification of vegetation types in the photos with observations on the ground.

The aerial photos and the oral histories thus are both *internally* valid and yet also provide tellingly different histories of forest change. The photos show that the areas that have improved the most are those closest to the villages but that overall forest cover has changed very little. This information, when compared to the oral histories that emphasise overall improvements, suggests that the accessible areas are of great importance and that villagers value these parts most. They can see tangible signs of improvement in the areas that are most accessible, and to them this *is* a dramatic change.

While this conclusion is perhaps not surprising, without the aerial photo work this would not have been obvious. When I asked people about which parts of the forest were most important they emphasised that the *whole* forest was important and were unwilling to separate it into different parts. Their answers shifted between describing the value of the forest in terms of the resources provided and the improvements made by community forestry due to their authority to exclude outsiders. Similarly, because some of these accessible areas are still off-limits for firewood collection, participant observation did not reveal the value of these parts. It was only by analyzing the incompatibilities between the photos (that showed minor change) and the histories (that insisted resources are much more accessible) that I was able to appreciate the inherent silences, and hence the partiality, of both methodologies. That is, the dramatic improvements talked about by the villagers were in reference to some places and *not others*, while the images captured by the aerial photos were devoid of local meaning.

But perhaps more significantly, the inconsistencies between the two "data sets" lent insight into the importance of control over the forest. As I mentioned above, some respondents did state that forest conditions had not improved much, but they were adamant that community forestry was better than state forestry management. Of course, these sentiments reflect not only the ecological conditions but also the political dimensions. If the villagers are seen to mismanage their land, it can be re-classified as national forest and turned over to the DFO again. By insisting on recent improvements in *their* forest, therefore, respondents were making ownership claims. The discrepancies between the aerial photo work and the oral histories helped to make it clear that they were referring to small areas of significant resource improvement in key places, and that resources and control over them were inseparable issues.

Conclusions: interrogating partial and situated knowledges

The epistemological issues raised by using different methods together highlight the relationship between research design and methods. I have used aerial photo interpretation in a non-positivist way by refusing to accept the maps generated as telling the 'real' story of forest change. This is not to reject the information produced by the photos, as this "data set" was critical for my analysis as outlined above. But, by comparing the inconsistencies rather than triangulating the histories and the photos, the objective, neutral image produced by the photos is challenged. In order to do this, however, the photos and the histories have to be used on equal terms and to investigate questions at roughly the same scale. A different research design might have used the photos merely to set the context for forest change and then used the histories to detail the cultural and political aspects of that change. Instead, by setting the data sets in relation to each other I have allowed for both to be acknowledged as partial and situated.

In doing so, I have problematized not only the notion that qualitative data provide a necessarily more 'authentic' resource, but also the authoritative status of quantitative methodologies in this specific context. Within policy circles and much of academia (particularly the natural sciences) the inherent neutrality and merits of remote sensing is largely unchallenged; indeed, it is the preferred type of data for managers. National and international environmental monitoring, evaluation and policy are increasingly dominated by remote sensed data. This is particularly true in Nepal where the rugged terrain and lack of transportation make travelling to remote areas difficult and time consuming and over the past several years, Maoist activity in the hills has made travel increasingly dangerous. These hegemonic representations are not views from "no-where;" they are views from a detached (increasingly from outer space) position and as such do represent a *situated, partial* view. It is therefore critical to demonstrate how and why remotely sensed data are partial and thus inadequate by itself for addressing social-ecological issues.

I want to end by urging feminist geographers to consider using multiple methods and data sources in their work as I have described. Feminist geographers have challenged the dominance of quantitative methods by engaging with their epistemological origins and questioning their relevance for particular kinds of research questions. I have noted here that mixing methods and analysing them in relation to each other provides another way to interrogate positivist science. Using different methods gives feminists an opportunity to demonstrate how hegemonic representations, such as remote sensed data for understanding land cover change, are insufficient. It is vital for scholars concerned with issues of power, positionality and hegemony to engage with this kind of work in order to exert influence within policy and development circles where positivist, statistical 'scientific' research continues to be dominant. Nature-society geography and political ecology often engage with physical science data to deconstruct it or provide a physical background (cf. Braun and Castree, 1998; Demeritt, 2001; Peet and Watts, 1996), yet as the case study presented here shows, physical data sets can be utilized in conjunction with other methods to probe human-environment issues in new ways. Qualitative, interpretive methods provide rich, thick results, but combined with other methods, these results can be richer and thicker, and we can demonstrate how fragmented and situated all knowledge is. What is at issue is not whether different methods, qualitative or quantitative, are feminist, but rather do they fully embrace the notion of *different* knowledges. When different kinds of knowledges are taken seriously and all are critically interrogated, richer results are generated, new interpretations emerge and the supremacy of any one kind of knowledge is challenged.

Acknowledgements

I wish to thank Verity Jones, Deborah Dixon and Sarah Jenkins for organising the session on feminist methodologies at the RGS/IBG meetings in 2002 and ensuring the papers stayed together as a collection. I am grateful to Deborah Dixon, Liz Bondi, Deborah Thien, Leila Harris, Caroline Desbiens, Susan

Frohlick, Pamela Moss and an anonymous reviewer for very helpful comments on earlier drafts. A special thanks to Dr. Mark Lindberg and the staff of the University of Minnesota Cartography Laboratory for doing the electronic mapping. Any errors that remain are my own. Thanks also to Eric Sheppard and Helga Leitner for inspiring my interests in methodology and epistemology and the people of Mugu for participating in my research. The fieldwork for this paper was supported by: the National Science Foundation under Grant No. 9900788, a Fulbright-Hays Doctoral Dissertation Award and a University of Minnesota Graduate School Special Grant. I am deeply grateful for the support they have provided. Finally, thanks to HMG Government of Nepal for allowing me to conduct research on several different occasions.

Notes

1 © Andrea Nightingale, 2003.
2 Here I am drawing a largely artificial distinction between human induced and so-called natural ecological changes (cf. Castree and Braun, 1998; Cronon, 1996). I prefer to retain this analytical distinction, however, in order to analyze changes in the physical environment that are not direct effects of human action (Nightingale, 2001, 2003).
3 While there is a large literature on the politics of mapping and map interpretation amongst cartographers, the need for critical analysis of remotely sensed images is rarely acknowledged by most remote sensing specialists who see the images as inherently neutral (see Curry, 1998; Harris and Weiner, 1998; and Sheppard, 2001 for some important exceptions in relation to GIS).
4 The original map itself is highly inaccurate, especially the topographic lines. The photos were taken for the forest department in 1978 and the topographic survey department in 1996. The politics of producing these images is another interesting and important issue, but outside the scope of this paper. See Demeritt (2001) for an analysis of the politics of picturing forest resources in the United States.

References

Arnold, J.E.M. 1998. *Managing Forests as Common Property*. Rep. 136, Food and Agriculture Organization of the United Nations, Rome.

Barnes, Trevor J. and Mathew Hannah. 2001. Guest Editorial: The Place of Numbers: Histories, Geographies, and Theories of Quantification. *Environment and Planning D: Society and Space* 19, 379–83.

Behar, Ruth. 1993. *Translated Woman: Crossing the Border with Esperanza's Story*. Boston: South End Press.

Bitter, Peter and Bhuban Shrestha. 2000. Orthophotographs to Assist Participatory Forest Management: Application in the Jhikhu Khola Watershed, Nepal. *Issues in Mountain Development*. Kathmandu, Nepal: International Centre for Integrated Mountain Development.

Bolstad, Paul V. 1992. Geometric Errors in Natural Resource GIS Data: Tilt and Terrain Effects in Aerial Photographs. *Forest Science* 38, 367–80.

Botkin, Daniel B. 1990. *Discordant Harmonies: A New Ecology for the Twenty-First Century*. New York: Oxford University Press.

Braun, Bruce and Noel Castree. 1998. *Remaking Reality: Nature at the Millennium*. New York: Routledge.

Castree, Noel and Bruce Braun. 1998. The Construction of Nature and the Nature of Construction: Analytical and Political Tools for Building Survivable Futures. In Bruce Braun and Noel Castree (eds.), *Remaking Reality: Nature at the Millennium*. New York: Routledge, 3–42.

Cope, Meghan. 2002. Feminist Epistemology in Geography. In, Pamela Moss (ed.), *Feminist Geography in Practice: Research and Methods*. Oxford: Blackwell Publishers, 43–55.

Cronon, William. 1996. The Trouble with Wilderness; or, Getting Back to the Wrong Nature. In William Cronon (ed.), *Uncommon Ground: Rethinking the Human Place in Nature*, New York: W.W. Norton and Company, 69–90.

Curry, Michael. 1998. *Digital Places: Living with Geographic Information Technologies*. London: Routledge.

Demeritt, David. 2001. Scientific Forest Conservation and the Statistical Picturing of Nature's Limits in the Progressive-Era United States. *Environment and Planning D: Society and Space* 19, 431–59.

Denzin, Norman K. and Yvonna S. Lincoln, (eds.), 1998. *Collecting and Interpreting Qualitative Materials*. Thousand Oaks, CA: Sage Publications.

Dyck, Isabel. 1993. Ethnography: A Feminist Method? *The Canadian Geographer* 37, 52–7.

Elwood, Sarah and Deborah Martin. 2000. "Placing" Interviews: Location and Scales of Power in Qualitative Research. *Professional Geographer* 52, 649–57.

Gibson-Graham, J. K. 1994. 'Stuffed If I Know': Reflections on Post-modern Feminist Social Research. *Gender, Place and Culture* 1, 205–24.

Gilmour, D. A. and R. J. Fisher. 1991. *Villagers, Forests and Foresters: The Philosophy, Processes and Practice of Community Forestry in Nepal*. Kathmandu: Sahayogi Press.

Graner, Elvira. 1997. *The Political Ecology of Community Forestry in Nepal*. Saarbrücken: Verl. für Entwicklungspolitik.

Hanson, Susan. 1992. Geography and Feminism: Worlds in Collision? *Annals of the Association of American Geographers* 82, 569–86.

Haraway, Donna. 1991. *Simians, Cyborgs and Women*. New York: Routledge.

Harris, Trevor and Daniel Weiner. 1998. Empowerment, Marginalization and 'Community-Integrated' GIS. *Cartography and Geographic Information Systems* 25, 67–76.

Katz, Cindi. 1996. The Expeditions of Conjurers: Ethnography, Power and Pretence. In Diane L. Wolf (ed.), *Feminist Dilemmas in Fieldwork*, Boulder: Westview Press, 170–83.

Kobayashi, Audrey. 1994. Coloring the Field: Gender, "Race," and the Politics of Fieldwork. *Professional Geographer* 46, 73–80.

Krueger, Richard A. 1998. *Analyzing and Reporting Focus Group Results*. Thousand Oaks: Sage Publications.

Kwan, Mei-Po. 2002. Quantitative Methods and Feminist Geographic Research. In Pamela Moss (ed.), *Feminist Geography in Practice: Research and Methods*, Oxford: Blackwell Publishers, 160–72.

Lawson, Victoria. 1995. The Politics of Difference: Examining the Quantitative/Qualitative Dualism in Post-Structuralist Feminist Research. *Professional Geographer* 47, 449–57.

Longhurst, Robyn. 2001. Geography and Gender: Looking Back, Looking Forward. *Progress in Human Geography* 25, 641–8.

Mattingly, Doreen J. and Karen Falconer Al-Hindi. 1995. Should Women Count? A Context for Debate. *Professional Geographer* 47, 427–435.

McDowell, Linda. 1991. Life without Father and Ford: The New Gender Order of Post-Fordism. *Institute of British Geographers Transactions* 16, 400–19.

McDowell, Linda. 1992. Doing Gender: Feminism, Feminists and Research Methods in Human Geography. *Transactions of the Institute of British Geographers* 17, 399–416.

McDowell, Linda. 1999. *Gender, Identity and Place: Understanding Feminist Geographies.* Minneapolis: University of Minnesota Press.

Moss, Pamela. 1995. Reflections on the 'Gap' as Part of the Politics of Research Design. *Antipode* 27, 282–90.

Moss, Pamela, (ed.). 2002. *Feminist Geography in Practice: Research and Methods.* Oxford: Blackwell Publishers.

Nagar, Richa. 1997. Exploring Methodological Borderlands through Oral Narratives. In John Paul Jones III, Heidi J. Nast and Susan M. Roberts (eds.), *Thresholds in Feminist Geography: Difference, Methodology, Representation*, Lanham, MD: Rowman and Littlefield.

Nightingale, Andrea J. 2001. *A 'Trialectic' of Community Forestry Management in Mugu District, Western Nepal: Power, Cultural Meanings and Ecology.* Doctoral Dissertation. University of Minnesota, Minneapolis.

Nightingale, Andrea J. 2002. *The Nature of Gender: Work, gender and environment.* Unpublished manuscript. School of GeoSciences, University of Edinburgh.

Nightingale, Andrea J. 2003. Nature-Society and Development: Cultural, social and ecological change in Nepal. *Geoforum* (forthcoming). Geoforum, Vol. 34, No. 4, 2003, p. 525–540.

Peet, Richard and Michael Watts. 1996. *Liberation Ecologies: Environment, Development, Social Movements.* New York: Routledge Press.

Pradhan, Pradmod S. and Basanta Shrestha. 1997. Geographic Information Systems and Remote-sensing Technologies for Mountain Development. *Issues in Mountain Development.* Kathmandu, Nepal: International Centre for Integrated Mountain Development, 97.

Professional Geographer, 1994. Women in the Field: Critical feminist methodologies and theoretical perspectives. 46, 54–102.

Professional Geographer, 1995. Should Women Count? The role of quantitative methodology in feminist geographic research, 47, 426–458.

Radcliffe, Sarah A. 1991. The Role of Gender in Peasant Migration: Conceptual Issues from the Peruvian Andes. *Review of Radical Political Economics* 23, 129–47.

Rocheleau, Dianne. 1995. Maps, Numbers, Text and Context: Mixing Methods in Feminist Political Ecology. *Professional Geographer* 47, 458–66.

Rose, Gillian. 1997. Situating Knowledges: Positionality, Reflexivities and Other Tactics. *Progress in Human Geography* 21, 305–20.

Safa, Helen I. 1981. Runaway Shops and Female Employment: The Search for Cheap Labor. *Signs* 7, 418–33.

Sassen-Koob, Saskia. 1984. Notes on the Incorporation of Third World Women into Wage-Labor Through Immigration and Off-Shore Production. *International Migration Review* 18, 1144–67.

Scoones, Ian. 1997. The Dynamics of Soil Fertility Change: historical perspectives on environmental transformation from Zimbabwe. *The Geographical Journal* 163, 161–9.

Sheppard, Eric. 2001. Quantitative Geography: Representations, Practices, and Possibilities. *Environment and Planning D: Society and Space* 19, 535–54.

Silverman, David. 2000. *Doing Qualitative Research: A Practical Handbook.* London: Sage Publications.

Soussan, J., B. K. Shrestha, and L.P. Uprety. 1995. *The Social Dynamics of Deforestation in Nepal.* New York: CRC Press-Parthenon Publishers.

Staeheli, Lynn A. and Victoria A. Lawson. 1995. Feminism, Praxis, and Human Geography. *Geographical Analysis* 27, 321–38.

Turner, B. L., William C. Clark, Robert W. Kates, John F. Richards, Jessica T. Mathews and William B. Meyer. 1990. *The Earth as Transformed by Human Action: Global and Regional Changes in the Biosphere over the Past 300 Years.* Cambridge: Cambridge University Press.

Wolf, Diane L. 1996. Situating Feminist Dilemmas in Fieldwork. In Diane L. Wolf (ed.), *Feminist Dilemmas in Fieldwork,* Boulder: Westview Press, pp. 1–55.

Yin, Robert. 1994. *Case Study Research: Design and Methods.* Thousand Oaks: Sage Publications.

Zimmerer, Karl. 1994. Human Geography and the "New Ecology": The Prospect and Promise of Integration. *Annals of the Association of American Geographers* 84, 108–25.

Zimmerer, Karl S. 2000. The Reworking of Conservation Geographies: Nonequilibrium Landscapes and Nature-Society Hybrids. *Annals of the Association of American Geographers* 90, 356–69.

7 The hidden side of group behaviour

A gender analysis of community forestry in South Asia

Bina Agarwal

Introduction

Community forestry groups, managing State or community-owned forest resources, represent one of the most rapidly growing forms of collective action in the developing world. They thus provide an especially useful study in how groups function. This chapter focuses on South Asian experience to illuminate how such groups, ostensibly set-up to operate on principles of cooperation, and meant to involve and benefit all members of the rural community, often effectively exclude significant sections, such as women. While seemingly participatory, equitable and efficient, they cloak substantial gender-related inequities and inefficiencies. The chapter also analyses what underlies such unfavourable outcomes and how the outcomes could be improved.

The interactive effect of one outcome on another is examined as well. For instance, excluding women (often the principal users of community forests) from a group's decision-making bodies could have a range of negative efficiency fallouts, such as the framing of inappropriate or inequitable rules that are difficult to enforce. I analyse these issues (which typically cut across class/caste divisions) mainly from a gender perspective. Where relevant, the interplay of class/caste with gender, in defining outcomes for different categories of women, is also outlined.

The chapter argues that the outcomes of group functioning are determined especially by rules, norms, and perceptions, in addition to the household and personal endowments and attributes of those affected. All of these factors can work to the disadvantage of women, both separately and interactively. To what extent they can be changed in women's favour will depend on women's bargaining power *vis-a-vis* the State, the community, and the family. The factors that are likely to affect women's bargaining power in these three arenas are spelt out as well. While the context here is community forestry, the conceptual framework would also be relevant to understanding the gendered dimensions of group functioning in a number of other contexts, including water users' associations.

The chapter is based largely on the fieldvisits and interviews I undertook during 1998–99 in eighty-seven community forestry sites across five states of India (Gujarat, Karnataka, Madhya Pradesh, Orissa, and the Uttar Pradesh hills) and two districts (Kaski and Dang) of Nepal.[1] These are supplemented by existing case studies and some earlier visits to selected sites. Information was obtained mostly through unstructured interviews with villagers, at times with women and men in separate groups, at other times with both sexes jointly, in addition to individual interviews with key informants, especially office bearers in the community forestry groups.

Background

In rural South Asia, forests and village commons have always been important sources of basic necessities and supplementary livelihoods, providing villagers with firewood, fodder, small timber, and various non-timber products. Especially for the poor and women who own little private land, they have been a critical element in survival. In India's semi-arid regions in the 1980s, for instance, the landless and land-poor procured over 90 per cent of their firewood and satisfied 69–89 per cent of their grazing needs from common pool resources (Jodha 1986). At that time, firewood alone provided 65–67 per cent of total domestic energy in the hills and desert areas of India and over 90 per cent in Nepal as a whole (Agarwal 1987). This situation remained largely unchanged even in the early 1990s when community forestry programmes were formally launched in both countries. Firewood was then still the single most important source (and for many the only source) of rural domestic energy in South Asia, and was still largely gathered, not bought. In 1992–93, for instance, 62 per cent of rural India's domestic energy came from firewood. In most Indian states over 80 per cent of rural households used some firewood as domestic fuel; and, taking an all-India average, only about 15 per cent of the firewood so used was purchased (Natrajan 1995).

Over time, however, people's ability to fulfil their needs had been eroding with the decline in communal resources, due both to degradation and to shifts in property rights away from communities to the State and to individuals. The push toward community forest management represented a small but notable reversal in these processes of statization and privatization, toward a re-establishment of greater community control over forests and village commons. Indeed there is now a mushrooming of community forestry groups (CFGs) in South Asia.[2]

In India, these CFGs include: (i) State-initiated groups formed under the Joint Forest Management (JFM) programme launched in 1990, in which village communities and the government share the responsibilities and benefits of regenerating degraded local forests; (ii) self-initiated groups, started autonomously by a village council, youth club or village elder and concentrated mainly in Bihar and Orissa; and (iii) groups with a mixed history, such as the *van panchayats* or forest councils in the hills of Uttar Pradesh (UP), initiated by the British in the 1930s. Some of these have survived or been revived by NGOs. JFM groups are the most widespread, both geographically and in terms of forest area. To date, virtually all Indian states have passed JFM resolutions. These allow participating villagers access to most non-timber forest products and to 25–50 per cent (varying by state) of any mature timber harvested. There were approximately 63,000 JFM groups in 2001, covering 14.2 million hectares (mha) or 18.6 per cent of the 76.5 mha administered as forest land (Sehgal 2001). In addition, there are a few thousand groups of the other types.

Nepal's community forestry programme launched in 1993 is largely State-initiated. Here the State transfers even good forest to a set of identified users who form a forest user group (FUG) and are entitled to all of the benefits.[3] In 2000, there were 9100 FUGs involving one million households and covering 0.66 mha or 11.4 per cent (the target being 61 per cent) of the country's 5.8 mha of total forest land (Government of Nepal 2000). In both India and Nepal, NGOs can act as catalysts or intermediaries in group formation and functioning.

Unlike the old systems of communal resource management which typically recognized the usufruct rights of all village residents, the new CFGs represent a more formalized system of rights. Typically these rights are based either on membership (as in the State-initiated groups), or on rules specified by selected (often self-selected) community members (as in the self-initiated groups). In other words, membership, or some other formal system, is replacing village citizenship as the defining criterion for establishing rights in the commons.

This raises some critical questions, such as: how are the CFGs performing in terms of participation, equity, and efficiency from the perspective of women, especially the poor? Are the emerging systems of rights in communal property inclusive and equitable, or are they replicating the patterns of élite and male centredness that characterize rights in private land? The section below focuses on these concerns.

Gendered outcomes: participation, equity, and efficiency

In terms of forest regeneration, many CFGs have had notable success. Typically what is involved is restriction of entry and protection, although in some cases replanting is also done. Even with simple protection, natural revival is often rapid if the rootstock is intact. Within five to seven years, many of the severely degraded tracts in semiarid India have been covered with young trees; and areas with some vegetation, but notably declining, show encouraging signs of regeneration. Hence in most ecological zones, as a result of the CFG initiatives, beneficial results are noted, and in a number of cases it has been reported that incomes and employment have increased,[4] seasonal outmigration has fallen,[5] the land's carrying capacity has improved, and biodiversity has been enhanced.[6] Some villages have even received awards for conservation (Shah and Shah 1995; my fieldvisits during 1998–99).

Viewed from a gender perspective (and especially the perspective of poor women), however, these results look less impressive on several important counts: effective participation; equity in the sharing of costs and benefits; and efficiency in functioning.

Participation

In both India and Nepal, the State-initiated groups generally have a two-tier organizational structure, consisting of a general body with members drawn from the whole village and an executive committee (EC) of some 9–15 persons. Typically the general body meets once or twice a year and the EC meets about once a month. Both bodies, interactively, define the rules for forest use and benefit sharing, the structure of fines for rule violation, the method of protection (e.g. guards, patrol groups, etc.), and so on. Which category of persons has a voice in the general body and the EC bears critically on how well these organizations function, and who gains or loses from them.

Women's effective participation in CFG decision-making would require not only that they become members of the group (the general body, the EC), but that they attend and speak up at group meetings, and (at least some of the time) can ensure that decisions are in their interest. Such participation is important both in itself, as an indicator of democratic institutional functioning, and for its effects on cost and benefit sharing and on efficiency. To what extent do women in general, and poor women in particular, so participate?

Participation in management

Women usually constitute less than 10 per cent of the general bodies in most JFM groups;[7] they are typically absent in the self-initiated groups;[8] and are few or none in the *van panchayats*.[9] Their presence in Nepal's FUGs is similarly sparse. A study of seven FUGs in eastern Nepal found that only 3.5 per cent of those recorded as users in the FUGs were women (Dahal 1994: 78).

In India, the eligibility criteria for membership in the JFM general body and the EC vary by state. Eight of the twenty-two states for which information is so far available restrict general body membership to only one person per household. This is almost always the male household head. In eight other states, due to amendments in the initial Orders, both spouses, or one man and one woman, can now be members, but this still excludes other household adults. Only three states have opened membership to all village adults. In the self-initiated autonomous CFGs, the customary exclusion of women from village decision-making bodies has been replicated. In Nepal's FUGs, again, the household is the unit of membership, and in male-headed households it is the man's name that is entered in the membership list (Seeley 1996).

Without being general body members, women usually hear little about what transpires at meetings. Many women complain:

> Our husbands don't tell us about meetings. They simply say they have a meeting and go, when the watchman brings around the notice for the meeting (woman to author, Five village mouza, Orissa, 1998).
>
> When we ask them what happened at the meeting, they say: what will you gain by knowing? (women to author, Five village mouza, Orissa, 1998).
>
> Typically men don't tell their wives what happens in meetings. Even if there is a dispute about something, they don't tell us; nor do they volunteer information about other matters (women to author, Khedipada village, Gujarat, 1999).

Women's representation in the ECs is also typically low, although there is some variation by region and context. In a study of twenty JFM groups in West Bengal (east India) 60 per cent had no women EC members, and only 8 per cent of the 180 EC members were women. Landless families, similarly, are little represented in the ECs (Sarin 1998). In many states, recent JFM resolutions require the inclusion of women in the EC, ranging from a minimum of two or three to one-third, but I found that the women so included were rarely chosen by other women as their representatives. Sometimes, to fill the mandatory slots, male EC members chose the women even without consulting them. Such women are seldom active or effective. In Nepal again, women have only a nominal presence in the ECs. Those who join are usually poorly informed about the activities of their FUG, and some are even unaware that they are EC members (Upadhyay and Jeddere-Fisher 1998; Moffatt 1998).

Whether from a lack of awareness, or the constraints discussed later, only a small percentage of the women who are general body or EC members usually attend the meetings. Those who attend rarely speak up, and if they do speak they say their opinions are given little weight.

> Men don't listen, except perhaps one or two. Men feel they should be the spokespersons (woman to author, Garbe Kuna village, Kaski district, Nepal, 1998).
>
> What is the point of going to meetings. We would only sit silently (women to author, Panasa Diha village, Orissa, 1998).
>
> I attend *van panchayat* meetings, but I only sign, I don't say much. Or I say I agree (woman *van panchayat* member to author, Sallarautela village, UP hills).

Having a voice in the EC is important since this is the site for discussions and decisions on many critical aspects of CFG functioning. As matters stand, women are not party to many crucial decisions. An analysis of JFM decision-making in five Gujarat villages revealed that all major decisions on forest protection, use, distribution of wood and grass, and future planning, were taken by men. The only joint decisions with women were those concerning tree nurseries

(Joshi 1998). Women are also often left out of the CFG teams that go on 'exposure' visits to other sites, or that receive technical training in new silviculture practices.

Although there are some contrasting examples of all-women CFGs and mixed CFGs with a high female presence, these remain atypical. All-women CFGs, for instance, are found especially in the UP hills and parts of Nepal where there is high male outmigration; and a scattering of them have also emerged in other regions, catalysed by a local NGO, a forest official, or an international donor.[10] Occasionally women initiate one themselves. There are no consolidated figures for India, but in Nepal, in 2000, all-women FUGs constituted less than 3.8 per cent of all FUGs, controlling 1.1 per cent of all FUG land. Half of the all-women FUGs have 10 ha or less, and virtually none have over 50 ha. Typically this is degraded land needing tree planting, while mixed FUGs commonly control a few hundred hectares and usually this is natural forest.[11] Similarly, mixed groups with a high female presence (say 30 per cent or even 50 per cent women in the general body) are found only in selected pockets, as in parts of Gujarat and West Bengal.[12]

Participation in protection

Despite their limited presence as formal members, many women play an active role in protection. In formal terms, the bounded forest area is usually protected either by employing a guard, with CFG members contributing the wage, or by forming a patrol group from among member households. A male guard or an all-male patrol characterized 45 and 18 per cent respectively of the eighty-seven sites I visited. Only a small percentage of patrols were constituted by both sexes, or by women alone, and there was a rare female guard. Occasionally, there are shifts from all-women to all-men patrols, and vice versa. More commonly, women patrol informally. In some villages of Gujarat and the UP hills, they have formed separate informal protection groups parallel to men's, because they feel men's formal patrolling is ineffective.

Women's informal vigilance improves protection in important ways. In most villages I visited, women told me that they had apprehended intruders both from other villages and from their own, and that when they caught women intruders they sought to dissuade them from breaking the rules. In fire fighting, likewise, women join in enthusiastically; and in several instances women's alertness alone has saved the forest.

On the one hand, therefore, most women are excluded from CFG membership and management; on the other hand, many women contribute notably to protection efforts, indicating their stake in forest regeneration. However, women's limited involvement in the decision-making process has implications for both equity and efficiency.

Equity

How equitable are the CFGs in the sharing of costs and benefits?

Cost bearing

The costs of forest closure are broadly of two types: those associated with protection and management and those associated with forgoing forest use (see Table 7.1). The former usually includes costs such as membership fees, the forest guard's pay, the opportunity cost of patrolling time, and so on. These types of costs are borne largely by men. The costs of forgoing forest use include the opportunity cost of time spent in finding alternative sites for essential items such as firewood and fodder, other costs (identified below) associated with firewood shortages, the loss of livelihoods based on non-timber forest products, and so on. These types of costs fall largely on women.

Consider the firewood-related effects in more detail. Firewood is an everyday need, obtaining which is mainly women's responsibility (compared with, say, small timber that is needed occasionally for agricultural implements, etc., and is mainly men's responsibility). Typically when protection begins, all human and animal entry is banned, especially in the semi-arid regions. Where the land was barren anyway this caused no extra hardship. But where earlier women could fulfill at least part of their firewood needs from the protected area, after closure they were forced to travel to neighbouring sites, involving additional

Table 7.1 Main potential costs and benefits of forest closure by gender

Mainly affecting women	*Mainly affecting men*
Costs	
Firewood shortages (more time and energy expended in collection and/or cooking; and adverse health effects)	Membership fee
Fodder shortages (more time and energy expended in collection)	Patrolling time/guard's pay
Increased time in stallfeeding animals	Fodder shortages (purchase)
Informal patrolling time	Loss of source for small timber
Erosion of some livelihoods: for example, firewood sellers, NTFP collectors	Erosion of some livelihoods: for example, blacksmiths using woodfuel in furnaces
Fines if caught stealing firewood	
Enhanced (late entry) membership fee	
Benefits	
Firewood supply (a few weeks, if forest is opened)	Small timber (if allowed)
Fodder supply (a few weeks, if forest is opened)	Housebuilding timber (if allowed)
NTFP (seasonal)	Cash (if distributed) from sale of forest products
	Use of collective fund

Note: This is a broad outline of the main *direct* costs and benefits. Not each of these need apply to every CFG. There may also be some *indirect* costs and benefits. For instance, a greater supply of firewood indirectly benefits the whole family.
NTFP = Non-timber forest products. These are largely collected by women.

time, energy, and (where those sites were also being protected) the risk of being caught and fined.[13] As a result, in the early years of JFM, Sarin (1995) noted that in some protected sites in Gujarat and West Bengal, women's collection time for a headload of firewood had increased from 1–2 hours to 4–5 hours or more, and journeys of half a kilometre had lengthened to 8–9. In many households, women were also compelled to take their daughters along, spending over six hours a day to walk five times farther, for the same quantity of fuelwood (Shah and Shah 1995). Over time this could negatively affect the girls' education. When neighbouring villages too started protecting, many women faced severe shortages. Most sought to make do with the limited firewood available from trees on their home fields, supplemented by inferior fuels which have other negative consequences (described below). But even with such shifts, some (especially the landless) were compelled to continue collecting clandestinely, risking fines and abuse if caught.

Such gendered consequences were widespread, causing considerable resentment among the women. For example, in Pingot village (Gujarat), women, when asked about the award for environmental conservation conferred on the village, responded with bitterness: 'What forest? We used to go [there] to pick fuelwood, but ever since the men have started protecting it they don't even allow us to look at it!' (Shah and Shah 1995: 80).

In most places, this picture has remained largely unchanged even after several years of protection. Of the eighty-seven CFGs I visited in 1998–99, firewood was available in eighty. Of these forty-five (60 per cent) still had a ban on firewood collection, with twenty-one not opening the forest at all and twenty-four opening it for a few days annually for drywood collection and/or cutback and cleaning operations. The remaining CFGs allowed some collection, usually only of fallen twigs and branches. Even after years of protection, women thus reported a persistence of firewood shortages in most of the villages I visited in Gujarat, the UP hills, Karnataka, parts of Madhya Pradesh bordering Gujarat, and in the Kaski and Dang districts of Nepal. The exceptions were some parts of Orissa, Karnataka and Madhya Pradesh with older forests, where collection was allowed on a regular basis from the start.

Some common responses by women are given below:

> We go in the morning and only return in the evening. Since the end of the rainy season, we have been going every day. I go myself and so does my daughter. Earlier too there was a shortage but not as acute (woman EC member to author, Kangod village, Karnataka, 1998).

> It is women who need the forest, they need firewood to cook. . . . Men preach to women about not cutting trees, but what can women do? They cannot cook food without firewood and they cannot collect firewood from other places (group discussion with women in Kabhre Palanchok, Nepal, cited in Hobley 1996: 147).

Where possible, women have tried to substitute other fuels. A few are able to switch to biogas (usually where there is an effective NGO programme), but most turn to twigs, dung cakes, agricultural waste, or even dry leaves. Fire from these latter fuels needs careful tending which increases cooking time and prevents women from simultaneously attending to other work. Moreover, these fuels cause more fumes, with negative health effects in poorly ventilated conditions. In several villages women also report economizing on fuel by forgoing a winter fire for space heating (even in the sub-zero temperatures of the Nepal hills), giving the animals cold feed, not heating winter bath water or heating it only for husbands, and so on.

Usually women from both middle and poor peasant households report such domestic energy problems, since even in the better-off homes firewood is typically gathered and not purchased. Most do not have many trees on their private land. Women of landless or landpoor households are, however, the worst off, since without private land they also have no crop waste or trees of their own, and few cattle for dung.[14] In fact, forest closure has forced many to sell-off parts of their animal stock. As a poor woman in Khut village (UP hills) told me: 'We don't know in the morning if we will be able to cook at night'. Another added: 'Our bahus (daughters-in-law) have to undertake a full day's journey to get a basket of grass and some firewood from the Reserve Forest.' Her daughter-in-law pitched in: 'But even in the Reserve Forest you can be caught by the forest guard. I paid Rs 20 as a fine to retrieve my axe, and all I was doing was cutting a fallen log.' Again in Nepal, in Tallo Goungonda village (Kaski district), a group of poor women told me: 'We go at night . . . Other women have gas and stoves, but we are poor, so we have to steal.'

Similarly, since grazing is usually banned, households with cattle have to procure fodder in other ways. Since cattle care is usually women's responsibility, if the household cannot afford to buy fodder women have to spend additional time looking for other sites to procure some. Moreover, animals now have to be stall-fed. In parts of Gujarat many women report an extra workload of 2–3 hours due to stall-feeding alone. Where some of the better-off households have replaced their goats with stall-fed milch cattle, it has further increased women's labour.

As the forests have regenerated, at best these hardships have been alleviated; they have not disappeared. Firewood shortages, for instance, continue to be reported even 8–10 years after protection in many regions, despite the quite large areas being protected. By one estimate, about 0.2 ha of forest is needed per household for meeting firewood and other subsistence needs, and many Gujarat villages have several times this area (Shah 1997). Certainly, in fifteen of my nineteen Gujarat fieldwork sites, the protected area per capita exceeded this norm (and nine of the nineteen villages were each protecting over 100 ha). Although specific estimates (such as 0.2 ha per household) might be contested by some, many agree that both in Gujarat and elsewhere more could be extracted sustainably than is currently being allowed. This could be supplemented by other measures, such as planning fuelwood plantations in a part of the forest, promoting and maintaining workable biogas plants, and so on. In many places, therefore, the scarcities that women are experiencing appear to have less to do with aggregate availability or a lack of potential solutions than with women's limited bargaining power, whereby their problems are seen as individual/ private rather than as warranting community attention.

Benefit sharing

There are also gender inequities in benefit sharing. In some cases the benefits are not distributed at all. Among Orissa's self-initiated groups, for example, a number of all-male youth clubs have completely banned forest entry and have been selling the wood (obtained from thinning and cleaning operations), as well as other forest produce. In many cases, the quite substantial funds so obtained have been spent on an annual religious festival (my fieldvisits, 1998), or on a clubhouse or club functions (Singh and Kumar 1993).

In other types of CFGs the money is normally put in a collective fund, to be used as the group deems fit. Women typically have little say in how it is used:

> The money obtained from grass and firewood is kept by them in their fund. We have not seen one penny of it. We buy grass, which is auctioned by bundles (women to author, Ghusra village, Dang district, Nepal, 1998).

Where the CFGs distribute the benefits, say in the form of firewood or grass, as in some of the JFM groups, women of non-member households usually receive none, since entitlements are typically linked to membership. Often these are poor households whose members have to migrate for work, or are out all day on wage labour and cannot easily contribute to patrolling or to the guard's wages.

Even in member households usually men alone can claim the benefits directly, either because only they are members, or because entitlements are on a household basis, so that even if both spouses are members they get only one share. Of course women can benefit indirectly in some degree, say if the benefits are in kind (such as firewood); or where member households continue to enjoy the right to collect dry wood or leaves from the protected area.[15] But where the CFGs distribute cash benefits, money given to men does not guarantee equal sharing, or even any sharing, within the family. In fact, outside the context of forest management there is substantial evidence of men in poor households spending a significant part of their incomes on personal items (tobacco, liquor, etc.), while women spend almost all of their incomes on basic household needs.[16] This pattern is found repeated in the context of CFGs. In many cases, men are found to spend the money on gambling, liquor, or personal items.[17]

Many women are aware that unless they receive a share *directly* (rather than through male members), they may get nothing. When asked their views on this at a meeting of three JFM villages in West Bengal, in which both women and men were present, all the women wanted equal and separate shares for husbands and wives (Sarin 1995). Being members in their own right would be one way in which women could benefit directly, provided that the individual and not the household was the unit of entitlement.

Inequities also arise because people differ in their needs, or in their ability to contribute or to pay. Broadly, three types of principles/norms can underlie the distribution of forest products: market-determined, contribution, and need. While seemingly neutral, these distributive principles have notable gender and class implications. The market principle (or willingness to pay) embodied in practices such as the auctioning of grass to the highest bidder, tends to be both unequal *and* inequitable, since those who cannot afford to pay have to do without. Given that rural women, even in rich households, tend to have less access to financial resources than men, auctions tend to be both anti-poor and anti-women. Distribution according to contribution, say by giving each household that contributes to protection an equal number of grass bundles, would be equal but inequitable for those more dependent on the commons for grass, such as the poorer households, and women in general. Moreover, women's ability to contribute may be circumscribed: for instance, even if they wanted to join patrol duty, they may be socially prevented from doing so by norms of seclusion. Only where distribution embodies some concept of economic need, such as where poor women are given rights to an additional grass patch, would the distribution be relatively more equitable, in that those most in need would get more.

In my fieldwork I found that contribution (in terms of membership fees, protection, etc.) was the most common criterion underlying distribution, with all contributing households having equal claims to the fuelwood or grass cut during the forest opening days. However, there were also occasional cases of auctioning, such as the auctioning of grass in the UP hills and Nepal, and of other forest produce by many of Orissa's self-initiated groups. Economic need seldom guided distribution. Hence for poor women in particular, the outcome tended to prove inequitable.

In recent collective action literature, questions of equity have been raised largely in terms of whether existing economic and social inequality affects the possibility of collective action and efficient institutional functioning.[18] There has been a relative neglect of whether or not the *outcomes* of collective action (in terms of, say, cost and benefit sharing) are equitable, and how those outcomes impinge on the sustainability of collective action. As argued above, equity of outcome is important, in itself, for evaluating institutions governing the commons, quite apart from the links between equity and efficiency (as between participation and efficiency) that are elaborated below.

Efficiency

Women's lack of participation in CFG decision-making, and gender inequities in the sharing of costs and benefits from protection, can have a range of efficiency implications. Some initiatives, for example, may fail to take-off at all;

others may not sustain the gains, or there may be a notable gap between the gains realized and those realizable (in terms of resource productivity and diversity, satisfying household needs, enhancing incomes, etc.). Inefficiencies can stem from one or more of the following problems (see also, Agarwal 2000*b*).

First, there are rule violations. In almost all the villages I visited there were at least some cases of rule violation, and at times this was a frequent occurrence. Violations by men usually involve timber for self-use or sale (the latter in areas with commercially valuable trees). Violations by women typically involve firewood. Where a CFG bans collection without consulting women or addressing their difficulties, many women are under great pressure to break the rules, given their daily need for fuelwood. Sometimes, in situations of acute need, women enter into persistent altercations with the guards.[19] In one Gujarat village the guard threatened to resign as a result. Only then did the EC address the issue and agree to open the forest for a few days annually. In Agrawal's study (1999) of a *van panchayat* village, women constituted 70–80 per cent of reported offenders between 1951 and 1991, many of them belonging to poor and low caste households.

A second source of inefficiency lies in inadequate information sharing with women. Information about the rules (especially membership rules), conflicts encountered, or other aspects of forest management, does not always reach women (my fieldvisits, 1998–99). Similarly, male forest officials seldom consult women or seek their feedback when preparing micro-plans for forest development. Some women hear about the plans through their husbands, others not at all (Guhathakurta and Bhatia 1992). Such communication problems can prove particularly acute in regions of high male outmigration.

Third, inefficiencies can arise if the male guard or patrol fails to notice resource depletion. During my 1995 fieldvisit to Gujarat, a women's informal patrol in Machipada village took me to the protected site, and, pointing out the illegal cuttings which the men had missed, noted: 'Men don't check carefully for illegal cuttings. Women keep a more careful lookout'. My subsequent fieldwork in 1998–99 revealed similar differences in several other field sites. This gender difference, at least in part, arises because women, as the main and most frequent collectors of forest products, are more familiar with the forest than men.

Fourth, and relatedly, there are problems in catching transgressors. In virtually all the regions I visited, all-male patrols or male guards were unable to deal effectively with women intruders because they risked being charged with sexual harassment or molestation. Threats to this effect were not uncommon when non-member women, or women from neighbouring villages, were caught. In some incidents, women and their families had even registered false police cases against patrol members, or beaten them up. Equally, however, women on their own find it difficult to do night patrolling or to confront aggressive male intruders. By all accounts, the most efficient solution appears to be a patrol team consisting of both sexes. Recognizing this, in some regions male patrollers have included some village women in their patrol, but this is atypical.

When women voluntarily set up informal patrols, even where there is a male guard or patrol, the efficiency of protection can improve notably. In their study of twelve *van panchayats,* Sharma and Sinha (1993) found that the four which could be deemed 'robust' all had active women's associations. However, in so far as women's groups are typically informal, they lack the authority to punish offenders who still have to be reported to the formal (typically all-male) committees. This separation of authority and responsibility introduces inefficiencies in functioning. For instance, sometimes male EC members fail to mete out punishments to the culprits women catch, causing women to abandon their efforts. I found several such cases in Karnataka, Gujarat and the UP hills. Also, when women catch intruders, they are seldom party to discussions or decisions on appropriate sanctions.

Fifth, and relatedly, efficient functioning requires effective methods of conflict resolution. This is made difficult with women's virtual exclusion from the formal committees, especially where the conflict involves women, as is not infrequently the case with firewood-related intrusions.

A sixth form of inefficiency stems from taking little account of women's knowledge of plants and species when preparing plans for forest regeneration. Women and men are often privy to different types of knowledge due to differences in the tasks they perform, and in their spatial domains. Women, as the main fuel and fodder collectors, can often better explain the attributes of trees than men (Pandey 1990), and can identify a large number of trees, shrubs, and grasses in the vicinity of fields and pastures (Chen 1993). In general, women are better informed about the local environment in which they gather and collect, and men about species found in distant areas (Gaul 1994). Women's systematic exclusion from decision-making and management of new planting programmes is thus likely to have negative efficiency implications, by failing to tap women's knowledge of diverse species for enhancing biodiversity.

A seventh form of inefficiency can arise from ignoring possible gender differences in preferences, say regarding when grass should be cut or which trees should be planted. I found that in the rare cases when women were consulted, they often came up with alternative, more suitable suggestions on when the forest should be opened for forest produce collection. Women are also known to usually prefer trees which have more domestic use value, such as for fuel and fodder, while men more typically opt for trees that bring in cash[20] (the exceptions being cases where fuel and fodder are ample, in which case women too might prefer commercial species: Chen 1993). Women's greater involvement in forest planning would thus better fulfil household needs and increase commitment to the initiative.

Basically, when examined from a gender perspective, it is clear that the CFGs are violating many of the conditions deemed by several scholars to be necessary for building enduring institutions for managing common pool resources. This includes conditions such as ensuring that those affected by the rules participate in framing and modifying the rules; that the rules are simple and fair; that there are effective mechanisms for monitoring the resource and resolving conflicts; and so on.[21]

Despite women's low involvement, forests might regenerate, but some of the initiatives might not sustain, and others might produce less than the full potential benefits.

What determines gendered outcomes?

The gender-related efficiency outcomes discussed above are in large part *secondary* outcomes, stemming from women's low participation in the CFGs and from inequities in the rules of forest use, benefit sharing, etc. Efficiency outcomes are therefore not discussed separately below. Rather, I focus on what underlies women's low participation and the inequities in cost and benefit sharing.

In broad terms, the degree of participation and the distribution of costs and benefits can be seen to depend especially on the following factors: rules, norms, perceptions, the person's individual endowments and attributes, and their household endowments and attributes (which define where they fall within the structural hierarchies of class, caste, etc).

Factors affecting women's participation

Rules

In formal CFGs, such as the JFM groups in India or the FUGs in Nepal, rules determine membership in the general body or the EC. As noted earlier, where the rule restricts membership to only one person per household, the male household head tends to join. The rule that allows one man and one woman per household is somewhat more inclusive; but full inclusiveness would require all adults to be allowed to join. This is rare.

In addition to the rules themselves, a lack of awareness of the rules, or of changes therein, can also constrain women's participation. In West Bengal, for instance, a study of nineteen CFGs showed that even four years after the state order was amended to allow women's inclusion, barely two-fifths of the members knew of the change (Raju 1997).

Among the self-initiated groups (who lack formal membership rules), long-standing conventions, which traditionally excluded women from public decision-making forums, also deny women entry into the CFGs.

Social norms

Even when membership rules are favourable and women join, they seldom attend or speak up at meetings because social norms place strictures on their visibility, mobility, and behaviour. These norms, whether internalized by women or imposed on them by threat of gossip, reprimand, or even violence, impinge directly on women's autonomy and ability to participate effectively in CFGs dominated by men.[22]

Some communities have quite strict female seclusion norms. But more pervasive is the subtle gendering of physical space and social behaviour. For instance, norms often dictate a gender segregation of public space. Women of 'good character' are expected to avoid village spaces where men congregate, such as tea stalls and the market place (Agarwal 1994). For older women, the restriction is generally less, but never fully absent. As a result, many women feel uncomfortable going to CFG meetings, unless explicitly invited by the men:

> The meetings are considered for men only. Women are never called. The men attend and their opinions or consent are taken as representing the whole family—it's understood (woman in a *van panchayat* village, UP hills, cited in Britt 1993: 148).

> Rural women and men can't sit together. But we convey our decisions to them (man to author, Chattipur village, Orissa, 1998).

The gender division of labour is another pernicious norm. The fact that women bear the main responsibility of childcare and housework, in addition to the load of agricultural work, cattle care, etc., makes for high work burdens and logistical constraints. This seriously restricts women's ability to attend lengthy meetings held at inconvenient times. As some women in Barde village (Karnataka, south India) told me in 1998: 'There are problems in attending meetings since we need to cook and serve the evening meal. The meeting is long. We also have to feed the cattle'. Men are usually reluctant to share not just domestic tasks and childcare, but even cattle care. Most women in the *van panchayat* villages she studied, told Mansingh (1991) that they did not have time to 'sit around for [the] four hours that it took to have a meeting in the middle of the day'. As a result women's attendance thinned out over time.

Norms also reduce women's participation by creating subtle gender hierarchies, such as by requiring women to sit on the floor while husbands and older men sit at a higher level on cots, or requiring women to sit at the back of the meeting space where they are less visible and less able to raise a point effectively. Moreover, where senior male family members are present, women either do not attend meetings, or do not oppose men publicly. The hierarchy that marks 'respectful' behaviour in the family also marks community gatherings.[23]

Social perceptions

Incorrect perceptions regarding women's abilities impinge on men's willingness to include women in the CFGs. Men often view women's involvement in CFGs as serving no useful purpose and tend to downplay their potential contributions. Some men's direct responses to my questions are indicative:

> There is no advantage in having women in the EC. We have been told by the forest officials that we must have two women in the committee, and that is why we have included them (male to author, Pathari village, Karnataka).

> Women can't make any helpful suggestions (man to author, Arjunpur village, Orissa, 1998).

> Women are illiterate. If they come to meetings, we men might as well stay at home (EC chairman to author, Ghusra village, Dang district, Nepal, 1998).

In some cases, I found that the men who were decrying my interviewing the women on the grounds that they were illiterate, were themselves illiterate!

Entrenched territorial claims

Men oppose women's inclusion much more strongly once their own claims are entrenched. For instance, where CFGs start out with only male members, or where men feel they have a prior claim to the land, they resist new claimants. Some young men in Basapur village (Karnataka) reacted to the idea of including women in CFGs as follows: 'Women have DWACRA,[24] they have savings groups, why don't you leave the CFGs to us men?' (author's fieldwork, 1998). Men in Asundriya village, Gujarat, strongly opposed NGO attempts to increase women's CFG membership, arguing: 'Why do we need women? What we are doing is ok' (author's fieldwork, 1999). In Kudamunda village, Orissa, when I asked the women who wanted their own separate patch for protection why they needed one, they responded:

> If we have our own forest, we would not need to ask the men each time for a bit of wood.

> They are not willing to give us even a patch to protect. Why would they be willing to give us a whole tree if we asked?

Personal endowments and attributes

Women's lesser access to personal property or political connections reduces the weight of their opinions. In addition, limited experience in public interaction undermines their effectiveness in public forums. Some of these disadvantages can partly be overcome if the women are older, married, have leadership qualities, and the self-confidence to speak up. In many CFGs, the few women members are widows, or older married women living in their parental homes, who often tend to be less socially constrained (Narain 1994; Britt 1993).

Household endowments and attributes

Finally, factors such as the class and caste position of the woman's household are likely to matter where the village is multi-caste and dominated by the upper-caste, or where the CFG is constituted of several villages that are caste/class homogeneous in themselves, but that differ hierarchically in this respect from other villages in the CFG.[25] But the caste factor works in complex ways. On the one hand, being low-caste and poor can adversely affect a person's ability to bargain for a better deal within a predominantly upper-caste community; and even low-caste men (like women in general) often hesitate to speak up at meetings in such contexts. On the other hand, low-caste women are less subject than upper-caste women to norms of seclusion, restricted mobility, and soft speech.

Factors affecting distributional equity

Similar *(but not identical)* factors affect gender inequitable outcomes in terms of costs and benefits. The principal factor underlying gender differences in cost sharing appears to be social norms governing the gender division of labour. As already discussed, women's primary responsibility for firewood and fodder means that women bear the bulk of the costs of forgoing forest use.

Benefit sharing is likely to be affected especially by five types of factors. One, there are the rules regarding entitlements to benefits. Here both entry rules and distribution rules matter. As noted earlier, access to some types of benefits is linked to membership. However, even if both spouses are members, the woman may not get a separate or additional share if the CFG has decided that the household rather than the individual will be the unit of distribution. In recent years, this has in fact proved to be a bottleneck in inducting women members in parts of Gujarat, where women are demanding shares on an individual basis as a condition for their joining. Hence while women's low participation in CFG decision-making affects equity of outcome through the distribution rules, inequitable distribution rules can, in turn, restrict women's participation.

Two, the norms/principles (willingness to pay, contribution, or need) underlying distribution affect equity of benefit sharing. At present (as noted earlier), contribution is the dominant criterion underlying distribution rules in most CFGs. Based on this, all those contributing get equal access to the resource or equal amounts of firewood/fodder when distributed. Auctions are undertaken in some cases, and distribution by economic need is rare.

Three, perceptions about need, contribution, and deservedness matter. Even if there were to be a shift from contribution to need as the defining principle, whether or not women get a better deal can still depend on whether they were *perceived* as deserving more (Agarwal 1997*b*; Sen 1990). There can be and often is a divergence between what a person actually contributes, needs, or is able to do, and perceptions about her/his contributions, needs, and abilities. Hence, for instance, women's contribution to household income is often undervalued, both by family members and by those implementing development programmes, because of the 'invisible' nature of many household tasks that rural women perform. These tasks (such as collecting firewood and fodder, stall-feeding animals, storing and processing grain, etc.) are often economically invisible since they usually do not bring in cash returns; and those done within the home compound are also rendered physically invisible. Hence, women *seen* to be participating in forest management would be better placed to claim equal benefits with men in that their contributions would be better recognized.

Four, whether or not the outcome is equitable depends on pre-existing personal endowments and attributes. Since women as a gender (even if not all women as individuals) have fewer private economic endowments, CFG shares given only to male members typically result in inequitable outcomes for women in both rich and poor households. Again, women's personal attributes such as age and marital status can affect intra-household distribution by influencing perceptions about deservedness.

Five, as we have noted, how acutely women are affected by forest closure or shortages is influenced by their household's economic endowments and social attributes, in particular by their household's class, caste, ethnicity, etc. However, in some respects, this can work in both directions. For instance, while women in upper-caste households that own land and animals can get some fuel and fodder from private assets, they are also likely to face greater social strictures on their mobility, which would limit their options with respect to alternative collection sites. Moreover, for fuelwood, except those able to afford cooking gas, the class difference may not be substantial, since many women even of middle peasant households have to depend mostly on what they themselves can gather.

Improving outcomes for women: the bargaining framework

In what ways can the factors noted above be acted upon to improve outcomes? Some factors predate the forestry programmes and have deep economic and social roots. The programmes could, however, either entrench them further or provide an opportunity for weakening them. Other factors, such as CFG rules, are part of institutional functioning. Both types of factors are constituted at several levels. Rules, for instance, are broadly made at two levels: the State and the community. Membership criteria under JFM are determined at the State level, but whether forest closure should be total or partial, or how different forest products should be distributed, is determined largely by the community. And social norms, social perceptions, and endowments, are constituted and contested at all levels—within the State, the community, the family, and various institutions of civil governance (including NGOs).

A promising analytical framework for examining the possibilities for change on all these counts is that of bargaining. Women's ability to change rules, norms, perceptions, and endowments in a gender-progressive direction would depend on their bargaining power—with the State, the community, and the family, as the case may be. What would affect women's ability to bargain effectively in these three arenas?

Bargaining: some conceptual issues[26]

The State

First, consider bargaining with the State. To begin with, the State too can be seen as an arena of bargaining at multiple levels. For instance, the State may formulate gender-progressive laws at the highest level, but it could face resistance in implementation from the local bureaucracy. Or some departments or ministries may pursue gender-progressive policies within an overall gender-retrogressive State structure (women's ministries are cases in point). Likewise, there are often some gender-progressive individuals within State departments who play key positive roles, typically but not only in response to demands made by interest groups.[27] In other words, the State is an arena of contestation between parties (such as policy-making and policy-implementing bodies), and/or between different regional elements of the State structure with varying commitments to gender equality.

The State might respond positively to demands by gender-progressive groups/NGOs because such groups could build up political pressure, say with the support of opposition parties and/or the media, with implications for voting patterns; or because of pressure from international aid agencies; or because the State recognizes the inefficacy both of market mechanisms and of its own machinery in implementing essential development programmes. In India, the State's attempts since the mid-1980s to enlist NGO support for various developmental projects, including that of community forestry, reflects this recognition.

We would expect women's bargaining strength with the State to depend on a complex set of factors, such as whether they function as a group or as individuals, and the cohesiveness and strength of the group. The bargaining power of such a group is likely to be higher the larger and more unified it is; the greater the political weight carried by the castes/classes of which it is composed; the greater its command over economic resources; the more the support it gets from NGOs, the media, academics, and international donors; and the more State officials are influenced by gender-progressive norms and perceptions.

The community

The second important arena of bargaining is the community. Implicit or explicit bargaining can occur between an individual (or a subset of individuals) and the community over the rules and norms governing, say, economic resource use, and social behaviour, and over the enforcement of those rules and norms. Non-compliance with CFG rules could be seen as a form of implicit bargaining.

As with the State, women's bargaining power within the community would be enhanced if they had support from external agents such as NGOs and the State. Group cohesiveness and strength is also important. For instance, an individual woman breaking seclusion norms could easily be penalized, say by casting aspersions on her character. Such reprisals are less possible if a group of women decide to transgress the norms.[28]

In addition, in a multi-caste/class-heterogeneous village, we would expect women's bargaining power to depend on the socio-economic composition of their group and their ability to command funds. In the sharing of communal resources, for instance, the negotiating strength of low-caste or poor peasant women, even if they formed a group, is likely to be weaker than that of high-caste or rich peasant women whose caste or class as a whole commands greater power in the village.

The family

The third major arena of bargaining is the family. Intra-family bargaining for a more equitable sharing of benefits or tasks, or for greater freedom to participate publicly, is perhaps the most complex aspect of bargaining. This complexity is spelt out in Agarwal (1994, 1997*b*), but broadly four types of factors are likely to impinge on a woman's bargaining power in the home: her personal endowments and attributes (educational level, whether or not she earns an income, property ownership, age, marital status, etc.); her ability to draw upon extra-household support from friends, relatives, women's groups in the village, gender-progressive NGOs outside the village, and the State; social norms (which might define who gets what, or who does what within the household); and social perceptions (say about deservedness).

Some of the common determinants of bargaining power in all three arenas discussed above are the following: support from external agents, social norms and perceptions, and group strength. Norms, perceptions, and group strength require some elaboration.

Social norms can affect bargaining power in both direct and indirect ways. For instance, norms that restrict women's presence in public spaces directly reduce women's ability to bargain for rule changes within CFGs. In addition, they do so indirectly by reducing women's ability to build contacts with NGOs or State officials. Social norms can also influence how bargaining is conducted: for example, covertly or overtly, aggressively or quietly. In cultures or contexts where social norms stifle explicit voice, women may be pushed into using covert forms of contestation within the family, such as persistent complaining or withdrawing into silence (Agarwal 1994). Moreover, attempts to change social norms can themselves constitute a bargaining process.

Social perceptions can affect women's bargaining power in so far as women's contributions and abilities diverge from perceptions about their contributions and abilities. As noted earlier, much of what women do is rendered invisible and therefore undervalued by both families and communities. To the extent that women internalize these perceptions, they can self-restrict their range of options or what they seek to change and bargain over. To enhance women's bargaining power, a necessary step would thus be to change women's own perceptions about their potential options and abilities, as well as the perceptions of their families, the community and the State regarding their abilities and the legitimacy of their claims.

Group strength can prove to be a critical factor at all levels of bargaining and in all forms of bargaining (including over social norms and perceptions). Here village women's group strength derives not merely from the number of women who would like, say, a change in rules and norms, but also from their willingness to act collectively in their common interest, an interest predicated on gender. In other words, it depends on whether gender is a basis of group identity, over and above the possible divisiveness of caste or class. The creation of such a group identity will thus need to be part of the process of improving outcomes for women.

Let us now consider ground experience in attempts to improve women's participation and distributional equity in CFGs. These experiences do not illustrate all elements of the bargaining framework spelt out above, but they reveal some key elements.

Bargaining: ground experience

The State

JFM experience indicates that successfully bargaining with the State for changing the initial rules of entry is not very difficult. Pressure from external agents such as gender-progressive NGOs and key individuals, for instance, has led a number of Indian states to make JFM membership rules more women-inclusive. Here village women did not have to explicitly bargain for changes, but the women's movement in South Asia has brought about a sufficient shift in perceptions regarding gender inequalities to make such issues easier to resolve with the State, through outside intervention. Village women, on this count, thus start from a position of some bargaining strength.

The community

Bargaining with the community to ensure that more women-inclusive membership rules are implemented, and to increase women's effective voice in CFGs, has proved more difficult. On the positive side, some of the gender-progressive NGOs, forest officials, and donors have used their bargaining power with the community to bring about changes in women's favour, sometimes on their own initiative, at other times when village women approached them.

For instance, some Indian NGOs have made high female membership in mixed groups a condition for forming the groups. In Gujarat, one NGO insists on 50 per cent women when starting new CFGs. Similarly, some state-level officials in India have increased women's membership in mixed groups, by stipulating that there should be at least 30 per cent women in the general body, or by refusing to start meetings unless the men also invite the women (Viegas and Menon 1993; Sarin 1998). For distributional equity, likewise, when the staff of a Gujarat-based NGO took up women's complaints about firewood shortages at a CFG meeting, it resulted in a shift from total closure of the forest to its opening for a few days annually. However, for a larger and sustained impact, an active input is required from women themselves.

Left to themselves, women have typically relied on covert forms of bargaining to change distributional rules, such as simply ignoring closure rules, challenging the authority of the patrol group or guard who catches them, persistently complaining, and so on. In some instances, this had led village committees to open the forest for short spells. However, complaining or breaking rules (with the risk of being caught and fined) are seldom the most effective ways of changing the rules. For effective change, women are likely to need more formal involvement in rule making and the bargaining power to ensure changes in their favour.

Ground experience suggests that to bring this about, for a start, a critical mass of vocal women is necessary. This can give women more voice in mixed forums, and help them challenge restrictive social norms and perceptions. As some women interviewed by Britt (1993: 146) in the UP hills stressed: 'without a good majority of women present, it is impossible to express opinions.' There is a growing consensus among gender-progressive NGOs and elements of the State apparatus that to build a mass of vocal women within CFGs will need, as a first step, the formation of separate women's groups. Maya Devi (a Nepalese grassroots activist with long experience in group organizing) put it to me emphatically:

> In mixed groups when women speak men make fun of them, so women need to learn to deal with this. . . . When women join a [separate] group they gradually lose their fear of making fools of themselves when speaking up. . . . Women need their own small groups. This is what I know from my 22 years of experience working with the government and NGOs.

There is less consensus, however, on what type of group this should be. Where all-women CFGs have been formed, many have done well in terms of protection and increasing women's self-confidence. However, so far, all-women CFGs (as noted earlier), have usually arisen in special circumstances, and are still marginal in terms of numbers and area protected. Also, they cannot solve the problem of women's low presence and lack of effective voice in the more typical all-male or mixed CFGs. For this, other kinds of efforts are needed. Toward this end, some rural NGOs have been forming all-women savings-and-credit groups, which, unlike CFGs, do not involve a resource over which there is a generalized community claim. In some regions, more multi-functional women's groups, such as *mahila mangal dals* in the UP hills, or *amma samuhs* in Nepal, are also doing well.

Such separate women's groups (organized around savings or some other issue) have helped build women's self-confidence and experience in collective functioning and promoted a sense of collective identity. They have also increased women's ability to deal with government agencies, improved male perceptions about women's capabilities, and brought about some change in social norms which earlier confined women to the domestic space. The response below is fairly typical:

> Men used to shut us up and say we shouldn't speak. Women learned to speak up in a *sangathan* (group). Earlier we couldn't speak up even at home. Now we can be more assertive and also go out. I am able to help other women gain confidence as well. (woman leader to author, Vejpur village, Gujarat, 1999)

These experiences are not dissimilar to those of many other rural women's groups across South Asia, which too indicate that group strength, external agency support, and activities that enable women to make a visible contribution (especially in monetary terms) can alter social norms and perceptions, and increase the social acceptance of women in public roles. But in many villages, separate women's groups have also sharpened gender segregation in collective functioning. Often women's savings groups are seen as 'women's groups' and the CFGs as 'men's groups'. Basically, working collectively in separate groups does not adequately challenge unequal gender relations or noticeably change the dynamics of *mixed* group functioning. In other words, forming separate women's groups appears to be a necessary condition but not a sufficient one for women's effective participation in the CFGs.

For effective integration, more concerted efforts appear necessary. In a few cases, NGOs working with both women and men have sought to integrate all-women groups with the CFG. An NGO in rural Karnataka, for instance, encourages women's savings groups to discuss CFG functioning, collect CFG membership dues, and persuade women to join the CFG. As a result, in several of its villages, some 80–90 per cent of the women in the savings groups are now in the CFG general body.[29] To bring this about, however, has taken many years of persistent effort and trust building between the NGO, the women, and the villagers.

An alternative approach (to my knowledge yet to be tried) could be to form a women's subgroup within each mixed CFG. Such a subgroup could first meet separately to discuss women's specific forest-related concerns, and then strategically place these concerns in the full CFG meeting. This could also enable female EC members to better represent women's interests within the CFG.

The family

Bargaining within the family, as noted, is one of the most complex issues to tackle, and rather few rural NGOs directly intervene in intra-household relations. Forming all-women groups can, however, have indirect positive effects. For instance, during my fieldvisits I found several cases where a women's group had supported individual women in their negotiations with husbands, or where joining a group had improved women's situation at home.

> There are one or two men who objected to their wives attending our meetings, and said you can't go. But when our women's association came to their aid, the men let their wives go (women to author in Almavadi village, Gujarat, 1998).

> My husband feels I contribute financially, take up employment, obtain credit for the home. This increases his respect for me (woman to author, Almavadi village, Gujarat, 1998).

In other words, group strength and women's visible contributions can help weaken restrictive social norms, and improve a man's view of his wife's deservedness. However, some norms, such as the gender division of domestic work, are particularly rigid. Also gender inequalities in economic endowments remain entrenched, putting women in a considerably weaker bargaining position in the family, relative to men (Agarwal 1994, 1997*b*).

Finally, any group, including a CFG, is likely to be affected not only by its immediate locale, but also by the wider context of structural and cultural inequalities within which it is located. For instance, both participation and distributional equity are affected by the pre-existing inequalities predicated on the caste and class of women's households, as well as on gender. These inequalities are unlikely to decline substantially within the parameters of CFG functioning. For instance, greater participation in CFGs alone is unlikely to notably improve the economic endowment position of women *vis-a-vis* men or of the poor *vis-a-vis* the rich. To change this would need more wide-ranging measures to enhance the access of women, and of poor and low-caste households in general, to land and other assets.

Conclusion

CFGs are a significant example of group functioning. While many have done quite well in regenerating the environment (at least in an immediate sense), they have been less successful in bringing about women's participation in CFG decision-making, or in ensuring gender equity in the sharing of costs and benefits from forest protection. As a result, they have also failed to tap the full potential of the collective effort. Improving participation and equity is thus important both in itself, and because it can prove complementary to (rather than, as usually assumed, in conflict with) efficiency.

This analysis shows that for more participatory, equitable, and efficient outcomes, it appears necessary that there are changes in factors such as rules, norms, and perceptions, and the pre-existing structural inequalities in endowments and attributes of women's households and of women themselves.

As argued here, it is useful to conceptualize such change within a bargaining framework, and to act on the factors that will strengthen women's bargaining power with the State, the community and the family. This has been achieved to some degree through the intervention of external agents, such as NGOs, forest officials and donors, who in some cases have acted both directly and indirectly, the latter especially by forming separate women's groups at the village level to enhance women's self-confidence and collective strength. At the same time, the analysis cautions that such separate women's groups can also lead to greater gender segregation, unless conscious steps are taken to integrate these women's groups with mixed CFGs. An alternative approach, which might work better, is to form women's subgroups within each CFG. In either case, these are only a few steps among the many that will be needed to transform mixed CFGs into more gender-egalitarian institutions.

Notes

This is a shorter, updated, and somewhat different version of a paper presented at the Conference on Group Behaviour and Development, held at UNU/WIDER, Helsinki, September 1999. The longer version has appeared as an UNU/WIDER working paper (Agarwal 2000*a*). I am most grateful to the conference participants for their responses, and especially to Judith Heyer for her comments on an earlier draft.

1 In India, the term 'state' relates to the biggest administrative divisions within the country and is not to be confused with 'State', used throughout this paper in the political economy sense of the word. In Nepal, the biggest administrative divisions are 'districts'. In India, districts are smaller divisions within states.
2 I will be using CFG as a general term to cover all types of community forestry groups.
3 The government, however, retains the right to reclaim any forests seen to be mismanaged by the FUGs.
4 See, e.g., Raju *et al.* (1993); Kant *et al.* (1991); and SPWD (1998).
5 See, Viegas and Menon (1993) and Chopra and Gulati (1997).
6 Raju *et al.* (1993); Arul and Poffenberger (1990); also my fieldvisits in 1995 and 1998–99.
7 Roy *et al.* (1992); Guhathakurta and Bhatia (1992); and Narain (1994); also my fieldvisits 1998–99.
8 Kant *et al.* (1991); Singh and Kumar (1993); and my fieldwork in 1998–99.
9 Sharma and Sinha (1993); Tata Energy Research Institute (TERI 1995); also my fieldwork in 1998–99. In the TERI study, out of the fifty *van panchayat* examined, only nine had any women members.
10 For example, Mukerjee and Roy (1993); Correa (1997); Adhikari *et al* (1991); Mansingh (1991); Regmi (1989); Singh and Burra (1993); and Raju (1997); also my fieldwork in 1998–99.
11 Calculated from figures given in Government of Nepal (2000).
12 Narain (1994); Viegas and Menon (1993); also my fieldvisits in 1998–99.
13 Sarin (1995); Agarwal (1997*a*): also my fieldwork in 1998–99.
14 See also Jodha (1986) on differences between landed and landpoor rural households in India, in their relative dependence on the commons for firewood and fodder.
15 Kant *et al.* (1991); ISO/Swedforest (1993); Arul and Poffenberger (1990); also my fieldwork in 1998–99.
16 See, Mencher (1988), and Noponen (1991) for India.
17 Guhathakurta and Bhatia (1992), and my fieldwork in 1998–99.

18 See e.g. Ostrom (1990); Bardhan (1993); Baland and Platteau (1996, 1999).
19 E.g. Shah and Shah (1995); Singh and Kumar (1993); Agarwal (1997*a*); also my field interviews during 1998–99.
20 See, e.g. Brara (1987) and Hobley (1996).
21 See, especially, Ostrom (1990) and Baland and Platteau (1996).
22 See also Stewart's (1996) more general discussion on the function of norms in hierarchical contexts.
23 See also, Raju (1997).
24 DWACRA: Development ofWomen and Children in Rural Areas. This is an anti-poverty programme of the Indian government under which, among other things, women's groups are given subsidized loans for income-generating activities.
25 My fieldwork in 1998–99. See also, Sarin (1998) and Hobley (1996).
26 For elaboration, see Agarwal (1997*b*).
27 See also, Sanyal (1991) and Agarwal (1994).
28 For elaboration and illustrative examples, see Agarwal (1994).
29 Personal communication in 1998 from Pratibha Mundergee, former worker in this NGO.

References

Adhikari, N., Yadav, G., Ray, S. B., and Kumar, S. (1991). Process Documentation of Women's Involvement in Forest Management at Maheshpur, Ranchi. In R. Singh (Ed): *Managing the Village Commons, Proceedings of the National Workshop*, December 15–16, Indian Institute of Forest Management, Bhopal, 118–123.

Agarwal, B. (1987). Under the Cooking Pot: The Political Economy of the Domestic Fuel Crisis in Rural South Asia. *IDS Bulletin*, 18 (1), 11–22.

Agarwal, B. (1994). *A Field of One's Own: Gender and Land Rights in South Asia.* Cambridge: Cambridge University Press.

Agarwal, B. (1997a). Environmental Action, Gender Equity and Women's Participation. *Development and Change*, 28 (1), 1–44.

Agarwal, B. (1997b). Bargaining and Gender Relations: Within and Beyond the Household. *Feminist Economics*, 3 (1), 1–51.

Agarwal, A. (1999). State Formation in Community Spaces: Control over Forests in the Kumaon Himalaya, India. Paper prepared for presentation at the *Workshop on Environmental Politics*, University of California, Berkeley, April 30.

Agarwal, B. (2000a). *Group Functioning and Community Forestry in South Asia: A Gender Analysis and Conceptual Framework*, Working Paper No. 172. Helsinki: The United Nations University and the World Institute for Development Economics Research.

Agarwal, B. (2000b). Conceptualizing Environmental Collective Action: Why Gender Matters. *Cambridge Journal of Economics*, 24 (3), 283–310.

Arul, N. J., and Poffenburger, M. (1990). FPC Case Studies. In R. S. Pathan, N. J. Arul, and M. Poffenburger (Eds), *Forest Protection Committees in Gujarat: Joint Management Initiative*, Working Paper No. 7. New Delhi: Ford Foundation, 13–25.

Baland, J. M., and Platteau, J. P. (1996). *Halting Degradation of Natural Resources: Is there a Role for Rural Communities?* Oxford: Clarendon Press.

Baland, J. M., and Platteau, J. P. (1999). The Ambiguous Impact of Inequality on Local Resource Management. *World Development*, 27 (5), 773–788.

Bardhan, P. (1993). Analytics of the Institutions of Informal Cooperation in Rural Development. *World Development*, 21 (4), 633–639.

Brara, R. (1987). *Shifting Sands: A Study of Rights in Common Pastures*. Mimeo, Institute of Development Studies, Jaipur.

Britt, C. (1993). *Out of the Wood? Local Institutions and Community Forest Management in Two Central Himalayan Villages*. Draft monograph, Cornell University, Ithaca.

Chen, M. (1993). Women and Wasteland Development in India: An Issue Paper. In A. Singh, and N. Burra (Eds), *Women and Wasteland Development in India*. Delhi: Sage Publishers, 21–90.

Chopra, K., and Gulati, S. C. (1997). Environmental Degradation and Population Movements: The Role of Property Rights. *Environment and Resource Economics*, 9 (4), 383–408.

Correa, M. (1997). *Gender and Joint Forest Planning and Management: A Research Study in Uttara Kannada District, Karnataka*. Dharwad: India Development Service.

Dahal, D. R. (1994). *A Review of Forest User Groups: Case Studies from Eastern Nepal*. Kathmandu: International Center for Integrated Mountain Development (ICIMOD).

Gaul, K. K. (1994). *Negotiated Positions and Shifting Terrains: Apprehension of Forest Resources in the Western Himalaya*. PhD dissertation, Department of Anthropology, University of Massachusetts, Amherst.

Government of Nepal (2000). *Forest User Group Statistics*. Kathmandu: Community and Private Forestry Division, Department of Forests.

Guhathakurta, P., and Bhatia, K. S. (1992). *A Case Study on Gender and Forest Resources in West Bengal*. Delhi: World Bank, June 16.

Hobley, M. (1996). *Participatory Forestry: The Process of Change in India and Nepal*. London: Overseas Development Institute.

ISO/Swedforest (1993). *Forests, People and Protection: Case Studies of Voluntary Forest Protection by Communities in Orissa*. New Delhi: Swedish International Development Agency (SIDA).

Jodha, N. S. (1986). Common Property Resources and the Rural Poor. *Economic and Political Weekly*, 21 (27), 1169–1181.

Joshi, S. (1998). *Report of the Workshop on JFM and Women*. Netrang, Gujarat: Agha Khan Rural Support Programme, September 14.

Kant, S., Singh, N. M., and Singh, K. K. (1991). *Community-Based Forest Management Systems: Case Studies from Orissa*. Delhi: Swedish International Development Agency (SIDA); Bhopal: Indian Institute of Forest Management; Delhi: ISO/Swedforest.

Mansingh, O. (1991). *Community Organization and Ecological Restoration: An Analysis of Strategic Options for NGOs in Central Himalaya, with Particular Reference to the Community Forestry Programme of the NGO Chirag*. MA dissertation in Rural Development, African and Asian Studies, University of Sussex.

Mencher, J. P. (1988). Women's Work and Poverty: Contributions to Household Maintenance in South India. In D. Dwyer and J. Bruce (Eds), *A Home Divided: Women and Income in the Third World*. Stanford: Stanford University Press, 99–119.

Moffatt, M. (1998). *A Gender Analysis of Community Forestry and Community Leasehold Forestry in Nepal with a Macro-Meso-Micro Framework*. MA dissertation in Development Policy Analysis, Department of Economics and Social Studies, University of Manchester.

Mukerjee, R., and Roy, S. B. (1993). *Influence of Social Institutions on Women's Participation in JFM: A Case Study from Sargarh, North Bengal*, Working Paper No. 17. Calcutta: IBRAD.

Narain, U. (1994). *Women's Involvement in Joint Forest Management: Analyzing the Issues*. Draft paper, Ford Foundation, New Delhi, May 6.

Natrajan, I. (1995). Trends in Firewood Consumption in Rural India. *Margin*, 28 (1), 41–45.

Noponen, H. (1991). The Dynamics of Work and Survival for Urban Poor: A Gender Analysis of Panel Data from Madras. *Development and Change*, 22 (2), 233–260.

Ostrom, E. (1990). *Governing the Commons*. Cambridge: Cambridge University Press.

Pandey, S. (1990). *Women in Hattidunde Forest Management in Dhading District, Nepal*, MPE Series No. 9. Kathmandu: International Center for Integrated Mountain Development (ICIMOD).

Raju, M. (1997). *Seeking Niches in Forest Canopy: An Enquiry into Women's Participation.* Mimeographed report, Ford Foundation, Delhi.

Raju, G., Vaghela, R., and Raju, M. S. (1993). *Development of People's Institutions for Management of Forests*. Ahemdabad: VIKSAT.

Regmi, S. C. (1989). *Female Participation in Forest Resource Management. A Case Study of a Women's Forest Committee in a Nepalese Village*. MSc dissertation in Social Development, Graduate School of the Ateneo de Manila University, Philippines.

Roy, S. B., Mukerjee, R., and Chatterjee, M. (1992). *Endogenous Development, Gender Role in Participatory Forest Management.* Calcutta: IBRAD.

Sanyal, B. (1991). Antagonistic Cooperation: A Caste Study of Non-Governmental Organizations, Government, and Donors Relationships in Income-Generating Projects in Bangladesh. *World Development*, 19 (10), 1367–1379.

Sarin, M. (1995). Regenerating India's Forest: Reconciling Gender Equity and Joint Forest Management. *IDS Bulletin*, 26 (1), 83–91.

Sarin, M. (1998) *Who is Gaining? Who is Losing? Gender and Equality Concerns in Joint Forest Management.* New Delhi: Society for Promotion of Wasteland Development.

Seeley, J. (1996). Who benefits from Participatory Forest Management. *Banko Janakari*, 6 (1), 38–39.

Sehgal, S. (2001). Joint Forest Management: A Decade and Beyond. Paper presented at the workshop on *Policy Implications of Knowledge with Respect to Common Pool Resources in India*, The Institute of Economic Growth, Delhi, September 14.

Sen, A. K. (1990). Gender and Cooperative Conflicts. In I. Tinker (Ed), *Persistent Inequalities: Women and World Development.* New York: Oxford University Press, 123–149.

Shah, A. (1997). Jurisdiction versus Equity: Tale of Two Villages. *Wastelands News*, February–April, 5863.

Shah, M. K., and Shah, P. (1995). Gender, Environment and Livelihood Security: An Alternative Viewpoint from India. *IDS Bulletin*, 26 (1), 75–82.

Sharma, A., and Sinha, A. (1993). *A Study of the Common Property Resources in the Project Area of the Central Himalaya Rural Action Group*. Mimeo, Bhopal, Indian Institute of Forest Management, Madhya Pradesh.

Singh, A., and Burra, N. (Eds) (1993). *Women and Wasteland Development in India.* New Delhi: Sage Publications.

Singh, N., and Kumar, K. (1993). *Community Initiatives to Protect and Manage Forests in Balangir and Sambalpur Districts*. Mimeo, Swedish International Development Agency (SIDA), New Delhi.

SPWD (1998). *Joint Forest Management Update, 1997*. New Delhi: Society for Promotion of Wastelands Development.

Stewart, F. (1996). Groups for Good or Ill. *Oxford Development Studies*, 24 (1), 9–25.

Tata Energy Research Institute (TERI) (1995). *Community Participation in Van Panchayats of Kumaon Region of Uttar Pradesh*, Paper No. 1, Part I. Delhi: TERI.

Upadhyay, S., and Jeddere-Fisher, K. (1998). *An Analysis of Community Forestry Characteristics in the Dhaulagiri Hills*, Report N/NUKCFP/09. Kathmandu: Nepal-UK Community Forestry Project.

Viegas, P., and Menon, G. (1993). Bringing Government and People Together: Forest Protection Committees of West Bengal – Role and Participation of Women. In A. Singh and N. Burra (Eds), *Women and Wasteland Development in India*. New Delhi: Sage Publications, 71–210.

Part III

Agroforestry, tenure, and conceptual framing in Africa

Africanist scholars have made a significant contribution to understanding how gender relations shape and are (re)produced through environmental management processes. At a micro scale, much of their work has built on and contributed to studies that depict the household as the "site of separable, often competing interests, rights and responsibilities" conditioned by gender norms and ideologies (Guyer and Peters 1987, p. 210; Folbre 1986; Guyer 1988; Udry 1996). Studies emerging from some African contexts have suggested that a fluid and shifting "conjugal contract" (Whitehead 1981) encodes spouses' expectations of one another,[1] defining the terms under which rights and responsibilities as well as access to resources and services are negotiated and exchanged among household members (Jackson 1995).

Many researchers have stressed the gendered asymmetries regarding tenure, labour, decision-making, and income opportunities in tree-based and agricultural systems (Carney and Watts 1991; Leach 1991a; Pearson and Jackson 1998; Schroeder 1999). As they have investigated these asymmetries, feminist political ecology emerged as a valuable (albeit loosely defined) approach by which to contextualize processes of environmental change across scales, through the lens of gendered power relations, and in historical perspective (Rocheleau *et al.* 1996; Elmhirst 2011). Moving beyond the micro scale of the household to examine how environmental management is shaped by macro processes in the political economy was a major contribution of the approach. Schroeder's works (Schroeder and Suryanata 1996; Schroeder 1999), including the one shared in this volume, represent some of the early influential pieces in this field.

The two chapters in Part III focus on gender and agroforestry. Schroeder demonstrates how shifts in international development tendencies differentially favoured women's and men's interests in two Gambian villages. Women's gardens prospered under the aegis of Women in Development (WID) initiatives even as donor interest in 'environmental stabilization' supported men's agroforestry projects on lands that had been allocated to women's gardens. The success of one production system threatened that of the other as female gardeners depended on usufruct rights and shade-free spaces to grow their vegetables on male-owned land. Men relied on women's labour to water their trees, which eventually shaded out women's production. Such experiences constitute vital reminders that purely technical solutions to forestry and agroforestry challenges are insufficient; we need to address gender and forests from a holistic perspective that also examines social relations and equity possibilities.

In this case, gendered property rights were renegotiated and reaffirmed as women planted trees, which men then removed, replanting others as their own. As men's trees shaded women's gardens, women were accused of sabotaging them. Trees thus became the physical locus of a material and ideological battle over gender rights and roles. Faced with the threat of eviction by the gardens' senior male landowners, women organized and collectively won the support of national authorities and international donors. Meanwhile, within the household, husbands sought some of their wives' garden profits and women assumed new financial responsibilities. In the face of change, the conjugal contract was renegotiated as "the fundamental determination of what it means to be a man or woman in Mandinka society (Carney and Watts 1991) and success or failure in negotiating the ecological challenges posed by persistent drought" were at stake (Schroeder 1993, p. 355 in original). The struggle to redefine gender roles and resources, which became explicit in local men's metaphor of the garden as women's 'second husbands', was thus closely linked to disputes over environmental resources and related economic outcomes.

Elias and Carney's work (this volume) shares a number of similarities with Schroeder's. They too contextualize environmental change within larger markets and international development tendencies. These contexts have created new opportunities for women in Burkina Faso to market shea butter (made from the nuts of *Vitellaria paradoxa*), a non-timber tree product. Moreover, they argue that agricultural policies have gender implications: new forms of mechanization using animals and tractors are dominated by men. And the new crops encouraged by such mechanization led to the removal of shea trees from cultivated fields. They further argue that women's need for income has drawn them progressively further into international markets for shea butter, often in new configurations as shea butter production moves to the collective space of cooperatives. As in the Schroeder case, Elias and Carney describe how gendered access rights to shea nuts and to income from their sale are reconfigured as development projects and the international shea value chain reach into Burkinabè agroforests.

Whereas Schroeder's study focuses on tree-rich gardens, Elias and Carney examine agroforestry parklands (managed landscapes of mixed vegetation where locally valued tree species are protected when fields are cleared for agriculture). Elias and Carney's attention is on how shea parklands reflect the cultures – and constitutive gendered knowledges, meanings, and social relations – with which they co-evolved and of which they are a part. They demonstrate that the "tangible and immaterial features of agroforestry systems represent a natural heritage" with "the shea tree, as well as the expertise and conventions that accompany its use, represent[ing] one such natural heritage that is integral to cultural and biodiversity conservation" (Elias and Carney 2007, p. 37, original). The knowledge required to process shea butter (re)produces female identity, and the physical features (size, type, and content of nut fat) of shea trees as well as their densities reflect gender-specific cultural preferences for tree specimens with specific traits. The authors highlight that new markets for shea butter require women to adopt standardized processing techniques that depart from the ethnically differentiated and place-specific methods they had practiced. These had responded to seasonal labour and resource shortages and resulted in butter with specific characteristics and flavours. International demands and policies thus reach down into "the very processes that have long formed the cultural heritage of shea" (p. 52, original).

Schroeder's and Elias and Carney's pieces – both primarily based in ethnographic methods with reliance on supplementary secondary sources and national statistics – draw attention to key issues in relation to gender and forests. They share a concern for women's labour, for the ways the meanings imbued in landscapes influence environmental outcomes, and for the gendered access rights that mediate women's and men's use of agroforestry resources. These themes have been and remain prominent in the Africanist gender and forests literature. Of particular interest in Schroeder's account is his sophisticated analysis of gendered negotiations around rights and resources at the household and community levels. These are manifest in the subtle and overt forms of everyday resistance and strategies local women and men used to capture benefits. Women's refusal to water men's seedlings or men's tactics for tapping into their wives' incomes are prime examples of this (see, similarly, Scott 1985; Carney and Watts 1990; Leach 1991b). Elias and Carney highlight gendered knowledge as constituent of female (in this case) identity and a female heritage. They clarify the manifestation of this knowledge in the physical features of the landscape, and show how these enhance our understanding of how gender is (re)produced in, and (re)produces, agroforests. They additionally explore how women negotiate between 'traditional' forms of cooperation in shea butter production and new production arrangements in formal associations emerging in a context of restructured global markets. Both studies flag that globally induced changes to local, historically rooted, and highly gender-specific environmental management strategies can provide not only challenges but also new opportunities for women. They show strong links between sociocultural and environmental outcomes, and the value of contextually and historically informed analyses across scales. They further demonstrate the importance of in-depth qualitative work that goes beyond what is seen at 'face value' to understand how gendered meanings and practices shape tree-based landscapes.

Note

1 Such expectations can also characterize and be complicated by generational differences, intersectionally.

References

Carney, J., and Watts, M. 1990. Manufacturing Dissent: Work Gender and the Politics of Meaning in a Peasant Society. *Africa* 60 (2): 207–241.

Carney, J., and Watts, M. 1991. Disciplining Women? Rice, Mechanization and the Evolution of Gender Relations in Senegambia. *Signs* 16 (4): 651–681.

Elias, M., and Carney, J. 2007. African Shea Butter: A Feminized Subsidy from Nature. *Africa* 77 (1): 37–62.

Elmhirst, R. 2011. Introducing New Feminist Political Ecologies. A special issue of *Geoforum* 42 (2): 129–132.

Folbre, N. 1986. Hearts and Spades. *World Development* 14 (2): 245–255.

Guyer, J. I. 1988. Dynamic Approaches to Domestic Budgeting: Cases and Methods from Africa. In D. Dwyer and J. Bruce, eds, *A Home Divided: Women and Income in the Third World*. Palo Alto, CA: Stanford University Press, 155–172.

Guyer, J., and Peters, P. E. 1987. Introduction. Conceptualizing the Household: Issues of Theory and Policy in Africa. A special issue of *Development and Change* 18 (2): 197–214.

Jackson, C. 1995. From Conjugal Contracts to Environmental Relations: Some Thoughts on Labour and Technology. *IDS Bulletin* 26 (1): 33–39.

Leach, M. 1991a. Locating Gendered Experience: An Anthropologist's View from a Sierra Leonean Village. *IDS Bulletin* 22 (1): 44–50.

Leach, M. 1991b. Engendered Environments: Understanding Natural Resource Management in the West African Forest Zone. *IDS Bulletin* 22 (3): 17–24.

Pearson, R., and Jackson, C. 1998. Introduction: Interrogating Development: Feminism, Gender and Policy. In C. Jackson and R. Pearson, eds, *Feminist Visions of Development: Gender Analysis and Policy*. New York: Routledge, 1–16.

Rocheleau, D., Thomas-Slayter, B., and Wangari, E., eds. 1996. *Feminist Political Ecology: Global Issues and Local Experiences*. New York: Routledge.

Schroeder, R. A. 1993. Shady Practice: Gender and the Political Ecology of Resource Stabilization in Gambian Garden/Orchards. *Economic Geography* 69 (4): 349–365.

Schroeder, R. 1999. *Shady Practices*. Berkeley, CA: University of California Press.

Schroeder, R., and Suryanata, K. 1996. Gender and Class Power in Agroforestry Systems: Case Studies from Indonesia and West Africa. In R. Peet and M. Watts, eds., *Liberation Ecologies: Environment, Development, Social Movements*. New York: Routledge, 188–204.

Scott, J. C. 1985. *Weapons of the Weak: Everyday Forms of Peasant Resistance*. New Haven and London: Yale University Press.

Udry, C. 1996. Gender, Agricultural Production, and the Theory of the Household. *Journal of Political Economy* 104 (5): 1010–1046.

Whitehead, A. 1981. "I'm Hungry Mum": The Politics of Domestic Budgeting. In K. Young, C. Wolkowitz, and R. McCullaugh, eds., *Of Marriage and the Market: Women's Subordination Internationally and its Lessons*. London: Routledge Kegan Paul, 93–116.

8 Shady practice

Gender and the political ecology of resource stabilization in Gambian garden/orchards

Richard A. Schroeder

> Rights over resources such as land or crops are inseparable from, indeed are isomorphic with, rights over people.
>
> (Watts 1992)

Since the first critiques of human ecology were presented by political economists in the late 1970s and early 1980s, political ecology has focused on identifying root causes of environmental degradation. This involves removing blame from victims and exposing the underlying political economic forces leading to resource deterioration.[1] While this effort continues,[2] recent developments suggest that the adoption of a parallel research agenda is in order. The popular outcry over such issues as global warming, ozone depletion, and acid rain, and the proliferation of "sustainability" motifs in development discourse, have prompted actors at all geographic scales to take unprecedented steps toward actively rejuvenating the resource base. A set of conflicts over reclamation and renewal—a politics of stabilization—has thus emerged, warranting critical attention.

Blaikie and Brookfield (1987) make one of the few attempts to explicitly theorize stabilization politics. They contend that "purposive," "long-lived" improvements to land resources ("landesque capital") often entail either the harnessing of slow-paced, regenerative natural processes or capital investments of a magnitude such that they can be recouped only over relatively long payback periods. Such conditions mitigate against "resource managers" investing either personal labor or private capital to reclaim resources. Instead, economic disincentives to stabilization efforts are often overcome by exploiting social and political advantages to coerce labor or capture outside subsidies.

This position is developed to some extent in the literature on agrarian change. Mann and Dickinson (1978) and S. Mann (1990) argue that low rates of return due to the inability to speed up natural processes have discouraged capitalist penetration of agriculture, perpetuating peasant and "family farm" forms of

Richard A. Schroeder, "Shady Practice: Gender and the Political Ecology of Resource Stabilization in Gambian Garden/Orchards," *Economic Geography* 69 (4), 349–365.

production that compensate through the exploitation of family labor. An extended debate has also highlighted the use of coercive labor relations to reclaim land through large-scale irrigation works.[3] It is thus important to recognize that environmental stabilization programs are no more politically "neutral" than many of the processes of degradation documented by political ecologists to date.

In the African context, gender and household-level social relations play key roles in organizing access to, and control over, productive resources (Berry 1989). The household unit is identified as "the site of separable, often competing, interests, rights and responsibilities" (Guyer and Peters 1987, 210). Familial social relations, once presumed to revolve around a natural core of "maternal altruism," are now viewed as governed by an ever-changing "conjugal contract" (Whitehead 1981, 88, 107; Harris 1981; Guyer 1984). Household subsistence budgets (Muntemba 1982; Folbre 1986; Guyer 1988; Dwyer and Bruce 1988), access to land (Roberts 1979; Dey 1981; Longhurst 1982; Davison 1988), and the allocation and remuneration of intrahousehold labor (Jones 1986; Roberts 1988; Carney 1988; Carney and Watts 1990, 1991) are seen as products of protracted negotiation and accommodation between men and women.

This paper explores the notion that household dynamics play a role in the new politics of environmental stabilization. Case materials are drawn from an ethnographic study conducted in 1989 and 1991 in The Gambia's North Bank Division (see Figure 8.1), where two decades of drought-related ecological changes have animated competition between male and female crop production systems over low-lying land and groundwater resources. Dozens of lucrative communal market gardens controlled by women are now under threat from largely male-dominated fruit orchards established in the same locations. By planting orchards (and wood-lots) directly on top of gardens that function as women's income-generating projects, male landholders reap a double benefit: from the subsidy paid by developers to install infrastructure (wells and fences) and from unpaid female labor, which the men manipulate to water their trees.[4]

On the face of it, this situation appears to be a fairly straightforward conflict between two social groups, each attempting to use commodity production as a strategy to withstand the shock of structural economic adjustment. Overlapping property rights to key resources and reciprocal claims on labor and income make the case compelling from the standpoint of gender and household studies (Schroeder and Watts 1991; Carney 1992). The main point of interest for this paper, however, is that the practice of tree-cropping in women's gardens has been constructed by environmentally conscious developers as a means to "stabilize" land resources. In many villages, funding for orchard projects now effectively replaces the equity-oriented women in development (WID) emphasis of the previous decade (Thoma 1989; R. Mann 1989; Worldview International Foundation 1990). In this regard, it is clear that the political ecology of one of The Gambia's premiere horticultural districts has become gendered to a remarkable degree.

The case study is divided into five sections. The first outlines the economic circumstances behind the boom in market gardening and traces their immediate

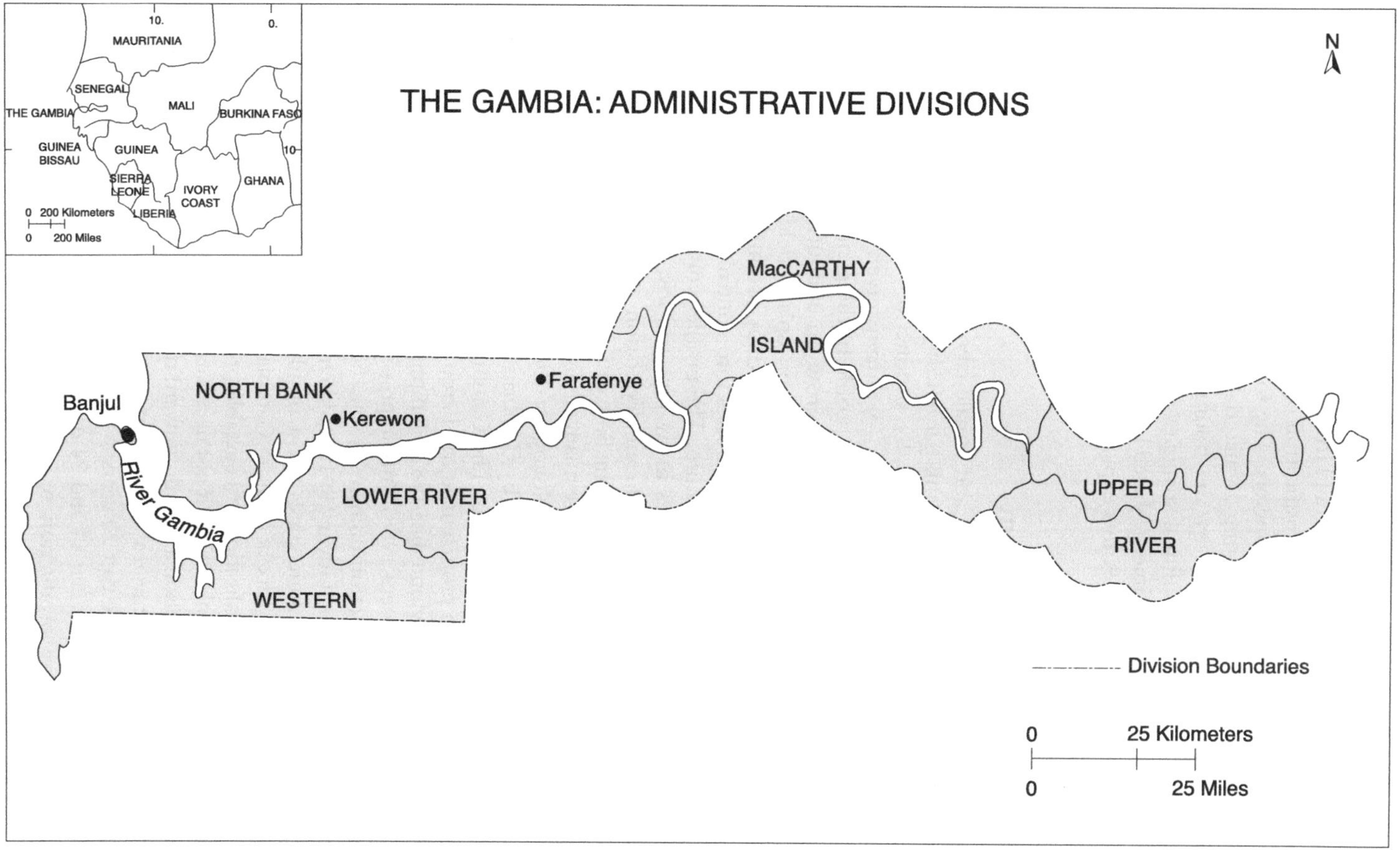

Figure 8.1 The Gambia: Administrative divisions.

effects on gardening practice in the Mandinka village of Kerewan. The second emphasizes opposing strategies adopted by husbands and wives to control the cash women earn from vegetable and fruit sales. The third links intrahousehold budget negotiations to the intensification of both gardening and tree cropping within communal garden perimeters. The fourth illustrates the critical significance of trees and tree planting as material and symbolic expressions of the gender dynamics produced by the boom. The paper concludes with the argument that parallel political struggles relating to the force of household budgetary obligation and lineage-based land use practices have created openings which development agencies now exploit to advance an agenda of environmental stabilization.

Roots of the garden boom

The primary allocation of male and female labor resources in the Mandinka villages comprising the North Bank garden district has, until recently, been constructed along lines designated by production: (1) of specific crops; (2) within particular spatial domains; (3) during given seasons; and (4) for returns of differing value. A simplified profile of the gender division of labor would indicate, accordingly, that men grow groundnuts (peanuts) and the coarse grains (millet, sorghum, and maize) on upland fields during the rainy season; and that their domination of groundnut production, the country's main source of foreign exchange, translates into control over most of the cash income generated through agriculture. The profile would also emphasize that women grow rice and vegetables in swamps and low-lying areas; that their gardens constitute the only significant dry-season activity in the traditional Mandinka labor calendar; and that the bulk of the produce they grow is strictly for home consumption (for a more nuanced discussion of this topic, see Carney and Watts 1991; Boughton and Novogratz 1989).

This picture is largely borne out by Gambian government statistics on national cropping patterns (Government of The Gambia 1991). It also conforms closely to the classic male cash crop/female food crop dichotomy exhibited elsewhere in Africa (Guyer 1980; Muntemba 1982; Hemmings-Gapihan 1982; MacCormack 1982; Schoepf and Schoepf 1988). Any impression that the division of labor into male and female production domains is somehow rigid and unchanging would be misleading, however (Guyer 1991; Daddieh 1989). Climate change, market fluctuations for groundnut sales, and a number of other factors have quite significantly altered the mix of crops in Gambian farming systems.

The rapid commercialization of vegetable production has taken place against the backdrop of an unambiguous (if not necessarily unprecedented; cf. Nicholson and Flohn 1980; Nicholson 1986) downward trend in rainfall averages. Norton et al. (1989) report that mean annual precipitation fell between 24 and 36 percent across the country in the past 20 years. Moreover, the length of the rainy season itself has decreased between 14 and 24 days over the same period. This has led farmers to increase production of a number of shorter-duration, drought-resistant, or cash-earning crops better suited for both ecological and economic reasons to the new growing conditions (Posner and Gilbert 1987). The adoption of shorter-duration rice varieties, for example, which can be harvested as much as two months earlier than the cultivars grown previously, has freed women to take up other endeavors. Instead of producing solely to meet joint household food needs, they now devote a significantly greater proportion of their labor to irrigated cash crops grown during the dry season.

These changes merely set the stage for several developments in the horticultural sector over the ensuing decade and a half. The establishment of weekly border markets and construction of a series of feeder roads transformed the economic geography of the North Bank in the early 1980s by linking important Gambian vegetable-growing areas to the large Senegalese market to the north (see Figure 8.2). At the same time, external funding began to flow into the sector as the effects of ideological shifts in development circles toward WID and environmental stabilization programming were felt locally for the first time. The first garden-enhancement grant received by Kerewan residents in 1978 initiated an expansion period that tripled the area under cultivation in ten years. Beginning in 1987, a second wave of enclosures carried out by entrepreneur-landholders interested in establishing orchards increased the size of Kerewan's garden enclave an additional 45 percent. The 22.5 hectares presently under cultivation serve 540 vegetable growers and encompass some 1,400 garden wells and 4,000 fruit trees, making the Kerewan area one of the most intensive vegetable-producing districts in the country.

In addition to inducements offered by funders and opportunities created by climate change and developmental shifts, motivation for women to expand their gardens also came in the form of sheer economic necessity. Several government policies in the 1980s combined to tighten the budgetary squeeze on rural households. Low groundnut prices prompted male heads of households to default on many of their traditional financial obligations to their families. Deregulation of staple food prices under the mandate of the IMF/World Bank Economic Recovery Program (ERP) caused rice prices to more than double in the period 1985–88 alone (Jabara 1990). A government retrenchment program and the deregulation of fertilizer prices exacerbated the rural economic situation further, and the pressure women felt to generate higher cash incomes increased proportionately.

In sum, the dramatic emergence of a female cash crop system in rural Gambia may be seen as the product of a conjuncture of several forces. Changes in the physical, social, and economic geography of the region have simultaneously forced and inspired women to assume unprecedented financial responsibilities within their families and communities. Gender is, consequently, a key factor in understanding The Gambia's ecological politics over the past two decades.

Commercialization and conjugal conflict

To establish the links between domestic social relations and ecological processes, it is helpful to examine the stakes involved in terms of income earned from gardening. Since no data exist for the period when most of the actual struggle

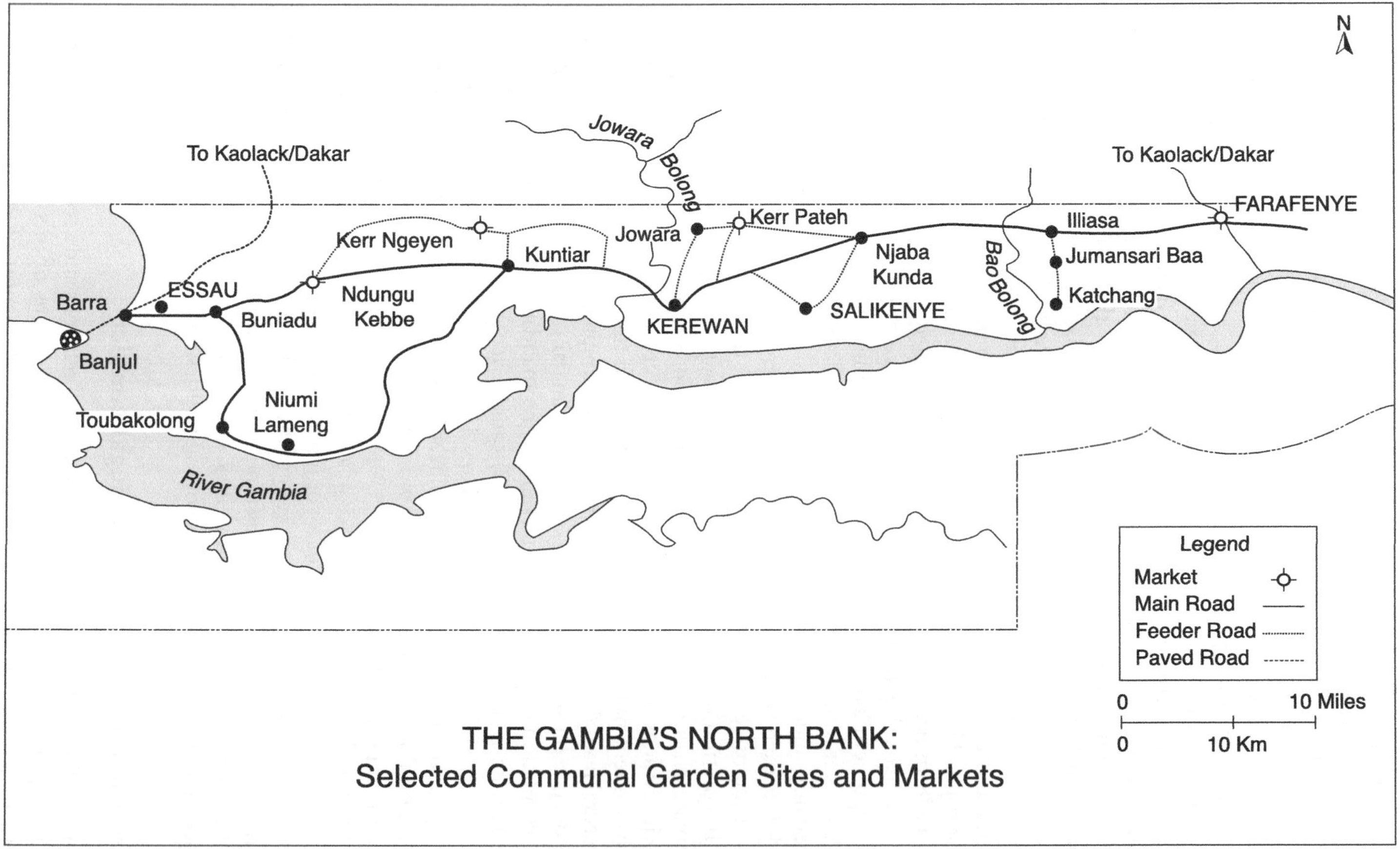

Figure 8.2 The Gambia's North Bank: Selected communal garden sites and markets.

over budgets, land, and trees took place, it is necessary to work backward from current data and formulate assumptions about the ways these conflicts have co-evolved since the onset of the boom in the mid-1970s.

The average gross incomes derived from vegetable sales by the top, middle, and lower thirds of a sample of 100 women gardeners in Kerewan in the dry season of 1991 were estimated at D1,845, D977, and D471, respectively.[5] The range of incomes for the same sample was D6,280 to D145. Thus, garden earnings on the North Bank approach the estimated rural per capita income of D1,500 ($180) (Jabara et al. 1991). A comparison of male and female cash crop incomes in two villages during the 1989–90 and 1990–91 cropping years amplifies these findings. Roughly 47 percent and 81 percent of women gardeners in each village, respectively, earned more money from their crop sales than their husbands did from groundnuts.[6]

By virtue of these cash incomes, women have taken on major new financial responsibilities within their households. Fifty-seven percent of women in the sample had purchased at least one bag of rice in 1991, at an estimated cost of D170–200, to supplement home-grown food supplies; 95 percent buy all their own clothes, 84 percent buy all their children's clothes, and 80 percent had purchased Islamic feast day clothes for at least one member of their family—all responsibilities borne either solely or primarily by men prior to the garden boom. Exactly how they took on these obligations bears close scrutiny, however. As Guyer notes, the makeup of household subsistence budgets in many parts of Africa is the product of negotiation:

> Male and female incomes come from different sources and are used for different purposes. If a couple's budget can be regarded as a single fund, it is not through the literal pooling of cash, but through the ongoing process of bargaining about the organization of interpersonal transfers and responsibilities under shifting conditions . . . Management of different responsibilities, with their different timing, has tended to be gender-specific in societies with a pronounced division into male and female spheres. But the specialization is never complete; it oscillates according to each sex's ability to cope with its own sphere, and its ability either to tap into the other or to shift the responsibilities.
>
> (Guyer 1988, 171–72; see also Roberts 1988; and Dwyer and Bruce 1988)

Thus, while all cash earned from vegetable sales is nominally controlled by women—male informants routinely claim total ignorance of their wives' incomes—growers' husbands have, nonetheless, devised a complex system of nearly a dozen separate tactics for alienating female earnings, or otherwise directing them toward ends of their own choosing (Schroeder, 1993). Informants alluded, for example, to several different ploys used by their peers for obtaining "loans" from their wives (e.g., using surrogates). Each carries its own measure of commitment to repayment, and its own underlying threat of reprisal if the loan is not forthcoming. Alternatively, men resort to "sweetness" (Mandinka: *diya*), which is intended to induce their wives to be generous with their money. Such behavior ranges from the husbands digging their wives' garden wells to simply maintaining an agreeable disposition. At the other extreme, some men reportedly steal their wives' cash outright.

In this setting, decisions made by growers regarding the expenditure of their garden incomes are also carefully strategized. For example, 38 percent of women give either lump sum or smaller, regular installments of cash to their husbands. These are clearly intended in many instances to be preemptive (cf. Jones 1986). Whereas a request for a loan cannot be flatly refused without straining the marriage or showing cause, the gifts allow growers somewhat greater latitude for controlling the terms of the cash transfers, while at the same time generating considerable goodwill. Several male informants admitted that they had once ridiculed women, joking that the gardens had become the women's "second husbands," whereas now they had "seen the benefit" and no longer opposed gardening in any way.

Another set of interrelated strategies widely employed by both growers and their husbands involves simple default. Men's main source of cash income in rural Gambia is groundnut production. The one-time sale of groundnuts typically takes place in December–January, through state or cooperative marketing services. Men are accordingly left with a single bulk sum of cash, from which they must subtract payments for debts incurred during the course of the year and with which they must devise a plan for meeting their family's subsistence needs over the June–August hungry season. Many men now fail to perform the latter responsibility. Default also becomes an option for women when the pressures exerted by their husbands' "loan-seeking" behavior become too great to withstand:

> What happens is, some men would like their wives to lend them money out of their garden proceeds. Now, many times women will give these credits, but most will never be refunded. Consequently, women gradually limit or refuse to grant the loans. We have a new tactic: whenever we go to market [to sell produce], we simply spend all of our money on things that we need, and come home with no money left to avoid the loan requests altogether.
>
> (Kerewan gardener; source: author's field notes)

This conflict may appear to be little more than petty marital bickering. Significant issues are at stake, however, including the fundamental determination of what it means to be a man or woman in Mandinka society (Carney and Watts 1991) and success or failure in negotiating the ecological challenges posed by persistent drought. The struggles over the spending of cash crop incomes are precisely the sort of politics to which Watts (1990) alludes in his critique of Blaikie and Brookfield's (1987) "regional political ecology" model. That budget conflicts turn so heavily on the seasonality of male and female income streams is especially relevant, insofar as they have added incentives to vegetable growers to alter the seasonality of their cropping strategies and expand production, as the following account of the effects of intensification demonstrates.

Intensification and demand for land

The technical innovations accompanying the garden boom included the replacement of poor-quality stick and thorn fences and hand-dug, unlined wells serving individual plots with communal wire-and-concrete structures that do not have to be replaced on an annual basis. These enhancements reduced prohibitive recurrent expenses, removed some of the threat of encroachment by grazing livestock, and improved access to groundwater. Adjustments within household units were as significant as any of these changes, however. The introduction of intensive crop production in the dry season meant a second major investment of labor power by women. In practical terms, the new work regime involved two trips to the garden daily, each of two to three hours duration, to water or otherwise tend to crops. Worse still, marketing took women out of villages altogether, sometimes for days at a time. That the gardening season coincided with a period when men expect enhanced domestic services only deepened male resentment.[7]

Much of the animosity at this stage may presumably be traced to the fact that gardens were not yet highly productive. Whereas promotional efforts by the state emphasized monocropping onions as part of a strategy of import substitution, onion sales were repeatedly undermined by market gluts. This left growers doubly vulnerable. A narrow selection of crops and relatively poor market returns meant that they were unable to meet their husbands' demands for either the expected complement of domestic services or greater financial support. Moreover, even as marginal increases were achieved, a strongly "pulsed" income stream left women vulnerable to the loan-seeking behavior described above. Growers consequently reverted to more complicated intercropping strategies that prolonged the market season and spread income over several months. By carefully mixing crops and planting them in particular sequences—i.e., by planting in series or double-cropping—women were able to increase returns without expanding plot area. Planting of fruit trees extended growers' use rights onto a vertical plane and improved the seasonality of the economic returns from gardens. Most papayas grown in Kerewan, for example, mature at precisely the time of year (November–December) when women need cash for recurrent expenses (e.g., redigging of wells, seed purchases).

Production of new crops such as cabbage, bitter tomatoes, and sweet peppers opened up sizable new markets, but this potential could only be met with an expansion of garden territory. Requests to enclose new areas for gardening purposes caused male landholders to reevaluate the long-term effects the garden boom was having on the prevailing tenure system. From the landholders' perspective, fruit production in the gardens only exacerbated the problem, insofar as tree planting confers a sense of permanence and legitimacy upon women's usufruct rights. At the same time, the trees provided an avenue for men to tap back into the low-lying land resources in ways that now threaten the garden boom in Kerewan altogether.

Tree counts and gendered tenure: measures of control in the gardens

The set of customary land tenure rules in the low-lying garden land in Kerewan may best be described as a nested system. First of all, customary land law among the Mandinka residents of Kerewan preserves a basic distinction between matrilineal and patrilineal land. Women's landholding rights are almost exclusively limited to swampland, where plots originally cleared by women are heritable property that passes from mother to daughter. Patrilineal land, by contrast, consists of virtually all arable upland, where men grow groundnuts, millet, and maize, and some swampland, where rice is grown by women for joint household consumption. The 12 communal garden perimeters established on the outskirts of the village have nearly all been founded on land controlled by relatively senior men in each of Kerewan's three founding lineages. The communal sites themselves range from a fraction of a hectare to nearly five hectares in size, and the garden groups that manage them have memberships varying from 20 to well over 200 growers, with many women holding plots in several locations at once.

Individual claims to plots in gardens may be advanced on any of several grounds. The most common is via a one-time cash payment (D5 to D30) to landholders, who ostensibly use the money to help pay for fence repairs or defray other expenses incidental to the garden's upkeep. These payments, known as *kumakaalu* in Mandinka, may be assessed when a new perimeter is founded, or levied individually by the landholder when use rights are transferred due to the death or relocation (e.g., change in marital status) of a prior occupant. In practice, no records are ever kept of these payments, and there is no way of holding landholders accountable for their use. In essence, then, they serve as a disguised rent, and mark a subtle shift toward land privatization and an emerging landlord-tenant relationship in the gardens.

Trees complicate this picture considerably. Generally, according to Mandinka customary law, trees belong to the person who plants them (Osborn 1989). In each of Kerewan's gardens, therefore, special consideration is given to the relationship between use rights to land and the right to plant trees. In most cases, landholders now specify not only whether tree planting is allowed, but which species are affected and to whom specific injunctions and dispensations apply. It is common for landholders to block women from planting taller, woody species such as mangoes or oranges, which represent a threat to the landholder's potential alternative uses and which, incidentally, pose the greatest shade problems for underlying vegetable crops, while allowing extensive cultivation of papayas and bananas on the same land. Alternatively, planting of "shade species" is allowed, but rights are reserved for the landholder himself.

Table 8.1 Tree-planting Density and Tenure Rights to Kerewan Gardens, 1991

	Site	*Area (ha)*	*Years Fenced*	*Mango*	*Orange*	*Other Species*
Group A: Mango and orange trees banned	1	4.3	16	6	12	572
	2	3.5	14	1	3	111
	3	0.5	10	0	4	258
	4	1.4	7	3	2	140
	5[a]	3.0	3	0	0	104
Group B: Full planting rights for women	6	3.6	15	64	20	225
Group C: Full planting rights for men and women	7	1.3	10	33	12	32
	8	0.2	0	na	na	na
Group D: Mango and orange planting restricted to men	9	2.2	4	90	14	168
	10	1.1	1	28	1	136
	11	0.4	1	45	13	48
	12[b]	1.0	1	100	40	600
Group E: North Bank site closed due to shade in 1988	13	4.4	8	32	108	2

Source: R. Schroeder, field notes

Note: Tree-planting density refers to trees/ha, at all stages of maturity.
a Tree counts taken on 0.6 ha prior to 1991 expansion.
b Tree counts taken on 0.1 ha prior to 1991 expansion.

Table 8.1 summarizes the situation for all 12 garden sites in Kerewan in 1991. For each site, the table indicates the size of the perimeter, the number of years since the garden was enclosed within a communal fence, and a measure of planting density for trees found in different size and shade categories. Counts for the shadiest species—mangoes and oranges—are listed separately. Trees grouped under the heading "Other Species" include the less "permanent," but numerous and economically beneficial, papaya and banana trees, as well as a small number of hardwood forest species, which for the most part predate the gardens themselves. Actual tree counts have been converted to measures of planting density for purposes of comparison. They also include trees at all stages of maturity, in an effort to reflect the potential as well as the current status of the shade threat posed in each setting.

The 12 sites have been numbered for reference purposes, and have been grouped according to the tree tenure system prevailing in each case. Landholders controlling gardens in Group A ban tall, shady trees altogether.[8] Group B includes the one site in Kerewan in which full and exclusive tree-planting rights are granted to women. In Group C, both growers and landholders have full tree-planting rights. Group D landholders reserve mango and orange planting privileges to themselves. And Group E consists of a single site in another North Bank community that closed its garden due to shade in 1988.

The data demonstrate that the bans imposed on shady species have largely been effective. At the same time, regardless of constraints stipulated or implied in use agreements, it is clear that women in the early stages of the garden boom deliberately sought to exploit the vertical dimension of their use rights. Where mangoes and oranges were banned—e.g., in gardens in Group A—the number of papayas and bananas is highest. The most striking feature of Table 8.1, however, is the clear trend toward increasing involvement in tree planting by men. There has also been a proliferation of mangoes, the species least compatible with an underlying vegetable crop. This is where the comparison with the garden in Group E, which was closed to vegetable production some five years after it was founded due to shade problems, is relevant. Comparable density figures for tree seedlings planted in garden/orchards controlled by men in Kerewan underscore the precarious nature of gardeners' tenure privileges; the likelihood of shade canopy closure during the next five to ten years is high. To get some sense of what may transpire, it will be necessary to look in greater detail at anecdotal evidence pertaining to spatial and shade problems in three of Kerewan's older gardens, where canopy closure is a real and present concern.

Shady contexts, shady practice

Intrahousehold social relations and a series of regional economic changes coincided in the early 1980s to exert a growing pressure for expansion of garden perimeters. Accordingly, when an expatriate volunteer was posted in Kerewan in 1983, local growers' organizations seized upon the opportunity to lobby for material support to expand two existing garden sites. Representatives of the growers' groups, extension agents from the civil service, and the volunteer met several times over a period of months to put together plans for the construction of several wells and an areal expansion sufficient to accommodate 479 women. Funding was to be provided by an outside donor, and the project was to be jointly administered by the volunteer and government extension agents. The Ministry of Agriculture would be responsible for surveying and allocating individual land parcels, and project materials would be temporarily kept in the Ministry of Agriculture's storage facilities until required by construction contractors.

These arrangements were subsequently challenged by one of the sites' landholders, referred to here as Al-Haji Abdou Sanyang. A backer of the project throughout the early stages of planning, Sanyang strongly objected to women signing a project agreement with the donor agency on their own behalf, arguing that such an act would seriously undermine his landholding rights. He had acted for a number of years as advisor to the group, and he felt he should sign the contract as project beneficiary. In addition, he asserted rights to store project materials in his private storerooms and award subcontracts to well diggers. When his demands were not met, he blocked construction efforts and threatened to evict growers from his land.

The growers' response was twofold. On the local front, several of the more militant women vowed they would be removed from the garden only by force, and designated a communal workday to tear down the old fence in spite of the landholder's opposition. This plan drew the attention of the police, who took three of the leaders of the garden group into custody. The detention prompted a protest demonstration on the part of several hundred growers, which resulted in the issuing of a temporary injunction against gardening on the site. Failing in their attempts to force a solution, leaders of the growers' group carried their case to national authorities. A deposition was sent to the National Women's Bureau, and project donors were lobbied to intervene on the group's behalf. Subsequent pressure on the North Bank divisional commissioner led him to invoke what amounted to a cooling-off period of several months, after which all principals to the dispute were called to air their grievances in court. In the final ruling on the case, nearly all of the substantive claims by the growers were upheld. The sole exception involved allegations made by the landholder that vegetable growers had planted dozens of fruit trees within the perimeter without authorization. His insistence that they be removed won the court's backing, and women were ordered to remove all trees at his request.

Within a day or two of the decision, Sanyang visited the garden and ordered several dozen trees removed. Then, in an action that foreshadowed much of what was to come in Kerewan's gardens, he immediately replanted several dozen of his own trees within the perimeter. By locating seedlings directly on top of garden beds already allocated to vegetable growers, his expectation was that water delivered by growers to the vegetable crop would support his trees until the ensuing rainy season. When the trees died for lack of adequate irrigation, he charged that they were deliberately abandoned, or otherwise sabotaged, by vegetable growers in retribution for his heavy-handedness during the previous year's controversy. Nonetheless, he was apparently satisfied that his landholding authority was no longer in question, since he took no further steps to establish an orchard on the site.

The controversy in Sanyang garden marks a watershed in the political ecology of gardening in Kerewan. Not only were several hundred women involved in the demonstration at the police station, but the case also received attention from politicians at the highest levels of government. As the affair became imbued with partisan political meanings, every step taken by Sanyang and every aspect of the women's claims to use rights were carefully scrutinized and debated throughout the village. This presumably forced other landholders to reappraise their own stance with respect to their management of low-lying land resources, and it set a precedent for landholders in the attempted use of female labor to establish private fruit tree orchards. Both effects are evident in the trends noted in the data in Table 8.1.

The ecological impact of these events is also apparent in terms of the relative shade problems experienced in the garden. The density figures recorded in Table 8.1 reveal the effects of the forced tree removal in Sanyang garden. There is a marked distinction between Site 1—Sanyang perimeter—and Site 6, which was established at roughly the same time and in which women gardeners enjoy full tree-planting rights. When growers in both gardens were asked in 1991 whether they experienced shade problems, 70 percent of the women in Site 6 complained that shade now had a significant impact on crop yields on their plots, as compared to only 30 percent of the growers in Sanyang garden. As one of the women's group leaders in Site 6 put it: "We are afraid of trees now. . . . You can have one [vegetables or fruit] or you can have the other, but you can't have both."

This comment encapsulates a growing dilemma facing Kerewan gardeners. Trees were once a means for somewhat surreptitiously extending use rights to land, a shady practice in its own right, given the cultural norms governing tree planting and land tenure in Kerewan (cf. Osborn 1989). By the mid-1980s, however, the relative economic benefits of tree planting and vegetable growing shifted decisively in favor of gardens, and trees became a threat to women. Growers began cutting back or chopping down trees in order to open up the shade canopy and expose their vegetable crops to sunlight. At the same time, landholders saw a new opportunity developing for themselves. Whereas male landholders like Sanyang had initially resisted tree planting on the grounds that it reduced their future land use options, the "capturing" of a female labor force to water trees, manure plots, and guard against livestock incursions within the fenced perimeters shifted the landholders' economic calculus *toward* fruit growing. Moreover, they quickly discovered that developers were eager to support tree planting under almost any condition. Site 7 is a case in point.

In 1983, Site 7 was founded immediately adjacent to Site 6, where women now experience the greatest shade problems. Given the land pressure at the time, many women from the older site took second plots in the new site. Under a somewhat novel arrangement between vegetable growers, the landholder, Forestry Department officials, and a donor agency, the garden was converted into a garden/orchard, with a dense stand of trees in a grid pattern over the entire area. The understanding was that ownership of the trees would be divided between the landholder and gardeners on an alternating basis; every other tree, in effect, belonged to the landholder. Within five or six years, however, shade problems began to appear. Gardeners had already determined that vegetables brought them a greater return than any harvest they could expect from their trees. Consequently, many of the maturing trees were either drastically trimmed or simply removed, including, apparently, many of the trees belonging to the landholder. In response, the landholder banned tree trimming in his garden, only to find his young trees still being destroyed as women burned crop residues to clear plots for each new planting season. While some of this destruction was doubtless accidental, the landholder claimed that growers also deliberately hung dry grass in tree branches so that fires set to clear plots would fatally damage trees.

Tree density figures for Site 7 reflect the fact that fully half of the original orchard no longer exists, so it is clear that vegetable growers have largely had their way on the site. In other gardens, landholders insist they will remain more vigilant and exert a greater degree of control. One man vowed: "We will sow the [tree] seedlings on their [garden] beds whether they will like it or not." And the town chief announced that any woman responsible for starting a bush fire that destroyed trees would be fined the equivalent of $200. At least one woman was evicted from her plot when a fire she started accidentally destroyed just one of the landholder's trees.

In sum, this comparison of three garden/orchards establishes that trees can be used as a means for claiming both material and symbolic control over garden lands. Tree planting on garden beds, moreover, is a mechanism for landholders to alienate surplus female labor and subsidies embodied in concrete-lined wells and permanent wire fences. At the same time, shade effects from tree planting threaten to undermine the productivity of gardeners, who now play key roles in providing for the subsistence needs of their families. This has brought about considerable resistance on the part of vegetable growers, who have demonstrated both individually and collectively their willingness to contest anything that they perceive as a threat to their new-found livelihoods. At the end of the 1991 rainy season, for example, gardeners chopped and burned almost at will in most of the village's perimeters. These actions suggest that, tougher rhetoric and recent clamp-down notwithstanding, the struggle to claim control over garden land is far from over.

Conclusion

> It is easy to invoke the environmental crisis and the poor people's energy crisis to open up new avenues for reductionist science and commodity production.
>
> (Shiva 1988)

Several attempts have been made in geography at deconstructing the core concepts of environmental theory.[9] O'Keefe, Westgate, and Wisner (1976), Waddell (1977), Torry (1979), Hewitt (1983), and Watts (1983a) combine forces to take the "natural-ness" and the "accidental" aspect out of natural disasters. Blaikie and Brookfield (1987) insist that environmental degradation be defined in explicit terms, by arguing that resources can only become "degraded" with respect to specific uses in particular geographic and historical circumstances. Watts (1991) challenges those employing a "lexicon of crisis" with respect to Africa, suggesting they abandon such "generic" theoretical pronouncements in favor of a more nuanced analysis of the "creative rupture(s)" occurring in agrarian systems across the continent (Watts 1991, 1–3).

In a similar vein, my premise here is that environmental stabilization is not a politically neutral process. Indeed, the ambiguity of the term "stabilization" begs the most basic of political economic questions: What is being "stable-ized"? On whose behalf? Who is in the position to define stability? And who determines when such a condition is achieved? To what extent does stabilization amount to a simple shift in resource access and control?

These questions could not be more germane to the case of The Gambia's garden boom. In each of the hundreds of garden perimeters springing up over the past two decades, the ecological and economic significance of wells, fences, soil improvements, and tree stands must be assessed in light of competing local, national, and international interests. Wells, fences, and soil improvements provide the necessary conditions for vegetable production and may thus be considered forms of "landesque capital" (Blaikie and Brookfield 1987) from the standpoint of both vegetable growers and their families, who depend heavily upon vegetable incomes. But such improvements also tie female labor to a specific spatial domain, thereby stabilizing conditions which allow landholders to establish orchards. The addition of the tree crop, in turn, negates the value of the infrastructure for gardeners, effectively *de*stabilizing their productive base once again, and actually compounding problems within a broader political economic context by enticing outside donors into the fray.

An overview of the case study from a slightly broader perspective reinforces the point. The loan-seeking behavior of men in Kerewan has forced their vegetable-growing wives to intensify horticultural production through expansion of fence enclosures and tree planting. Landholders have finessed the issue of enclosure in a way that allows them to control women's labor and capture subsidies intended for the construction of garden infrastructure. Nongovernmental donor agencies use landholders' leverage over vegetable growers to meet their own objectives of land stabilization via tree planting (Norton Staal 1991; Worldview International Foundation 1990; R. Mann 1989; Thoma 1989; Lawry 1988). And the state and multilateral donors build on NGO successes to meet national goals in environmental stabilization, agricultural diversification, and full-scale economic readjustment (Government of The Gambia n.d.; Government of The Gambia 1990; USAID 1991; Thoma 1989; Agroprogress International 1990; Thiesen et al. 1989). This implies, quite simply, that developers at all levels have pinned their hopes, indeed have staked their very legitimacy in some cases,[10] on the continued mobilization of unpaid female labor.

It is important to realize that the Gambian case, as compelling as it may be in its own right, may be part of a larger pattern. Witness, for example, the international notoriety of Kenya's Green Belt and India's Chipko movements, both led by women (*UNESCO Courier* 1992). Without denigrating in any way either the impressive accomplishments or the principles of these movements, there are, nonetheless, grounds for skepticism about the fact that they are widely touted as models of environmental stabilization and conservation (see the growing debate among ecofeminists in Biehl 1991; Merchant 1992; de Oliveira and Corral 1992; and Runyan 1992). Momsen (1991) reports that 35 African countries have now initiated national tree-planting campaigns that in some sense emulate the Green Belt movement. In the original Green Belt program, sponsored by the National Council of Women of Kenya (NCWK), women workers were paid for their efforts in propagating and transplanting tree seedlings; the Gambian case demonstrates that this arrangement has not necessarily been replicated elsewhere. This suggests that developers have a blind spot with respect to the fundamentally exploitative nature of many of the work processes into which women are being incorporated across the continent. While circumstances doubtless vary from one case to the next, it appears that women are often featured not so much because they have any natural affinity for nature per se, but because men refuse to perform tree-watering tasks themselves.

Equally troubling is the basic lack of recognition of the critical factor motivating women to undertake land improvements in the first place, namely the preservation of livelihoods. Shiva (1988, 90), writing on the struggles of women in the Chipko movement to preserve ecological diversity and economic livelihoods in India's forests, decries the "reductionist" scientific paradigm that offers as a "universal solution" to ecological breakdown the practice of planting trees. Instead of stabilizing resources, official afforestation programs "destroy the forest as a diverse and self-reproducing system, *and* destroy it as commons, shared by a diversity of social groups with the smallest having rights, access and entitlements" (original emphasis, Shiva 1988; 70; cf. Barrett and Browne 1991). Faced with such reductionist logic, political ecologists can only feel compelled to expand the scope of their critical inquiry to include both the peculiar problems of resource stabilization and the new ways in which patterns of gender exploitation are being factored into environmental initiatives.

Notes

* The author wishes to acknowledge the following research grants which funded fieldwork in 1991: the Fulbright-Hays Doctoral Dissertation Research Award, the Social Science Research Council/American Council of Learned Societies International Doctoral Research Fellowship for Africa, the Rocca Memorial Scholarship Award for Advanced African Studies, and the National Science Foundation Fellowship in Geography and Regional Science. A research trip in 1989 was funded by the Rockefeller Foundation. Special thanks to research assistants Malaine Hydara and Borang Danjo and the staff of Save the Children/The Gambia. Thanks also to Michael Watts, Gillian Hart, Louise Fortmann, Judith Mayer, Vinay Gidwani, and Tad Mutersbaugh, who read the manuscript and offered advice and encouragement.

1 The debate in geography originally centered around human ecological approaches to natural hazards, which emphasized adaptation, environmental perception, rational choice theory, and cybernetic ecosystemic modeling (Kates 1971; White 1974; Burton, Kates, and White 1978). Critics of these efforts stressed the structural vulnerability of many disaster victims and the political economic constraints bounding their choices and perception (O'Keefe and Wisner 1977; Waddell 1977; Torry 1979; Hewitt 1983; Watts 1983b). Zimmerer (forthcoming) traces the origins of the term "political ecology" to an anthropological symposium that linked struggles over land use practices to property relations (Wolf 1972). Blaikie and Brookfield (1987) give the term currency in geography.

2 Zimmerer (forthcoming), Bryant (1992), Watts (1990), Butzer (1989), and Bassett (1988) review the relevant geographic literature.

3 See the discussion of the controversy surrounding Wittfogel's (1957) thesis on "oriental despotism" in Worster (1985).

4 The term "developer" is used here to refer to nongovernmental and private voluntary agencies, as well as branches of the state and multilateral donors.

5 Data for this sample were recorded over an 18-week marketing period. Yield figures reflect the efforts of production units of varying makeup—"mother and adolescent daughter(s)," "daughter and elderly mother," and "unassisted individual woman" being the most common types. During the period these data were collected, the Gambian dalasi (D) was exchanged at a rate of roughly D7.5 = U.S. $1.00.

6 This comparison is based on earnings of 36 couples from Niumi Lameng and 75 couples from Kerewan. Niumi Lameng vegetable growers provided data for the March–April 1989 period; Kerewan women were surveyed from February to June 1991; and men in both groups provided data on the June–October 1990 groundnut season. Given the sporadic nature of the data collection, these findings should be interpreted with caution. In comparison, Jabara et al.'s (1991) estimates for per capita incomes from all sources (including imputed values for "own-produced food") for three North Bank villages show that vegetable income averages 12% of total household income. This compares with 21% from groundnuts, 17% from cereals, 15% from gifts/remittances, and 14% from "Business."

7 The dry season is generally a more relaxed time than during the rains. Food stocks are relatively plentiful, and there is more time for meal preparation and child care. Family members spend a greater amount of time in and around family compounds, and men in particular are in a position to observe and more directly organize compound affairs. Consequently, they expect a greater measure of personal service than they get at other times of the year. With intensification of vegetable production, these services have been reduced.

8 As of 1992, only three sites in the area were reserved exclusively for gardening purposes. The landholder in Site 3 converted his garden into an orchard. And a large portion of Site 5 was confiscated by managers of the Gambian affiliate of a Norwegian NGO. As fieldwork drew to a close, plots had been reassigned and trees were being planted as part of a strategy for the agency to generate recurrent operational expenses locally. The terms of access to the newly enlarged and fenced garden/orchard stipulate that women water the project trees or forfeit their usufruct privileges.

9 Pred's (1989, 1990) studies of linguistic practice accomplish the same purpose for "gendered languages."

10 The striking emphasis on women and the use of commercial vegetable production to justify funding was not lost on a team of environmental experts touring The Gambia in 1987. The group of USAID analysts was assembled under a mandate from a U.S. Congress grown weary of African disaster narratives. Their mission was to find and retrieve "success stories" in natural resource management on the Sahel (Shaikh et al. 1988). By the time they reached the North Bank for a tour of Save the Children-sponsored projects, under my supervision, they had been on the road for weeks and were nearing the end of the Gambian leg of their research trip. In discussing the day's proposed itinerary, one of the leaders of the group commented, "I don't care what we see, just don't show us another garden project." He elaborated that, at virtually every stop over the course of the team's two or three day Gambian tour, they had been shown trees planted within communal gardens as evidence of the host agency's commitment to environmental stabilization.

References

Agroprogress International. 1990. Project preparation consultancy for an integrated development programme [European Community] for the North Bank Division. Bonn: Agroprogress International.

Barrett, H. R., and Browne, A. W. 1991. Environmental and economic sustainability: Women's horticultural production in The Gambia. *Geography* 76 (332):241–48.

Bassett, T. J. 1988. The political ecology of peasant-herder conflicts in the northern Ivory Coast. *Annals of the Association of American Geographers* 78 (3):453–72.

Berry, S. 1989. Social institutions and access to resources. *Africa* 59 (1):41–55.

Biehl, J. 1991. *Rethinking ecofeminist politics*. Boston: South End Press.

Blaikie, P., and Brookfield, H. 1987. *Land degradation and society*. London: Methuen.

Boughton, D., and Novogratz, J. 1989. The Gambia: Women in development project, agricultural component. Banjul: Government of The Gambia/World Bank.

Bryant, R. 1992. Political ecology: An emerging research agenda in Third World studies. *Political Geography* 11 (1):12–36.

Burton, I.; Kates, R; and White, G. 1978. *The environment as hazard*. New York: Oxford University Press.

Butzer, K. 1989. Cultural ecology. In *Geography in America*, ed. G. L. Gaile and C. Wilmott, 192–208. Columbus, Ohio: Merrill.

Carney, J. 1992. Peasant women and economic transformation in The Gambia. *Development and Change* 23 (2):67–90.

——. 1988. Struggles over crop rights and labour within contract farming households in a Gambian irrigated rice project. *Journal of Peasant Studies* 15 (3):334–49.

Carney, J., and Watts, M. 1991. Disciplining women? Rice, mechanization and the evolution of gender relations in Senegambia. *Signs* 16 (4):651–81.

——. 1990. Manufacturing dissent: Work, gender and the politics of meaning in a peasant society. *Africa* 60 (2):207–41.

Daddieh, C. K. 1989. Production and reproduction: Women and agricultural resurgence in sub-Saharan Africa. In *Women and development in Africa: Comparative perspectives*, ed. J. Parpart, 165–93. New York: Dalhousie University Press.

Davison, J., ed. 1988. *Agriculture, women and land*. Boulder, Colo.: Westview Press.

de Oliveira, R. D., and Corral, T., eds. 1992. *Terra femina*. Rio de Janeiro: IDAC/REDEH.

Dey, J. 1981. Gambian women: Unequal partners in rice development projects? *Journal of Development Studies* 17 (3):109–22.

Dwyer, D., and Bruce, J., eds. 1988. *A home divided: Women and income in the Third World*. Stanford: Stanford University Press.

Folbre, N. 1986. Hearts and spades: Paradigms of household economics. *World Development* 14 (2):245–55.

Government of The Gambia. 1990. National natural resource policy. Banjul.

——. 1991. *National agricultural statistical survey, 1990*. Banjul: Ministry of Agriculture, Department of Planning.

——. n.d. *Executive summary. Programme for sustained development. Sectoral strategies*. Banjul: Gambia Round Table Conference.

Guyer, J. 1980. Food, cocoa and the division of labor by sex. *Comparative Studies in Society and History* 22:355–73.

——. 1984. Naturalism in models of African production. *Man* 19:371–88.

——. 1988. Dynamic approaches to domestic budgeting: Cases and methods from Africa. In *A home divided*, ed. D. Dwyer and J. Bruce, 155–72. Stanford: Stanford University Press.

——. 1991. Female farming in anthropology and African history. In *Gender at the crossroads of knowledge: Feminist anthropology in the postmodern era*, ed. M. di Leonardo, 257–77. Berkeley: University of California Press.

Guyer, J., and Peters, P., eds. 1987. Special issue: Conceptualizing the household: Theory and policy in Africa. *Development and Change* 18 (2).

Harris, O. 1981. Households as natural units. In *Of marriage and the market*, ed. K. Young, C. Wolkowitz, and R. McCullagh, 49–68. London: CSE Books.

Hemmings-Gapihan, G. 1982. International development and the evolution of women's economic roles: A case study from Northern Gulma, Upper Volta. In *Women and work in Africa,* ed. E. Bay, 171–90. Boulder, Colo.: Westview Press.

Hewitt, K., ed. 1983. *Interpretations of calamity (from the viewpoint of human ecology).* Boston: Allen and Unwin.

Jabara, C. L. 1990. *Economic reform and poverty in The Gambia: A survey of pre- and post-ERP experience.* Ithaca, N.Y.: Cornell Food and Nutrition Policy Program, Monograph 8.

Jabara, C. L.; Tolvanen, M.; Lundberg, M. K. A.; and Wadda, R., 1991. Incomes, nutrition, and poverty in The Gambia: Results from the CFNPP household survey. Washington, D.C.: Cornell Food and Nutrition Policy Program.

Jones, C. 1986. Intra-household bargaining in response to the introduction of new crops: A case study from North Cameroon. In *Understanding Africa's rural households and farming systems,* ed. J. Moock, 105–23. Boulder, Colo.: Westview Press.

Kates, R. 1971. Natural hazard in human ecological perspective: Hypotheses and models. *Economic Geography* 47 (3):438–51.

Lawry, S. 1988. Report on Land Tenure Center mission to The Gambia. Madison: University of Wisconsin Land Tenure Center.

Longhurst, R. 1982. Resource allocation and the sexual division of labor: A case study of a Moslem Hausa village. In *Women and development: The sexual division of labor in rural economies*, ed. L. Beneria, 95–118. Geneva: ILO.

MacCormack, C. 1982. Control of land, labor and capital in rural southern Sierra Leone. In *Women and work in Africa*, ed. E. Bay, 35–54. Boulder, Colo.: Westview Press.

Mann, R. D. 1989. Africa: The urgent need for tree-planting. Manuscript. Methodist Church Overseas Division.

Mann, S. 1990. *Agrarian capitalism in theory and practice.* Chapel Hill: University of North Carolina Press.

Mann, S., and Dickinson, J. 1978. Obstacles to the development of capitalist agriculture. *Journal of Peasant Studies* 5:466–81.

Merchant, C. 1992. *Radical ecology: The search for a livable world.* New York: Routledge.

Momsen, J. H. 1991. *Women and development in the Third World.* London: Routledge.

Muntemba, M. S. 1982. Women and agricultural change in the railway region of Zambia: Dispossession and counterstrategies, 1930–1970. In *Women and work in Africa,* ed. E. Bay, 83–104. Boulder, Colo.: Westview Press.

Nicholson, S. 1986. Climate, drought, and famine in Africa. In *Food in sub-Saharan Africa,* ed. A. Hansen and D. McMillan, 107–28. Boulder, Colo.: Lynne Rienner.

Nicholson, S., and Flohn, H. 1980. African environmental and climatic changes and the general atmospheric circulation in Late Pleistocene and Holocene. *Climate Change* 2:313–48.

Norton, G.; Mills, B.; Gilbert, E.; Sompo-Ceesay, M. S.; and Rowe, J. 1989. Analysis of agricultural research priorities in The Gambia. Manuscript. Banjul: Department of Agricultural Research, Ministry of Agriculture.

Norton-Staal, S. 1991. Women and their role in the agriculture and natural resource sector in The Gambia. Banjul: U.S. Agency for International Development.

O'Keefe, P.; Westgate, K.; and Wisner, B. 1976. Taking the naturalness out of natural disaster. *Nature* 260:566–69.

O'Keefe, P., and Wisner, B. 1977. *Land use and development.* London: International African Institute.

Osborn, E. 1989. Tree tenure: The distribution of rights and responsibilities in two Mandinka villages. Master's thesis, University of California, Berkeley.

Posner, J., and Gilbert, E. 1987. *District agricultural profile of Central Baddibu, North Bank Division*. Gambian Agricultural Research Papers No. 2. Banjul: Department of Agricultural Research.

Pred, A. 1989. The locally spoken word and local struggles. *Environment and Planning D: Society and Space* 7:211–33.

——. 1990. In other wor(l)ds: Fragmented and integrated observations on gendered languages, gendered spaces and local transformation. *Antipode* 22:33–52.

Roberts, P. 1979. "The integration of women into the development process": Some conceptual problems. *IDS Bulletin* 10 (3):60–66.

——. 1988. Rural women's access to labor in West Africa. In *Patriarchy and class,* ed. S. Stichter and J. Parpart, 97–114. Boulder, Colo.: Westview Press.

Runyan, A. S. 1992. The "state" of nature: A garden unfit for women and other living things. In *Gendered states: Feminist (re)visions of international relations theory,* ed. V. S. Peterson, 123–40. Boulder, Colo.: Lynne Rienner.

Schoepf, B. G., and Schoepf, C. 1988. Land, gender and food security in Eastern Kivu, Zaire. In *Agriculture, women and land,* ed. J. Davison, 106–30. Boulder, Colo.: West-view Press.

Schroeder, R. 1993. Shady practice: Gender and the political ecology of resource stabilization in Gambian garden/orchards. Ph.D. diss., University of California, Berkeley.

Schroeder, R., and Watts, M. 1991. Struggling over strategies, fighting over food: Adjusting to food commercialization among Mandinka peasants in The Gambia. In *Research in rural sociology and development.* Vol. 5, *Household strategies,* ed. H. Schwarzweller and D. Clay, 45–72. Greenwich, Conn.: JAI Press.

Shaikh, A.; Arnould, E.; Christophersen, K.; Hagen, R.; Tabor, J.; and Warshall, P. 1988. *Opportunities for sustained development, successful natural resources management in the Sahel.* Vols. 1 and 2. Washington, D.C.: USAID.

Shiva, V. 1988. *Staying alive: Women, ecology and development.* London: Zed Press.

Thiesen, A.; Jallow, S.; Nittler, J.; and Philippon, D. 1989. African food systems initiative, project document. The Gambia: U.S. Peace Corps.

Thoma, W. 1989. *Possibilities of introducing community forestry in The Gambia,* pt. 1. Gambia-German Forestry Project, Deutsche Gesellschaft für Technische Zusammenarbeit (GTZ). Feldkirchen, Germany: Deutsche Forstservice.

Torry, W. 1979. Hazards, hazes and holes: A critique of *Environment as Hazard* and general reflections on disaster research. *Canadian Geographer* 23:368–83.

UNESCO Courier. 1992. Women speak out on the environment (special issue). March.

U.S. Agency for International Development (USAID) 1991. Agricultural and natural resource program, program assistance initial proposal. Banjul.

Waddell, E. 1977. The hazards of scientism: A review article. *Human Ecology* 3 (4):249–73.

Watts, M. 1983a. On the poverty of theory: Natural hazards research in context. In *Interpretations of calamity,* ed. K. Hewitt, 231–62. Boston: Allen and Unwin.

——. 1983b. *Silent violence: Food, famine and peasantry in Northern Nigeria.* Berkeley: University of California Press.

——. 1990. Review of Harold Brookfield and Piers Blaikie, *Land degradation and society. Capitalism, Nature, Socialism* 4:123–31.

——. 1991. The agrarian crisis in Africa: Debating the crisis. *Progress in Human Geography* 13 (1):1–42.

Watts, M. 1992. Idioms of land and labor. In *Land tenure in Africa,* ed. T. Bassett, 159–96. Madison: University of Wisconsin Press.

White, G., ed. 1974. *Natural hazards: Local, national, global.* New York: Oxford University Press.

Whitehead, A. 1981. "I'm hungry mum": The politics of domestic budgeting. In *Of marriage and the market,* ed. K. Young, C. Wolkowitz, and R. McCullagh, 88–111. London: CSE Books.

Wittfogel, K. 1957. *Oriental despotism: A comparative study of total power.* New Haven: Yale University Press.

Wolf, E. 1972. Ownership and political ecology. *Anthropological Quarterly* 45 (3): 201–5.

Worldview International Foundation. 1990. *WIF Newsletter* 3 (1):4.

Worster, D. 1985. *Rivers of empire: Water, aridity and the growth of the American West.* New York: Pantheon.

Zimmerer, K. Forthcoming. Ecology. In *Conceptual thinking in human geography,* ed. C. Earle and M. Kinzer. Boston: Routledge.

9 African shea butter

A feminized subsidy from nature

Marlène Elias and Judith Carney

Biodiversity preservation is an age-old phenomenon. For millennia, peoples worldwide have selected for, managed and conserved species that fulfil nutritional, medicinal, cultural and spiritual functions. Amid the rapid global decline in fauna and flora, the urgency of preserving natural resources has increased. Environmentalists are now placing their hopes on traditional agro-forestry systems that can point the way towards the sustainable use and management of forest resources (Berkes 1999; Depommier and Ramakrishnan 2002).

Agro-forestry systems comprise biophysical features as well as the interrelated knowledges, meanings and social relations that mediate the ways societies and individuals relate to the physical landscape. These complex systems are an integral part of the very cultures with which they co-evolved (Berkes *et al.* 2000). The tangible and immaterial features of agro-forestry systems represent a natural heritage that is passed down and even developed, from one generation to the next. In the African savanna the shea tree, as well as the expertise and conventions that accompany its use, represent one such natural heritage that is integral to cultural and biodiversity conservation.

African shea butter is becoming increasingly familiar in the West. This vegetal oil has emerged from obscurity to prominence as a favourite ingredient in natural lines of cosmetics. Over the past fifteen years, shea – or *karité* as it is known in French – has become the focus of many development initiatives because it is one of the few economic commodities in the region under the control of women. Part of a *filière feminine* (a female commodity chain), shea has long been collected, processed and traded by women. The current global market demand extends the shea commodity chain, linking African women producers to Western female consumers.

Africa's shea tree (*Vitellaria paradoxa* C.F. Gaertn.) grows naturally in eighteen countries along a 5,000-kilometre expanse of the semi-arid Sahel and Guinean savanna woodlands (Figure 9.1).[1] Major shea-producing countries are among the poorest in the world. As a result of women-in-development (WID) projects

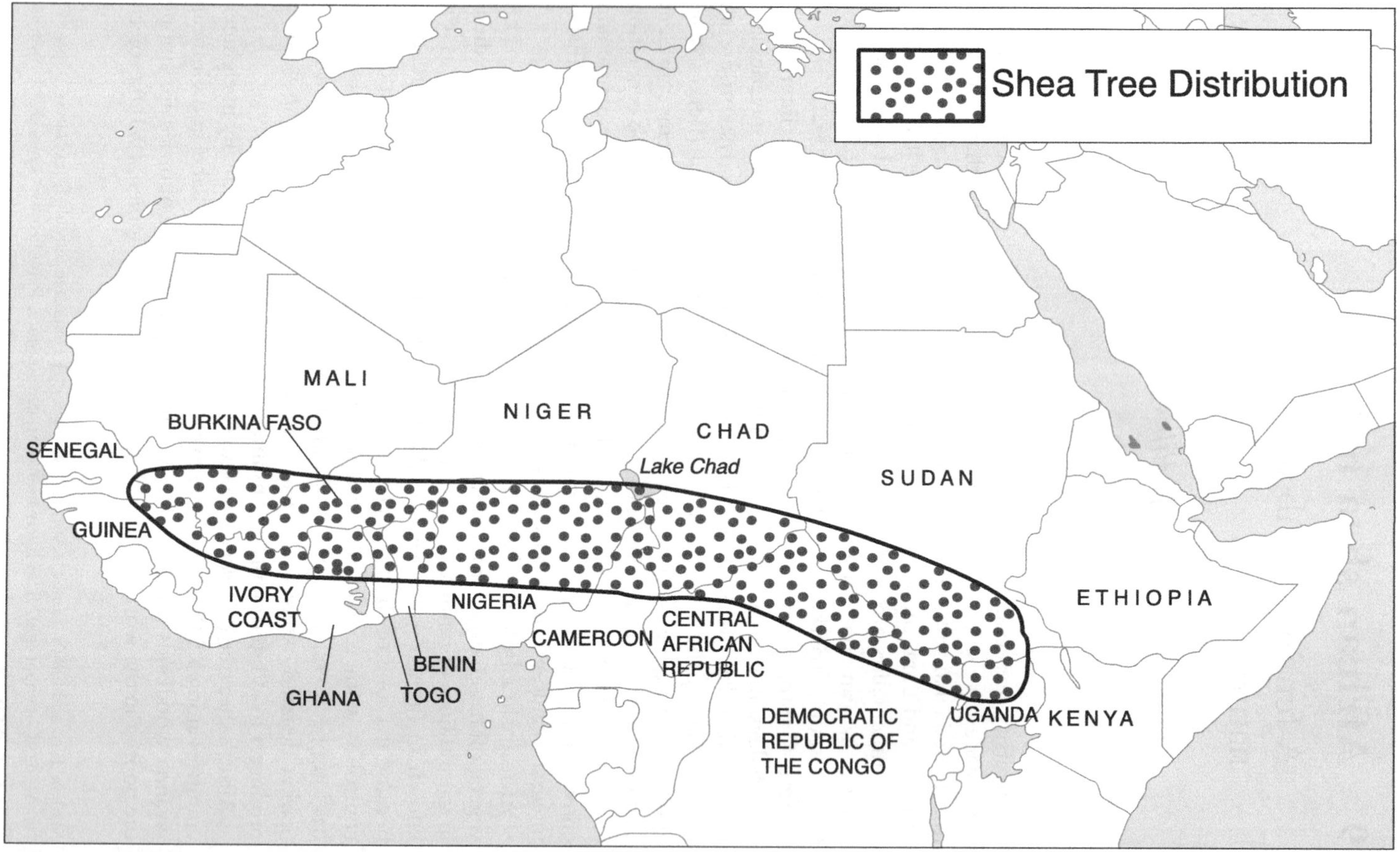

Figure 9.1 Shea tree distribution.

supported by the UN, bilateral aid agencies and NGOs, shea butter exports from Africa have increased since the 1990s. Donors have sponsored fair-trade initiatives that promise to offer higher prices to female producers through direct contracts with firms promoting natural skin products. The current wave of shea commercialization, however, is prompting changes in traditional resource processing and management as well as agro-forestry practices in countries like Burkina Faso, West Africa's largest exporter.

Based on fieldwork in Burkina Faso's provinces of Boulgou (2001) and Sissili (2005), this article examines the role of shea as a female heritage in Burkina Faso.[2] Emphasis is placed on the knowledge systems that inform the transformation of nuts into butter and the tree's management in farmed fields. The objective is to illuminate the cultural and botanical heritage of shea as well as the tree's role in biodiversity protection, African natural heritages and female knowledge systems.

Divided into four sections, the article begins with an overview of the female knowledge informing the preparation of shea butter and perceived product quality differences that affect its regional trade. The next section considers the traditional land-use practices and local agronomic knowledge shaping shea parklands and fostering the tree's conservation. The article then turns to contemporary markets for shea products, identifying preliminary trends of shea commercialization on female tenure rights and biodiversity conservation. Issues surrounding the marketing of shea as a 'heritage product' are finally considered, with emphasis on the ways standardized shea-processing techniques are breaking from the specialized and diverse traditions related to the resource.

Traditional shea butter production and trade

Preparation of shea butter

The shea belt crosses many sub-Saharan countries and ethnicities. Yet, in areas where the tree is found, women have long been the ones to collect and process shea nuts (Lewicki 1974). The process of rendering butter from shea nuts represents an ancient knowledge system that has been passed on generationally from mother to daughter. There are many ways to process the butter, however, and female producers and buyers of the product recognize the differences in quality that are associated with distinctive methods in specific geographical regions. The preparation of shea butter represents a cultural heritage of many different ethnic groups in the sub-Saharan shea zone. The female shea commodity chain thus reaches deep into the environmental knowledge systems of Africa and the techniques developed by African women over the centuries. An overview of the ways shea is collected and transformed into the product now esteemed in Western skin care elucidates this point.

The collection and processing of shea nuts involves a great deal of work. Shea trees come into production at the onset of the rainy season and bear fruit throughout most of the agricultural period. At this time, women are especially burdened with farm work. Butter making thus produces an intensification of women's workloads during the rains. The preparation of shea butter requires large quantities of firewood and water, which women collect. The production of a single kilogram of shea butter demands between 8.5 to 10 kilograms of fuelwood (Hyman 1991: 1250).

Shea nut processing also occurs during the dry season, when women's workloads are lessened. In some areas of Burkina Faso, however, this is not possible because village wells seasonally run dry. As seasonal water shortages increase the distance to permanent water sources, Lobi women in southwestern Burkina Faso produce shea butter during the rainy season (Crélerot 1995: 116). In recent decades, the steady fall of farm commodity prices and deepening impoverishment of rural society are increasing the need for money. The result is that many shea producers now make butter year-round for sale.

Despite its importance to female incomes, there is little research to date on the diversity of nut-processing techniques developed by rural women. These techniques, however, represent a significant cultural heritage passed down through generations of female producers. The steps of the butter-making process and key techniques are summarized below for Burkina Faso, the shea tree's probable centre of domestication (Maranz and Wiesman 2003).

The butter-making process begins when the shea tree bears its fruits. This corresponds with the end of the dry season and continues for several months into the rains. Women and children collect the fallen fruits, canvassing an area within a radius that extends between one and three kilometres from the household. The nuts are head-carried to the homestead for processing. The initial stage of butter production involves pulping the fruits to remove the nuts. In order to ferment the pulp and extract the nuts, shea fruits are sometimes buried in underground pits for at least twelve days. Nuts are then boiled (if it is raining) or left to dry in the sun for approximately two days – a process that prevents them from germinating. The de-pulping stage concludes with roasting or smoking the nuts over a fire for three to four days (Hyman 1991). Upon completion of these preliminary steps, nuts can be stored for up to nine months, until a woman is ready to convert them into butter. The specialized techniques women use for preservation and storage prolong the product's 'shelf-life' and ensure nut availability throughout the year (Howard 2003: 13).

When the time for making shea butter arrives, shea nut shells are cracked and removed. Nuts may then be roasted or smoked over a stove for 24 hours prior to being crushed one by one with a stone on the ground. They are thereafter warmed in a cauldron (Figure 9.2) and pounded in a mortar with a pestle. This yields a coarse brown batter, which is placed on a large stone and ground, with a smaller stone, into a finer-grained paste.

The following stages typically involve several women sharing the workload. Water is added to the paste and the mixture is kneaded. Two or three women jointly reach into the thick shea batter to beat the paste so the foam floats to the surface. Every few minutes, they relieve each other of the work to reduce fatigue from the labour. As the kneading motion is rhythmic, those waiting

Figure 9.2 Woman stirring shea nuts as they are warmed over a fire. Photo by Marlene Elias

their turn raise the spirits of the ones working by singing and clapping to the tempo of the kneading (Elias 2003). For women, the preparation of shea butter is a social process.

The foam is then transferred to a bucket of water, where it is 'washed' by hand, with women spinning the mixture in basins of water to eliminate unwanted residues (Figure 9.3). Subsequent washings – repeated as many as four times – yield progressively whiter foam, which is then boiled for many hours. The top layer is skimmed, or clarified, and upon cooling becomes the white butter so desired in international markets. The conversion of ten kilograms of shea nuts into butter typically demands eight to ten hours of an individual woman's labour (Faucon *et al.* 2001; Crélerot 1995; Elias 2003).

Differences in methods of butter making were recorded as early as the eighteenth century by Scottish explorer Mungo Park, as he searched for the source of the River Niger. On his journey through the Bambara landscape, Park witnessed shea nut collection, processing and trade. He observed the drying and roasting of nuts during the wet season. Park's village hosts informed him that the nut transformation method he witnessed produced the best shea butter:

> In one corner . . . was constructed a kiln for drying the fruit of the Shea trees: it contained about half a cart-load of fruit, under which was kept up a clear wood fire. I was informed, that in three days the fruit would be

> ready for pounding and boiling; and that the butter thus manufactured, is preferable to that which is prepared from fruit dried in the sun; especially in the rainy season; when the process by insolation is always tedious, and oftentimes ineffectual.
>
> (Park 2000: 215)

Park commended the taste of shea butter, writing that the product had the 'advantage of its keeping the whole year without salt'. He also noted that it 'is whiter, firmer, and to my palate, of a richer flavour, than the best butter I ever tasted made from cow's milk' (Park 2000: 201).

Variations in nut processing do indeed yield butters with different qualities. Regional shea markets recognize the quality differences associated with specific localities. This same point has been made with the African locust bean tree (*Parkia biglobosa*) or *néré*, another valued parkland species, from whose pod women prepare a paste used as a sauce-flavouring agent. Women in Benin base their market purchases of *néré* on tangible differences in product characteristics between localities (Gutierrez and Juhé-Beaulaton 2002: 468). These differences reflect distinctive methods of locust bean preparation. Female shea butter makers similarly contend that the manner in which the nuts are prepared results in product differences. Some appreciate the partially fermented product

Figure 9.3 Women washing shea. Source: Carol J. Pierce Colfer

obtained from nuts stored in underground pits (Hyman 1991), while others prefer the butter that results from boiling the nuts (Crélerot 1995; Elias 2003).

Many of the techniques women employ in butter preparation are related to water, firewood and labour availability. Smoking, roasting or boiling the nuts – and even the number of times the paste is rinsed – account for perceived differences in quality and taste. For instance, while the taste of butter derived from smoked nuts is enjoyed in certain communities, it is unappreciated in others. As with the preparation of *néré*, the diversity of product offerings derives from cultural repositories of knowledge that reside in specific communities.

Women develop the rich knowledge of shea nut processing at a young age. This knowledge is imbued with meanings that draw on the sociological, cosmological and ritual realms (Appadurai 1986). Shea butter production is a gendered identity marker, as well as a way rural women cement their social ties. Women are recognized for the quality of their butter and skilled butter makers take great pride in their reputation. In south-western Burkina Faso, butter producers offer their finest shea butter as a gift at births and weddings and as a gesture of gratitude for acts of kindness (Crélerot 1995). Similar practices are reported elsewhere – as in central Anatolia, Turkey, where women maintain their social networks through gifts of wild plants they collect (Ertuğ 2003). The shea butter remaining after gift offerings is consumed within the household or commercialized.

Regional trade in shea

Trade in shea butter is the principal economic activity of rural Burkinabé women (Crélerot 1995). Nut collectors and butter makers sell shea products out of their houses, in local markets, or to bulk buyers who purchase and transport the goods to regional outlets. While commerce in shea is active year-round, product prices nearly double during the dry season. The economic value of shea nuts and butter is lowest between June and September, when shea fruits come to maturity and their by-products abound in local markets.[3] Nonetheless, considerable amounts of nuts and butter are sold at this time. This is because household grain reserves are at their lowest in the pre-harvest agricultural period, and women need cash to purchase critical food items (Gosso 1996). One way to raise the value of shea nuts is to convert them into butter, but this requires women to exert even more labour when they are already burdened with agricultural work.

Buyers (principally women) take advantage of low wet season prices to purchase nuts cheaply from female collectors. These buyers then stock the nuts until their values rise. The nuts may be sold to merchants, wholesalers, local food dealers and, at times, even resold to the very females whose cash needs originally forced them to part with the nuts so cheaply (Terpend 1982; Audette 1995). Terpend (1982) estimates that a rural Burkinabè woman gathers between 560 kg and 650 kg of nuts in a typical year. If she transforms all her nuts into butter and sells the entire lot in the dry season, her earnings will average 50,000 to 58,500 FCFA (US$91–106). Such figures, however, overestimate the real value earned by women because most of the gathered nuts are retained for household consumption rather than sold, and because women must often sell the product when prices are seasonally lower.

Attempting to estimate the percentage of nuts actually consumed by rural Burkinabè households, Boffa *et al.* (1996) calculated that 60 to 90 per cent of a woman's collection is processed to feed her family. The remainder is marketed as raw nuts or butter. A more realistic estimate of female shea earnings takes this point into account and considers the 450–650 hours each woman typically labours to produce butter for household subsistence and petty cash sales. From this perspective, a rural woman earns only between 12,500 and 14,600 FCFA annually, or between US$23 and $27 per year from petty sales (Elias 2003).[4] While this may seem a pittance, the income from shea marketing still makes a contribution to the region's otherwise depressed farm incomes (Boffa *et al.* 1996).

Traders specializing in shea acquire the butter or raw nuts at local markets or directly from females selling from their homes or along the roadside. Chalfin (2001: 217) reports that female shea traders in one north-eastern Ghanaian town profit from proximity to commercial networks in south-east Burkina Faso and Niger. They typically buy in single transactions three to five calabashes of butter, weighing about 20 kg each, for US$15–20. They then sell to more capitalized merchant middlemen and wholesalers, typically men, who distribute the butter regionally or on the export market.

Research on northern Ghana's female shea traders also reveals buyer recognition of discernible differences in nut and butter quality (Chalfin 2000). At times this is the result of storage and preparation methods that result in nuts of low oil content. Ghanaian market women also show considerable savvy in noting regional and ethnic differences in the quality and taste of shea butter. In bringing together in one place the products from many localities and diverse ethnic groups, regional markets operate as sites for profiling butters of different quality. Near the border between Ghana and Burkina Faso, shea producers from the same region occupy different parts of the market. Their butter is moulded into different shapes and sizes depending upon their geographical origin and ethnic background. These provisions render distinctions in shea butter characteristics and in butter traders' origins obvious to customers (Chalfin 2001).

The assortment of shea butters sold in markets reflects the place-based and culturally embedded heritage associated with butter production. But this cultural repository also includes the very management of the shea landscape, which involves the selection and management of shea trees. Indigenous agroforestry practices contribute to the maintenance of shea biodiversity and have assured the conservation of the species across generations.

The shea landscape and biodiversity conservation

Shea is a slow-growing tree. It can take as many as 15 years to produce flowers, and fruit production peaks between 45 and 50 years (Boffa 1999). The tree bears fruit from the end of the dry season into the rains (in Burkina Faso, from May until mid-September), at a time when farmers are busy with field preparation and farming (Ruyssen 1957; Terpend 1982; Schreckenberg 1996). A mature shea tree produces an average of twenty kilograms of fresh fruit annually, but the quantity and quality of the fruit can vary unpredictably over short-term cycles (Chalfin 2000: 992). Climatic, biophysical and human practices affect tree yields.[5] Local populations eat the fruit, while nuts are retained for butter making. With traditional techniques of production, 20 kg of fruit typically yield about 4 kg of dried nuts and between 0.7 and 1.5 kg of butter (Terpend 1982).

V. paradoxa is found on over one million square kilometres of savanna south of the Sahel, where annual rainfall ranges between 500 and 1,400 mm (Hyman 1991: 1248) (Figure 9.1). The current biogeographical range of the species extends nearly to the Atlantic Coast in The Gambia. Domestic animals as well as wild elephants, birds, ungulates, primates and bats contribute to its long-distance seed dispersal (Burkill 1985; Hall *et al.* 1996). So do human beings. The diffusion of the shea tree to The Gambia, for instance, occurred with the migration of ethnic groups from the west African interior who deliberately established the culturally valued specimen (Maranz and Wiesman 2003). *V. paradoxa* comprises two subspecies: subsp. *paradoxa* for the tree present in the western Sudano-Sahelian zone and subsp. *nilotica* (Kotschy) A. N. Henry *et al.* for its eastern counterpart (Hall *et al.* 1996).

In Burkina Faso, *V. paradoxa* extends across the country from the semi-arid north-east to the south-west, which respectively receive an average of 500 mm and 1,200 mm of rain per annum. Shea trees are estimated to occupy 6.5 million hectares of savanna woodlands in Burkina Faso, or one quarter of the country's total land area (Kessler and Geerling 1994). In addition to rainfall patterns, human land use regulates the density and distribution of the species. One study from the late colonial period in Burkina Faso showed average per hectare densities of 55 trees in the country's south-west, 25 in the densely settled central region, and 35 in the north (IHRO, cited in Terpend 1982). However, recent studies indicate that the number of standing shea trees has considerably decreased as a result of the shift to orchard crops in the country's south-west, the intensification of agriculture in the central region, and the expansion of cotton monocultures, ox ploughing and fuelwood scarcity elsewhere (Saul *et al.* 2003: 159). These surveys report densities between six and nineteen trees per hectare, a figure consistent with neighbouring Mali, where the per hectare average is fifteen (Boffa 1991; Maranz and Wiesman 2003).

Traditional **V. paradoxa** *conservation*

Traditional agro-forestry techniques do not involve deliberate establishment of the shea tree. Shea's prolonged growth and maturation periods, unreliable production, recalcitrant seeds and out-crossing breeding system favour other methods (Lovett and Haq 2000).[6] Nonetheless, groves were established in The

Gambia during colonial rule and pilot plantations are under development in Burkina Faso (Saul *et al.* 2003; Carney and Elias, forthcoming).

Instead, the existence of *V. paradoxa* parklands results in considerable part from anthropogenic management and preservation of the species (Boffa 1999; Maranz and Wiesman 2003). Owing to their valued products, there exist many local taboos against cutting down *Vitellaria* trees (Lovett and Haq 2000). The Bobo in the western part of the country have long prohibited the cutting of valuable shea trees during the rainy season when they bear fruit. In other West African regions, customary law interdicts collection of shea nuts during certain periods (Lovett and Haq 2000; Boffa 1999). Bans on tree products with economically valuable environmental resources are typical of indigenous conservation measures reported in many areas of West Africa (Freudenberger *et al.* 1997). These cultural mechanisms promote the preservation and regeneration of valued species.

Managed landscapes of mixed vegetation and protected trees are known as agro-forestry parklands. They are the result of a calculated land management system among specific ethnic groups, which protects arboreal species providing desirable products when fields are cleared and burned for agriculture. The dominant tree species of parklands reflect local climatic, agricultural and cultural factors (Pélissier 1980). Livestock production may also be a significant component of these systems (Boffa 1999).

Along with *néré* (*Parkia biglobosa*) and gum arabic (*Faidherbia albida*, syn. *Acacia albida*), shea is found on cultivated and fallow lands throughout the West African Sudano-Sahelian zone (Breman and Kessler 1995; Boffa 1999). Shea and *néré* parklands are associated with agricultural societies that rely upon shea butter as their primary source of fat and cooking oil (Schreckenberg 1999). In contrast, protection of *V. paradoxa* is often not a priority among Fulani herders, whose diet is based on animal fat. They rely instead on *Acacia albida*, a leguminous tree that bears its leaves in the dry season, thereby providing shade and fodder for their animals (Seignobos 1982).

Among west African farming societies, shea tree conservation dates back to antiquity (Harlan 1992). At the end of the eighteenth century, Mungo Park (2000: 201) noted that in Mali's shea-dominated landscapes, shea trees were the only ones spared when forested land was cleared for cultivation. Recent palynological evidence from Burkina Faso suggests that the practice of preserving shea in cultivated fields was already occurring by 1000 AD (Neumann *et al.* 1998). Landscapes dominated by aged, slow-growing shea trees serve as testimony to longstanding *V. paradoxa* selection and parkland management.

While farmers preserve shea trees, they cull other species in cultivated fields. This increases the relative abundance of shea with respect to other parkland species. In southern Burkina Faso, Boffa (1995) observes that the relative occurrence of shea trees in farmed fields is five times greater than that found in uncultivated savanna. In one area he compared, shea accounted for 83 percent of the woody individuals on farmed land, but only 16 percent of those in uncultivated bush. In Benin, well over 70 percent of the trees encountered in farmed parklands are shea (Agbahungba and Depommier 1989), while in northern Ghana, *V. paradoxa* trees constitute more than 80 percent of the wooded farmland vegetation (Lovett and Haq 2000).

Protection of the shea tree serves vital ecological functions. Its extensive, shallow root system preserves soil structure and drainage where the tree grows (Bonkoungou 1987; Boussim and Guinko 1993; Gray 2003) while its canopy buffers parklands from wind and soil erosion (Kessler 1992). The tree's resistance to fire also helps prevent devastation in the savanna woodlands when fields are burned for agriculture or pasture grass regeneration (Burkill 1985). The ecological importance of shea is underscored by the fact that the tree figures among the few savanna species whose physical presence is used as a vegetative descriptor throughout the Sudano-Sahelian parklands (Harlan 1992; Hall *et al.* 1996).

Local agronomic knowledge

Traditional management practices additionally shape the relatively large size of shea trees in parklands compared to those on uncultivated areas. Local agronomic knowledge guides the selection of robust shea trees that appear best adapted to local growing conditions. Those with undesirable characteristics are culled. Due to this practice, and to the enhanced growing conditions found in cultivated fields, shea trees found on farmed lands typically attain a diameter double those of the same age growing in uncultivated areas (Boffa 1995).

Local agronomic knowledge also affects butter quality. In a study comparing shea pulp traits from trees in Mali and Burkina Faso, Maranz and Wiesman (2003: 1507) show how strong local selection for desired fruit and nut traits led to selective tree preservation and the culling of specimens with undesirable traits. As shea is exploited principally for pulp and fat, three economically valued characteristics were investigated: pulp sweetness, which is desirable because fruits are eaten, as well as fat content of the seed and type of fat in the kernel, which bear upon shea butter processing and quality. Shea populations in central Burkina Faso displayed the highest kernel fat and saturated fatty acid content. The percentage of stearic acid is a measure of fat hardness, which makes the butter retain a solid state at temperatures that exceed 40 degrees Celsius. In most Burkinabè nut populations, there is a slightly higher percentage of oleic acid to stearic acid. However, the percentage is reversed in shea kernels from the Moose (Mossi) Plateau of central Burkina Faso, where stearic acid dominates (Maranz and Wiesman 2003). Such factors affect shea butter quality in foreign markets. A lower ratio of stearic acid results in soggy butter that does not hold its form as solid pats. The interplay between shea nut traits and processing methods in different geographical regions thus influences the quality of the butter produced. Both traits and processing methods are in continuous evolution. Shea tree management methods highlight the way cultural preferences and practices preserve individuals with distinct traits in different parkland environments. This also has implications for shea butter commercialization.

Commercialization of shea production

Trade in African shea butter dates back at least to the fourteenth century, when Muslim travellers first recorded the practice (Lewicki 1974). The overseas export market for shea developed with the imposition of colonial rule in the nineteenth century. Demand grew with the use of shea as a cocoa butter equivalent in the manufacture of chocolate and margarine. While most of the colonial trade focused on nuts, butter exports steadily increased between 1932 and 1947. In 1937, at the height of the world depression, Burkina Faso continued to export shea: 8,451 tons of shea kernels and 2,927 tons of butter were collected from the key producing areas of Ouagadougou, Bobo and Gaoua (in the central and western parts of the country) and exported to France (Massa 1995).

Following the country's independence in 1960, shea exports grew (Pehaut 1976; FAOSTAT 2006). By the 1970s, shea nuts and butter had become the country's third largest foreign-exchange earner (Saul *et al.* 2003). But the international shea market's volatility was one cause of a decline in exports in the early 1990s (World Bank 1989; Saul *et al.* 2003) (Table 9.1). Poor tree productivity, as well as the disorganization of the commercial shea sector, the artisanal nature of butter production and trade, and unreliable statistics for national output further account for the export fluctuations depicted in Table 9.1 (UNCTAD 2006).

Despite annual fluctuations, the export demand for shea nuts continues to increase. In the two years between 1995 and 1997, nut exports from Ghana leaped from 15,000 to 32,000 tons, which represented an increase from two to seven million dollars in foreign exchange revenues (Chalfin 2000). Burkina Faso's shea nut exports have likewise increased in the past decade, with average annual nut exports rising from 10,000 tons over the ten-year period 1984–94 to 15,000 tons in 1994–2004 (FAOSTAT 2006). Shea butter exports from Burkina Faso also climbed over the past decade, averaging 630 tons per year between 1997 and 2001 (ONAC 2001) (Table 9.1).

The contemporary shea market continues to be characterized by raw nut rather than butter exports (Conti 1979). Two reasons primarily account for this. First, many European companies prefer processing imported nuts with modern technologies that guarantee butter of a desired quality. Second, low producer remuneration confers on women little incentive to engage in arduous nut transformation activities for the export market. Thus, while millions of African women produce shea butter for household consumption and local trade, most of their production remains in Africa (Hyman 1991; Boffa 1995). The dominance of raw nut over butter exports perpetuates the pattern established with colonialism, where value-added processing activities occur outside the producing area.

International demand for shea derives from cosmetics and food industries in the North, which are mostly headquartered in France, Great Britain, Scandinavia, Japan and North America (Pehaut 1976). The chocolate industry still accounts for 90 percent of the international demand for shea, used as a

Table 9.1 Shea exports from Burkina Faso, selected years (1961–2004)

Year	*Nuts (metric tons)*	*Butter (metric tons)*
1961	2,891	75
1962	2,572	681
1963	3,196	316
1964	6,681	773
1965	4,340	1,154
1966	11,611	1,142
1967	3,366	
1968	15,084	1,185
1969	12,342	1,024
1970	14,280	
1971	7,667	
1972	10,648	
1973	3,856	
1974	8,762	
1975	11,597	
1976	40,489	
1977	30,613	
1978	21,516	
1979	23,697	
1980	34,700	
1981	43,622	
1982	23,543	
1983	26,051	
1984	41,079	
1985	11,005	
1986	6,298	
1987	4,240	
1988	2,676	
1989	3,072	
1990	17,222	
1991	3,314	
1992	5,000	
1993	5,000	
1994	14,657	
1995	7,263	
1996	10,004	
1997	9,964	53
1998	20,663	2,367
1999	7,930	271
2000	11,575	190
2001	17,980	269
2002	34,975	
2003	26,686	
2004	11,891	

Sources: Péhaut (1976: 1319) (shea butter data 1961–9); FAOSTAT (shea nut data 1961–2004); ONAC (shea butter data 1997–2001)

cocoa butter equivalent (CBE) (UNIFEM 1997). The market for shea is thus closely linked to that for cocoa.[7] Shea sales are favoured in years of low cocoa yields and high cocoa prices, while the contrary is true of years of low cocoa prices. The availability of a handful of other alternatives to cocoa butter further drives down both shea and cocoa producer prices. In 2000, the European Union ruled in favour of allowing up to 5 percent of cocoa butter substitutes into chocolate. This legislation has stimulated the demand for shea and other vegetal oils within the chocolate industry (Fold 2000). Yet, the labour involved in making shea butter and the low prices offered by agro-food industries do not provide female African producers attractive income opportunities under conventional marketing arrangements.[8]

Current market demand for shea butter by global cosmetics firms, along with fair trade contracts, is creating a demand for export-quality butter and shifting more of the product into the export trade. The popularity of shea butter results from its emergence over the past fifteen years as a key ingredient in lines of 'natural' cosmetics (Compaoré 2000). In this new market niche, West African producers could enjoy a potentially more advantageous position with buyers. However, the food-processing industry is also poised to produce the shea butter demanded by cosmetics firms through its raw nut imports and advanced refining technology. At the turn of the twenty-first century, about half the shea butter used in the cosmetics sector was supplied by European food-processing firms. The latter processed the nuts and sold shea butter to the cosmetics industry at double the market price for its use in food applications (Boffa 1999).

While the 1999 import demand for shea by the cosmetics industry was estimated at only 200 tons (a figure well below the tonnage produced by Burkina Faso alone that year), the potential for integrating the butter in cosmetics products worldwide is estimated to reach 1,500 tons annually (Boffa 1999). With that in mind, international women-in-development (WID) projects are promoting new technologies to improve export butter quality by West African women's groups. The idea is to encourage cosmetics firms to purchase shea butter directly from the female producers, thereby building a marketing relationship based on 'fair' trade. With the middleman's profits eliminated, women's incomes are expected to rise. Contracts have already been negotiated with global cosmetics firms, such as The Body Shop and L'Occitane, willing to pay female producers superior prices for their product. In 2001 the French company purchased 60 tons of shea butter in Burkina Faso and planned to increase imports by another 90 tons in 2002 (Harsch 2001). The Body Shop is involved in similar contracts with producer groups in Ghana (TBS 1997). Such contracts have earned African women more than twice the prevailing market value per kilogram of shea butter (ANDINES 2002; Elias, fieldwork, 2005).

Those convinced that female income opportunities rest on the production of shea butter, rather than raw nut exports, have high hopes that women will benefit from new niche markets. Wholesale trade figures from Burkina Faso reveal the potential. In 1997 one ton of unprocessed shea nuts sold domestically for FCFA 70,000 and externally for FCFA 100,000 while the same ton processed into shea butter obtained FCFA 148,000 (Harsch 2001: 6).

In providing new technologies to producer cooperatives, donor projects help to ease the labour and natural resource demands associated with shea processing and to improve product quality. Many Burkinabè cooperatives already have access to mechanical mills that grind shea nuts and some have acquired equipment to assist them in the nut crushing, heating and kneading steps. While these cooperatives have a greater capacity to turn out large quantities of quality butter and reduce fuelwood demand for processing, the technologies are not without some drawbacks. They demand costly fuel inputs and spare parts. When shea presses break down, cooperatives are left shouldering large debts to repay the initial capital investment and maintenance costs (Compaoré 2000).

For the time being, fair-trade shea projects have offered thousands of Burkinabè women a singular economic opportunity to earn more from butter preparation. Even though the bulk of shea continues to be traded conventionally as nuts, the market share of direct shea butter sales to cosmetics companies is rising. As new markets link female butter makers with consumers of the product across geographic space, the tentacles of the female commodity chain now reach to the very processes that have long formed the cultural heritage of shea.

Changing property rights with shea commercialization

Female income opportunities with shea butter depend fundamentally on access to the tree's nuts. As the shea tree is not deliberately planted, its distribution on different types of land confers varying access rights. Shea trees are found on household landholdings as well as on unclaimed land used by villages for pasture, fuelwood and the gathering of medicinals. In open-access forests, women collect nuts on a first-come-first-served basis. On cultivated or fallow fields, only women with privileged access hold the rights to gather the nuts.

Burkinabè rural households follow a tenure pattern typical of Sahelian common property systems (Carney 1988), dividing their landholding into personal and family fields. If a shea tree grows on her personal field, a woman is entitled to collect the fruits and the value of marketed butter (Terpend 1982). If the tree grows on family fields, the male family head grants female members of the household the right to gather shea fruits (Ruyssen 1957; Boffa *et al.* 1996). The decision to cut or leave shea in parklands is ultimately made by the household head and reflects existing policy measures, economic incentives, and the value of the tree and its by-products. For instance, the shade from a shea tree can reduce the yield of other crops such as sorghum by 44 percent and millet by 60 percent (Kater *et al.* 1992). Nut availability is jeopardized every time a male household head decides to fell shea trees on family land. Longstanding customary village taboos reduce the incidence of cutting down shea trees, and some West African countries legislate specifically against their removal, even if forestry departments are too often unable to enforce such laws (Schreckenberg 1999; Wezel and Haigis 2000).

Recent research suggests that shea commercialization is changing women's rights over nuts and the value derived from making butter. In southwestern

Burkina Faso, near regional export markets in Côte d'Ivoire, nut collection remains a female activity. However, women are being made to share shea revenues with male household heads. Boffa *et al.* (1996) report this development in a quarter of households they surveyed in a village known for its shea production. In an additional 7 percent of the households, the male family head claimed the entire value of female shea nut sales (Boffa *et al.* 1996). The decline in cocoa commodity prices in Côte d'Ivoire has further led young men to enter into competition with women in collecting shea nuts. The men do not process the nuts, but sell them directly to wholesalers. These male nut collectors do not select high quality nuts nor do they put their harvest through the necessary pre-transformation steps to prevent nut germination. The deepening demand for nuts has thus adversely affected nut and butter quality (Bliss and Gaesing 1992, in Boffa 1999).

The extent of men's longer-term involvement in the shea commodity chain remains to be seen. History has shown that female-controlled crops can become the purview of men upon an increase in economic value, all the while maintaining women's role in their labour-intensive processing steps. Such was the case, for instance, with palm oil in Nigeria at the beginning of the twentieth century. As the product's export value increased, men gained control of the associated profits while assuming only superficial palm oil processing tasks. Meanwhile, women retained their traditional fruit transport duties and the bulk of oil processing responsibilities. In Martin's words (1984: 419),

> The entry of men into palm production made a difference to women mainly in that it deprived them of their right to initiate and control the production process and to control the use of the resulting oil. By the early twentieth century men were well established as the owners of palm fruit and of palm oil. Women were rewarded for their role in oil production by being allowed to keep some oil for cooking as well as the by-products of oil processing . . . which had no major local use.

Will shea replicate the history of palm oil, where men captured the market for palm oil as well as the female labour required to process it?

Market mechanisms and biodiversity conservation

The longer-term impacts of shea markets and current land-use patterns on *V. paradoxa* conservation cannot yet be evaluated fully. However, a few trends are evident. High prices for shea nuts and butter – associated with fair trade and WID contracts – are encouraging the selection and conservation of shea trees on agricultural land. In contrast, shea trees are felled when alternative land use is more valued or the price and need for fuelwood exceeds that of shea products. The density of shea trees on farmed fields thus remains closely linked with the economic value of the tree's products to the household economy (Schreckenberg 2004a).

Gender issues also appear crucial in shea conservation strategies. Preliminary evidence suggests that more shea trees per hectare are found on personal fields controlled by women (Boffa, 1995). In Thiougou, male household heads maintain densities of 20 shea trees per hectare on their personal fields while female-managed areas averaged shea densities of 27 trees per hectare. Burkinabè income streams are gendered, and men do not customarily control the products of shea, as they do other parkland species. They may thus opt to cut down *Vitellaria* individuals to the detriment of the women who harvest the nuts (Boffa 1999). Tree felling may occur even though overall revenues from a typical parkland in Burkina Faso with five to ten shea trees and two or three *néré* trees is of the order of 8000 FCFA per hectare (Saul *et al.* 2003: 131).

Agricultural policies and extension packages that support draught animal traction and mechanized ploughing also adversely affect the incidence of shea trees (Kessler 1992; Boffa 1999). Tree removal facilitates the unimpeded movement of draught animals and avoids striking tree roots. As a result, such policies cause a drastic decline in tree densities on cultivated fields. In Thiougou, the average shea density on land farmed with hand cultivation techniques was 31 per hectare. Land ploughed with donkey or oxen reduced densities to 21–25 trees per hectare (Boffa 1995). Even fewer shea trees are maintained on land mechanically ploughed in northern Ghana (Lovett and Haq 2000). While farmers are encouraged to replant trees in non-ploughed areas, they opt for exotic fruit trees or fast-growing pole and timber species at the expense of slow-growing indigenous species whose value is associated with women (Schreckenberg 1999: 288).

Mechanical ploughing, however, should be seen as separate from the process of agricultural intensification. Agricultural intensification does not always result in the destruction of shea trees and a reduction in parkland biodiversity. In southwest Burkina Faso, land is becoming scarce and fallow periods are diminishing. Gray (2003) observes that the shortage of farmland has increased the worries of borrowers, predominantly migrants, who fear losing their fields or the privilege to farm if they leave fields fallow. Instead of clear-cutting the arboreal vegetation, vulnerable households are responding to the lack of permanent land rights and soil degradation with traditional agro-forestry practices. In the very area where landholdings are least secure, borrowing households are investing in soil quality to strengthen user rights to land to which they have few formal claims. A key component of their land-use strategy is to preserve specific tree species (shea, *néré* and acacia) for their economic and ecological value. The trees reduce soil erosion and water run-off as well as supplying the farmland with organic matter through leaf litter and root decay (Gray 2003). They also stand as sentinels, preserving the cultural identity of migrants – who rely upon shea butter – in their new environment.

Heritage product: shea and geographical indicators as a marketing strategy

In its burgeoning cosmetics niche, shea's cultural heritage is explicitly valorized as a marketing strategy. African-American retail outlets as well as global firms such as The Body Shop emphasize shea butter's African origins. Their publicity describes the inherited skills African women employ in the commodity's production. Shea's central role in Sudano-Sahelian lifestyles and traditions renders it an ideal addition to the growing number of 'heritage products' linking consumers to 'authentic' place-based heritages (Bessière 1998). The novel forms shea butter assumes within elaborate Western cosmetics combine these African traditions with modern innovations.

As international interest in culinary heritages grows, does shea's cultural heritage hold additional promise within the food industry? Predominantly within the European Union, accreditation of food products by origin, preparation and identity is prevalent. Ranging from labels specifically designed to highlight sound environmental or labour practices in food production to a product's geographical provenance, these markers foster consumer confidence in goods bearing a tradition of quality. Such is the case, for instance, for France's *produits du terroir*. This distinguished niche market draws upon the notion that some of France's regions are specialized in the preparation of particular foods with distinct characteristics (Bessière 1998; Roussel and Verdeaux, this volume). These distinctive culinary traditions are the result of *in situ* cultural heritages similar to those exemplified by shea in Africa. They result from different methods of food preparation, which affect product characteristics. The *Appellation d'Origine Contrôlée* (AOC) label highlights the regional origins of these *produits du terroir* within France to consumers (Bessière 1998). While an analogous tradition exists for shea in Africa, what are the prospects for a comparable market valuing shea butter's place-based culinary traditions?

The esteem the international community places on shea's traditional function as cooking oil is uncertain. Even within West Africa, the consumption of shea butter is waning where alternatives such as sesame, groundnut, cottonseed and palm oil are found (Hall *et al.* 1996; Schreckenberg 2004a, 2004b). While shea butter is often the most affordable local cooking oil, many Africans prefer the taste of its alternatives, with shea retaining a specialized use only in commemorative meals. Internationally, there is no evidence that expatriate African populations in Europe or North America place the same value on shea butter in cooking as they do on palm oil or *néré*. The foreign market for it as a food product thus appears unviable, unless its role in chocolate manufacture can be linked to current fair trade initiatives.

There is, nonetheless, a current attempt in West Africa to link the chemical characteristics or 'signature' of shea nuts and butter to their geographic origins. Sponsored by the Common Fund for Commodities, the Dutch government and the Food and Agriculture Organization (FAO), the Pro*Karité* project aims to document the distinct traits of shea nuts and butter originating in specific regions. The aim is to improve the 'trace-ability' of shea products along the shea commodity chain should a specialized market for the product emerge (Pro*Karité* 2004; Masters 2005).

Rather than focusing on regional production specialties, however, the current trend for international shea butter sales is one of standardization. To secure international markets for shea butter, producers must comply with high quality standards. Shea butter must be of reliable quality and reflect a low percentage of impurities, humidity and acidity. In Burkina Faso, WID projects advise women on the production of a quality product meeting strict international demands. Workshops given to female producer associations call for a standardization of processing techniques – the very indigenous methods that have long resulted in the different types of butter produced regionally and by diverse ethnic groups. For instance, all butter producers are encouraged to boil their nuts rather than to bury them. Women are similarly dissuaded from smoking their nuts and advised to subject the shea paste to multiple washings. This yields a product with fewer impurities but demands considerable amounts of water.

The advent of new technologies further standardizes the production process and final product quality. As steps are mechanized, the cultural heritages embedded in traditional shea nut transformation are erased. As such, these technologies reduce the need for the specialized gendered knowledges that inform processing practices (Biquard 1992). In standardizing the butter-making process, new technologies open the door to marketing entrepreneurs who wish to capitalize on the products of women's labour and emerging market opportunities (Biquard 1992).

Official certification by the Fairtrade Labelling Organization (FLO) International will provide additional momentum to the standardization of shea butter production. Fair-trade shea butter is not yet officially certified by FLO. Yet the organization is currently establishing criteria to regulate the way butter sold on the fair-trade market should be produced and traded.[9] Producer associations desiring certification need to demonstrate compliance with established social and ecological regulations, including biodiversity and sustainable harvesting requirements related to shea nut collection and processing. Certified importers, similarly, are required to respect a set of trade criteria, including a minimum price per kilogram of shea paid to producer associations. Official labelling of fair-trade shea butter in North America and Europe is likely to increase product pricing and foster this market niche by building consumer awareness and confidence in the product. The economic returns and ecological requirements of fair-trade shea butter also promise to assist local communities in protecting the natural heritage passed down from their ancestors. Yet, to comply with global product norms, producers must adopt standardized processing techniques that stray from the distinct and specialized cultural traditions developed regionally by butter makers over the centuries.

Conclusion

As this article has sought to demonstrate, the shea tree constitutes a botanical and cultural African heritage. Owing to its myriad functions and valued by-products, the species plays an integral role in African subsistence and is embedded with cultural meaning. The tree's local significance ensures its anthropogenic selection and preservation in what have become shea parklands. For more than a thousand years, successive Sudano-Sahelian farming communities have managed and shaped these parklands. Now globalization has reached these areas. Expanding shea markets, new European Union legislation concerning the use of CBEs, international women-in-development advocacy groups, and agricultural extension programmes are affecting the very management systems of these ancient parklands. As changes sweep over the socio-cultural shea landscape, what will be the future outcome of this African natural heritage? Who will be the beneficiaries as shea parklands are transformed?

While both men and women use, value and protect *Vitellaria* trees, shea is above all a female heritage. Across the Sudano-Sahelian zone, the species is a marker of gender identity. Women acquire knowledge of shea trees, nut collection and processing at a young age; over their own life cycle they bequeath this knowledge to their daughters, who in turn renew the tradition. The significance of shea to female identity and livelihoods is commemorated in a song sung by butter makers in Mali (Biquard 1992: 173–4).

> Shea is women's wealth . . . shea is women's wisdom.
>
> Confide in it before speaking to your husband.
>
> Wild shea, born of this very earth;
>
> This shea has stayed with the village, has given it all its riches and protection.
>
> Shea was already present when the Ancestors founded this place, composing with nature the harmony we must each reproduce.[10]

As this refrain tells us, the roots of the shea tree reach deep into culture and nature. With each year, the transformation of nature's subsidy into shea butter also roots the identity of Sudano-Sahelian women into place.

Notes

1 The natural range of shea includes Benin, Burkina Faso, Cameroon, Central African Republic, Chad, Côte d'Ivoire, Ethiopia, Guinea, Mali, Niger, Nigeria, Senegal, Sudan and Togo. Over the past centuries, the tree has spread to Gambia (Maranz and Wiesman 2003), the Democratic Republic of the Congo, Ghana and Uganda due to human influence (Hall *et al.* 1996). The subspecies *paradoxa* dominates in the western Sudan while ssp. *nilotica* is found in the Democratic Republic of the Congo, Ethiopia, Uganda and the Sudan.

2 Elias carried out primary data collection for the study over two months in each fieldwork period. Interviews were conducted with male and female farmers, NGO personnel, academics and government officials working in the shea sector. More in-depth participant observation occurred with members of the Laafi women's group and the Union des Productrices de Produits Karité de la Sissili et du Ziro. Primary and secondary sources obtained in Burkina Faso supplemented the collected information.

3 A weighted average producer price of 500 FCFA per kilogram of butter is calculated from local Burkinabè markets, based upon annual means for the last decade (1990–2000) (ANDINES 2002; Elias 2003).

4 Figures are based upon an exchange rate of 550 FCFA to US$1 (2006).

5 Terpend (1982) discusses a three-year cycle in annual production, wherein production is good one year, poor the next, and mediocre the third. Seasonal variation in the harmattan – the arid, cool, dry season wind that blows south from the Sahara across the Sahel – also affects fruit yield. The tree's flowers are torn in years of pronounced winds, and subsequent production is reduced. Precipitation further influences yields, with high rains leading to increased subsequent production. Diseases, parasites and predators additionally decrease productivity. Anthropogenic factors responsible for lower yields include the setting of fires to clear land for agriculture during the shea tree's flowering period, which adversely affects the timing and quantity of flowers produced (Abbiw 1990).

6 'Recalcitrant seeds' refers to the fact that seed viability drops very rapidly. For *V. paradoxa*, viability declines within a week of seed removal from the fruit and is completely lost within three to six weeks (Hall *et al.* 1996). Out-crossing complicates the selection for 'superior', 'true-to-type' individuals stemming from heterozygous parents.

7 Statistics on the international shea market for the food industry are difficult to obtain because the few firms that dominate production conduct their activities in a secretive manner. Four large European importers – Aarhus, Karlshamns, Unilever and Van Dermoortele – dominate the international shea market.

8 Producer prices hover around 300 FCFA (US$0.60) per kilogram of shea butter (Elias, fieldwork, 2005). This price is inordinately low in light of the fact that production of one kilogram of butter demands nearly 10 hours of female labour (Crélerot 1995; Elias and Carney 2005).

9 Inaugurated in 1997, FLO (Fairtrade Labelling Organizations International) is an umbrella organization that regroups 17 fair-trade labelling initiatives throughout the world. FLO sets global standards for, and offers third-party certification of, fair-trade products. For more information on the organization, see: <http://www.fairtrade.net/>.

10 Authors' translation.

References

Abbiw, D. K. 1990. *Useful Plants of Ghana: West African uses of wild and cultivated plants*. London: Intermediate Technology Publications and The Royal Botanic Gardens, Kew.

Agbahungba, G. and D. Depommier. 1989. 'Aspects du parc à karité-nérés (*Vitellaria paradoxa* Gaertn. F. *Parkia biglobosa* Jacq. Benth.) dans le sud de Borgou (Bénin)', *Bois et Forêts des Tropiques* 222: 41–54.

ANDINES. 2002. *Beurre de karité: Groupe LAAFI (Burkina Faso) et ANDINES (France)*, Ile Saint Denis: ANDINES SA.

Appadurai, A. 1986. 'Commodities and the politics of value' in A. Appadurai (ed.), *The Social Life of Things: commodities in cultural perspective*. Cambridge: Cambridge University Press.

Audette, R. 1995. *Étude de la filière karité. Programme Micro-Réalisations Burkina Faso*. ACDI, Projet 960/16010.

Berkes, F. 1999. *Sacred Ecology: traditional ecological knowledge and resource management*. Philadelphia, PA: Taylor and Francis.

Berkes, F., J. Colding and C. Folke. 2000. 'Rediscovery of traditional ecological knowledge as adaptive management', *Ecological Applications* 10: 1251–62.

Bessière, J. 1998. 'Local development and heritage: traditional food and cuisine as tourist attractions in rural areas', *Sociologia Ruralis* 38 (1): 21–34.

Biquard, A. 1992. 'Femmes et innovations technologique: pertes sans profit. L'exemple du beurre de karité (Mali)' in J. Bissiliat, F. Pinton and M. Lecarme (eds), *Relations de genre et développement: femmes et sociétés*. Paris: ORSTOM.

Bliss, F. and K. Gaesing. 1992. *Möglichkeiten der Einbeziehung von Frauen in Massnahmen der ressourcenschonenden Nutzung von Baumbeständen. Forschungsberichte des Bundesministeriums für Wirtschaftliche Zusammenarbeit.* Munich: Weltforum Verlag.

Boffa J-M. 1991. 'Land and Tree Management and Tenure Policies in Burkina Faso's Agroforestry System'. MS thesis, Purdue University, West Lafayette, Indiana.

—— 1995. 'Productivity and Management of Agroforestry Parklands in the Sudan Zone of Burkina Faso'. PhD dissertation, Purdue University, West Lafayette, Indiana.

—— 1999. *Agroforestry Parklands in Sub-Saharan Africa.* FAO Conservation Guide No. 34. Rome: FAO.

Boffa, J-M., G. Yaméogo, P. Nikiéma and D. M. Knudson. 1996. 'Shea nut (*Viellaria paradoxa*) production and collection in agroforestry parklands of Burkina Faso' in R. R. B. Leakey *et al.* (eds), *Domestication and Commercialization of Non-Timber Forest Products in Agroforestry Systems: non-wood forest products.* Rome: FAO.

Bonkoungou, E. G. 1987. *Monographie du Karité,* Butyrospermum paradoxum *(Gaertn. F.) Hepper, espèce agroforestière a usages multiples.* Ouagadougou: Institut de Recherche en Biologie et Écologie Tropicale.

Boussim, I. J. and G. S. A. S. Guinko. 1993. 'Tapinantus parasite du karité au Burkina Faso, première partie', *Bois et Forêts des Tropiques,* 238: 45–7.

Breman, H. and J. J. Kessler. 1995. *Woody Plants in Agro-Ecosystems of Semi-Arid Regions, with an Emphasis on the Sahelian Countries.* Berlin: Springer Verlag.

Burkill, H. M. 1985. *The Useful Plants of West Tropical Africa, Volume 4.* Kew: Royal Botanic Gardens.

Carney, J. 1988. 'Struggles over crop rights and labour within contract farming households in a Gambian irrigated rice project', *Journal of Peasant Studies* 50 (3): 334–49.

Carney, J. and M. Elias. Forthcoming. 'Revealing gendered landscapes: indigenous female knowledge and agroforestry of African shea', *Canadian Journal of African Studies.*

Chalfin, B. 2000. 'Risky business: economic uncertainty, market reforms and female livelihoods in northeast Ghana', *Development and Change* 31: 987–1008.

——2001. 'Border zone trade and the economic boundaries of the state in north-east Ghana', *Africa* 71 (2): 202–24.

Compaoré, P. N. 2000. 'Femmes, Développement et Transfert de Technologies. Le Cas des Presses à Karité au Burkina Faso'. PhD dissertation, University of Montreal.

Crélerot, F. 1995. 'Importance of Shea Nuts for Women's Activities and Young Child Nutrition in Burkina Faso: the case of the Lobi'. PhD dissertation, University of Wisconsin-Madison.

Depommier, D. and P. S. Ramakrishnan (eds). 2002. *Traditional Ecological Knowledge, Conservation of Biodiversity and Sustainable Development.* Pondicherry: Institut Francais de Pondichery.

Elias, M. 2003. 'Globalization and Female Production of African Shea Butter in Rural Burkina Faso'. Masters thesis, University of California, Los Angeles.

Ertuğ, F. 2003. 'Gendering the tradition of plant gathering in Central Anatolia (Turkey)' in P. L. Howard (ed.), *Women and Plants.* London: Zed Books.

FAOSTAT. Trade statistics 1961–2004, Accessed online 10 February 2006, at: http://apps.fao.org/faostat/form?collection=Trade.CropsLivestockProducts&Domain=Trade&servlet=1&hasbulk=0&version=ext&language=EN

Faucon, M., A. Sauvageau and S. Bahl. 2001. 'Coût de production pour le beurre de karité du groupement Laafi'. Unpublished data, Centre d' Études et de Coopération Internationale (CECI), Ouagadougou.

Fold, N. 2000. 'A matter of good taste? Quality and the construction of standards for chocolate products in the European Union', *Cahiers d'Economie et Sociologie Rurales* 55–56: 91–110.

Freudenberger, M., J. Carney and A. R. Lebbie. 1997. 'Resilience and change in common property regimes in West Africa: the case of the Tongo in The Gambia, Guinea and Sierra Leone', *Society and Natural Resources* 10: 383–402.

Gosso, D. 1996. 'Étude de l'importance socio-économique du karité et des dangers qui menacent sa survie'. Thèse pour Diplôme de Technicien Supérieur (DTS), Ouagadougou.

Gray, L. 2003. 'Investing in soil quality: farmer responses to land scarcity in southwestern Burkina Faso' in T. Bassett and D. Crummey (eds), *African Savannas: global narratives and local knowledge of envrionmental change*. Oxford and Portsmouth, NH: James Currey and Heinemann.

Gutierrez, M.-L. and D. Juhé-Beaulaton. 2002. 'Histoire du parc à Néré (*Parkia biglobosa* Jacqu. Benth.) sur le plateau d'Abomey (Bénin): de sa conservation pour la production et la commercialisation d'un condiment, l'afitin', *Les Cahiers d'Outre-Mer* 220: 453–74.

Hall, J. B., D. P. Aebischer, H. F. Tomlinson, E. Osei-Amaning and J. R. Hindle. 1996. *Vitellaria paradoxa: a monograph*. Bangor: School of Agricultural and Forest Sciences, University of Wales, Publication No. 8.

Harlan, J. R. 1992. *Crops and Man*. Madison, WI: American Society of Agronomy/ Crop Science Society of America, second edition.

Harsch, E. 2001. 'Making trade work for poor women', *Africa Recovery* 15 (4): 6.

Howard, P. L. 2003. 'Women and the plant world: an exploration' in P. L. Howard (ed.), *Women and Plants*. London: Zed Books.

Hyman, E. L. 1991. 'A comparison of labor-saving technologies for processing shea nut butter in Mali', *World Development* 19 (9): 1247–68.

Kater, L. J. M., S. Kante and A. Budelman. 1992. 'Karité (*Vitellaria paradoxa*) and néré (*Parkia biglobosa*) associated with crops in South Mali', *Agroforestry Systems* 18: 89–105.

Kessler, J. J. 1992. 'The influence of karité (*Vitellaria paradoxa*) and néré (*Parkia biglobosa*) trees on sorghum production in Burkina Faso', *Agroforestry Systems* 17: 97–118.

Kessler, J. J. and C. Geerling. 1994. *Profil environnemental du Burkina Faso*. Wageningen: University of Agronomics.

Lewicki, T. 1974. *West African Food in the Middle Ages*. New York: Cambridge University Press.

Lovett, P. N. and N. Haq. 2000. 'Evidence for anthropic selection of the Sheanut tree (*Vitellaria paradoxa*)', *Agroforestry Systems* 48: 273–88.

Maranz, S. and Wiesman, Z. 2003. 'Evidence for indigenous selection and distribution of the shea tree, *Vitellaria paradoxa*, and its potential significance to prevailing parkland tree patterns in sub-Saharan Africa north of the equator', *Journal of Biogeography* 30 (10); 1505–16.

Martin, S. 1984. 'Gender and innovation: farming, cooking and oil palm processing in the Ngwa region, south-eastern Nigeria, 1900–1930', *Journal of African History* 25 (4): 411–27.

Massa, G. 1995. 'Introduction' in G. Massa and Y. G. Madiéga (eds), *La Haute Volta Coloniale: témoignages, recherches, regards*. Paris: Karthala.

Masters, E. Personal communication, 8 July 2005, Ouagadougou.

Neumann, K., S. Kahlheber and D. Uebel. 1998. 'Remains of woody plants from Saouga, a mediaeval West African village', *Vegetation History and Archaeobotany* 7: 55–77.

ONAC (Office National du Commerce). Trade statistics, 1997–2001, Ouagadougou, Burkina Faso.

Park, M. 2000 [1799]. *Travels in the Interior Districts of Africa*. Durham, NC: Duke University Press.

Pehaut, Y. 1976. *Les Oléagineux dans les pays d'Afrique Occidentale associés au marché commun*. Paris: Éditions Honoré Champion, 2 vols.

Pélissier, P. 1980. 'L'arbre dans les paysages agraires de l'Afrique noire', *Cahiers ORSTOM, Série Sciences Humaines* 17: 131–6.

ProKarité. 2004. 'ProKarité'. Accessed online 12 February 2006, at: http://www.prokarite.org/index-eng.html.

Roussel, B. and F. Verdeaux 'Natural patrimony and local communities in Ethiopia: advantages and limitations of a system of geographical indications', in this issue.

Ruyssen, B. 1957. 'Le karité au Soudan', *Agronomie Tropicale* 12: 143–72, 279–306, 415–40.

Saul, M., J-M. Ouadba and O. Bognounou. 2003. 'The wild vegetation cover of western Burkina Faso: colonial policy and post-colonial development' in T. Bassett and D. Crummey (eds), *African Savannas: global narratives and local knowledge of envrionmental change*. Oxford and Portsmouth, NH: James Currey and Heinemann.

Schreckenberg, K. 1996. 'Forests, Fields, and Markets: a study of indigenous tree productions in the woody savannahs of the Bassila region, Benin'. PhD dissertation, University of London.

—— 1999. 'Products of a managed landscape: non-timber forest products in the parklands of the Bassila region, Benin', *Global Ecology and Biogeography* 8 (3–4): 279–89.

—— 2004a. 'The contribution of shea butter (*V. paradoxa* C.F. Gaertner) to livelihoods in Benin' in T. Sunderland and O. Ndoye (eds), *Forest Products, Livelihoods and Conservation: case studies on non-timber forest product systems*. Bogor Barat, Indonesia: CIFOR.

—— 2004b. 'Shea butter: from cooking fat to cosmetics and chocolate' in C. Lopez and P. Shanley (eds), *Riches of the Forest: for health, life and spirit in Africa*. SMK Desa Putera, Indonesia: CIFOR.

Seignobos, C. 1982. 'Végétations anthropiques dans la zone soudano-sahélienne: la problématiques des parcs', *Revue de Géographie du Cameroun* 1–23.

Terpend, M. N. 1982. *La filière karité: produit de cueillette, produit de luxe*. Paris: Les dossiers faim-développement.

The Body Shop (TBS). 1997. Promotional literature in brochure form. London: TBS.

UNCTAD (United Nations Conference on Trade and Development). 'Information de marché dans le secteur des produits de base.' Accessed online 12 February 2006, at: <http://r0.unctad.org/infocomm/francais/karite/ plan.htm>

UNIFEM (United Nations Development Fund for Women). 1997. *Le karité, l'or blanc des africaines*. Dakar, Sénégal : Bureau Régional de UNIFEM.

Wezel, A. and J. Haigis. 2000. 'Farmers' perception of vegetation changes in semi-arid Niger', *Land Degradation and Development* 11 (6): 523–34.

World Bank 1989. 'Burkina Faso', in World Bank (ed.), *Trends in Developing Economies*. Washington, DC: World Bank.

Part IV

Conjugal relations, migration, and forest-based livelihoods in Southeast Asia

As was noted in Chapter 1, there have been increasing calls for more attention to home life and to power relations between men and women. We begin with Li's (1998) contribution, which provides an informative comparison with Part III in its use of methods developed in Africa on conjugal bargaining. Li's contribution includes discussion of both city and forest, and its insights have relevance for marital relations in either context. Her examination of divorce proceedings to ground her ethnographic observations in material impacts on women and men is clever and useful methodologically.

One of her important concerns is "the renegotiation of cultural practices in the context of changing material conditions and shifting fields of power" (Li 1998, p. 675 in original) – a very relevant and timely concern for gender and forests. She also argues against some of the common dichotomies (as in Chapter 1) that have been used in analyses of the gendered division of labour in Africa and elsewhere.

She summarizes key features of many Southeast Asian societies and cultures that contribute to women's relative equality before turning to property relations linked to inequitable political power and prestige. To do this, she focuses first on divorce proceedings and conjugal property disputes. Here, she brings in the distinction from the African literature between sex-sequential and sex-segregated labour processes (see Glossary), with differing implications in different contexts for women in property disputes. In both settings, the concept of personhood is important for both men and women.

Although Li's findings in urban Singapore are not directly applicable in many of today's forests, what she describes – the recognition among women that their labour can be considered a commodity – *is* likely to be the case in the forests of the global North (though the sense of individual autonomy seems greater in Southeast Asia). Singaporean women describe their work at home as a gift, conceptually striving to remove it from the world of commerce and commodities. Yet divorce exposes the clear vulnerability of women's position. This analysis brings home the points made in Chapter 1 regarding the 'reproductive sphere' and domestic work, and the degree to which these are devalued, invisible, and/or ignored, to the disadvantage of women.

In those forests of Sulawesi where swidden agriculture is the economic base, again each person – male or female, adult or child – is seen to be autonomous, and his/her property is linked to the labour invested. Men's dominance in public affairs, their greater strength needed for land clearing, and their access to cash via wage labour give them advantages in establishing property rights that women do not have. Governmental policies also favour land ownership by men (see Elmhirst, this volume). Yet Li's analysis, like those of Arora-Jonsson (2011) and Reed (this volume), consistently recognizes the agency of women, not their victimhood.

Elmhirst, writing about Lampung, a province on the island of Sumatra, uses the concept of a 'political forest' (from Peluso and Vandergeest 2001), which refers to "ideas, practices and institutions that seek to regulate peoples' access to resources, providing recognition and legitimacy to some, whilst excluding and criminalizing others" (Elmhirst 2011, p. 173 in original). She analyzes the gender implications of the concept of the ideal citizen – including glorification of heteronormative marriage – promulgated within ex-president Suharto's regime. Like Rocheleau and Ross (this volume), Elmhirst shares a feminist political ecological orientation; but by incorporating elements from queer theory, she highlights an issue long ignored within the world of forestry: the role of the state's (and others') conceptions of the family in allocating forest benefits.

Her attention to the experience of migrants – so common in tropical forests – is welcome as well, given the tendency to focus on those who have a place-based affinity for their forested environments. She graphically portrays the risks to the livelihoods and farms of migrants whose presence is technically illegal. They are under threat of forced eviction, particularly from protected areas.

Elmhirst traces the impacts of the various governmental policies throughout the twentieth century and into the twenty-first century. She presents telling archival material about, for instance, the criteria for selection in the Dutch transmigration program. These show the view of unmarried men as 'unruly' and of families as necessary for stability, harmony, and order – views later reinforced and replicated by Suharto's regime. Single women seem not to have been included at all. The power of state discourses about masculinity, femininity, and citizenship and their impacts on people's lives come across clearly. Men were seen as breadwinners (and received the direct government benefits to transmigrants, including title to land); women were seen as homemakers and socializers of children, dependent on their husbands. As local land-based livelihoods became more precarious with increasing land-related conflict in the countryside in recent years, both young men and women have migrated for work, some overseas – leaving the elderly (or the remaining spouse) to mind the farm.

Elmhirst's paper emphasizes the Javanese orientation of governmental policy in the late twentieth century and beyond. In the case of Javanese transmigrants, the governmental assumptions about marriage and family are familiar. However, the strength of the pressure on them to conform is clear. Such pressure is magnified when the population in question is from a different ethnic group (e.g., the matrilineal Minangkabau or the more egalitarian Kenyah) with quite different views of marriage and the family. Similar kinds of policy pressure, from politically and numerically dominant ethnic groups on more marginalized ethnic groups, are replicated in many forest regions (e.g., Bose, 2011, on India; Sijapati, 2008, on Nepal; Paulson, this volume, on the Andes).

Both of these authors bring to bear the emphasis on harmony and cooperation that characterize many Southeast Asian cultures while also conveying the means and reality of conflict and negotiation.

References

Arora-Jonsson, S. 2011. Virtue and Vulnerability: Discourses on Women, Gender and Climate Change. *Global Environmental Change* 21 (1): 744–751.

Bose, P. 2011. Forest Tenure Reform: Exclusion of Tribal Women's Rights in Semi-Arid Rajasthan, India. *International Forestry Review* 13 (2): 220–232.

Elmhirst, R. 2011. Migrant Pathways to Resource Access in Lampung's Political Forest: Gender, Citizenship and Creative Conjugality. *Geoforum* 42 (2): 173–183.

Li, T. M. 1998. Working Separately but Eating Together: Personhood, Property, and Power in Conjugal Relations. *American Ethnologist* 25 (4): 675–694.

Peluso, N. L., and Vandergeest, P. 2001. Genealogies of the Political Forest and Customary Rights in Indonesia, Malaysia and Thailand. *Journal of Asian Studies* 60 (3): 761–812.

Sijapati, B. 2008. *Gender, Institutions and Development Intervention: A Case Study of Community Forestry in Nepal.* Unpublished PhD Dissertation, London School of Economics and Political Science.

10 Working separately but eating together

Personhood, property, and power in conjugal relations

Tania Murray Li

Spurred initially by a recognition that "the household" could not serve as a unit of analysis in many of the ethnographic contexts in which they worked, Africanists have given considerable analytical attention to gendered divisions of labor and reward mediated through the conjugal relationship. These issues have been less thoroughly explored in Southeast Asia, where a cultural emphasis on harmony and complementarity has sometimes deflected attention from the contractual and potentially unequal basis of conjugal exchange.

In this article I draw on the theoretical insights of feminist anthropologists working in Africa and apply the conceptual repertoire they have developed to the analysis of personhood, property, and conjugal relations in Southeast Asia. While I am aware of the risks entailed in lifting analytical constructs from one context and deploying them in another, there are also, as Edward Said points out, many insights to be gained when theory "travels" (1983:242). In this case, as I hope to demonstrate, the gains are twofold: a fresh perspective on gender issues in Southeast Asia which highlights a regional ethnographic area of inquiry hitherto underexplored in the literature, and some enrichment of the body of theory itself, as the comparison helps to clarify the origins of theory within the time and place from which it emerged and to which it responds, and opens it up toward other realities (Said 1983:242).

Cultural repertoires always carry within them a range of possible understandings with diverse implications for practice, although the outcomes are not simply a matter of choice. A particular strength of the Africanist literature to which I allude in this article is the sustained attention to questions of agency, and the attempt to situate the renegotiation of cultural practices in the context of changing material conditions and shifting fields of power. Constructs of personhood and the capacity of differently located persons to control property are treated not as constants, but instead as domains of struggle. The effort to comprehend the nature and limits of "struggles over meaning" (Berry 1988b), and the directions of change that both contribute to and emerge from these struggles, lie at the heart of feminist research on questions of personhood and property in Africa.[1] It is the exploration of the capacity of the theories and concepts developed in that context to produce fresh insights in another, rather than a comparison of abstract models of African and Southeast Asian kinship, that drives my analytical agenda here.

The analytical framework within which my comparative exercise takes shape is one that emphasizes the role of agency in deriving unique (and partial) resolutions to the structural dilemmas associated with a particular time and place. I begin with a review of the conceptual repertoire developed by Africanists. This is followed by an overview of concepts of personhood, property, and conjugal relations drawn from the literature on Southeast Asia. I then examine these issues in detail in urban Singapore and upland Sulawesi, to explore the cultural understandings, practices, and strategies that emerge at two distinct sites. My data for Singapore derive from household interviews and participant-observation undertaken in the context of my doctoral research from 1982–84. For Sulawesi, I draw on field research totalling eight months and conducted between 1990–96 in several mountain hamlets. I end my article by revisiting the Africanist formulations and drawing some conclusions focused around the issue of power.

The conceptual framework in my analysis adds an important dimension to my search for the elusive "difference that makes a difference" (Atkinson 1990:90). This "difference" accounts for men's greater power and prestige in the region despite the relatively high status and nominal equality of women in many contexts. It is not a framework that begins or ends with the victimization of women by men, by the state, or by capitalist market pressures. Nor does it highlight moments of dramatic resistance. Rather, it addresses the issue of power through a focus on gendered rhythms of work and exchange. It brings into view the everyday discourse and practice through which women and men reshape the social world, engaging and redirecting state agendas while reworking the cultural meanings of things and relationships and reconfiguring the more intimate domains of conjugal commitment and generosity.

Personhood, property, and the conjugal contract: feminist theorizing about Africa

Personhood and property

Ann Whitehead's important contribution to feminist theorizing about personhood and property in rural Africa is framed as an intervention in classic social theory. She observes that classic discussions of the relationship between personhood and property were founded on evolutionary presumptions and relied on a dichotomous contrast between types of society. Marx's analysis of commodity fetishism hinges on the distinction between the Western (capitalist or bourgeois) concept of property which embodies "the legal separation of subject and object, and . . . the legal separation of subject from subject in his or her capacity to have control over the disposal of a thing which has been designated his or her property," and its "historical opposite," in which these characteristics are reversed (Whitehead 1984:180). A similar dichotomy underlies Mauss's distinction between archaic (gift) societies and modern (commodity) societies in which people "transact freely with one another" and objects are "alienated commodities, separate from the giver and the recipient" (Carrier 1992:200; Gregory 1982:10–24).

Critics have argued that the contrast posed by these models is overdrawn. It obscures "matter out of place" (Carrier 1992:204): morally burdened relationships articulated through giftlike transactions in market-dominated societies, and commodity-like transactions that occur in times or places in which, according to the model, they are not expected (see also Appadurai 1986a; Parry 1986; Parry and Bloch 1989).[2] Reformulated, however, and shorn of their evolutionary associations, these classic discussions indicate a significant area for empirical inquiry. This is the issue to which Whitehead (1984) draws our attention in her essay on women and property:

> I find this suggestion, that bound up with the concepts of property are concepts of the person, and that these concepts of the person in turn can be characterised by their degree of individuation within social relations, an extraordinarily powerful one to begin thinking more generally about women (and men) and property.
>
> [1984:180]

Raising the possibility that degrees of individuation are variable by gender, she argues further:

> the issue raised is the extent to which forms of conjugal, familial and kinship relations allow [a woman] an independent existence so that she can assert rights as an individual against individuals. In many societies a woman's capacity to act in this way may be severely curtailed compared to a man's.
>
> [1984:189]

The ethnographic work of Whitehead and other feminist scholars who have examined gender issues in African contexts highlights the difference between men and women in their capacity to assert themselves as persons under changing cultural, economic, and political conditions. This body of work makes a distinction between women's nominal right to own property (which exists to some degree in most African societies) and their capacity to operationalize these rights in practice. Among the structural limitations that women in Africa commonly encounter are dependence on husbands, sons, or male kin to provide access to patrilineal land; lack of access to labor power, including their own, when they must meet labor obligations to husband and household before engaging in their own enterprises; and lack of access to capital in those cases when they are burdened with the responsibility for household provisioning or required by a doctrine of maternal altruism to put family needs above personal accumulation or investment (Berry 1988a, 1988b; Moore 1988; Stamp 1989; Whitehead 1984:185).

Women's capacity to own property depends also on the fields of power within which they make and pursue their claims. Whitehead observes that rural women in many African societies are less-than-equal persons in the political realm, not permitted to represent themselves and barely represented in customary courts even in "woman cases" (Whitehead 1984:188). They are therefore disadvantaged when defending their claims to property against others (husband, other wives, sharecroppers), especially when claims overlap. To work around these limitations, women in some of the agrarian contexts examined by Berry (1988a) base their investment decisions on their capacity to collect, preferring sectors in which the payoffs are quick and not dependent on maintaining long-term control over assets.

Divisions of labor and the conjugal contract

The conjugal relationship is one of several relationships in which women experience the embeddedness in kinship relations that potentially limits their individuation and personhood. It is especially important as a focus for feminist research as a site through which gendered divisions of labor and reward are negotiated. Feminist scholars working in Africa have investigated this relationship through the concept of the conjugal contract, defined as the "terms on which husbands and wives exchange goods, incomes, and services, including labor, within the household" (Whitehead 1981:93). They have emphasized the shifting nature of this contract as partners negotiate rights and responsibilities over long periods of time under changing material conditions (Berry 1988a:156). They have also pointed out the ways in which room for maneuver is constrained (but not eliminated) by official legal systems that bring the power of the state to bear on the definition of the conjugal contract and on the interpretation of "tradition" (Carney and Watts 1991).

Within Africa, it has been noted that there is much variation in the extent to which the economic terms of the conjugal contract are specified and handled overtly. In some instances the contractual aspects are quite explicit and may include remuneration for goods and services among household members (Whitehead 1985:45). In other cases expectations are left implicit, or clothed in a language of cooperation that mutes or masks the economic aspects of marital transactions. In either case, in order to assess the effective terms of the contract, it is necessary to investigate "resource flows themselves (for example, the mobilisation of women's labor for men's crops) and focus questions on the forms of interdependence and local understandings that enable them" (Leach 1991 a:50).

A key arena of struggle within the conjugal relationship is the meaning and value to be attached to gendered labor. When women press claims for a greater degree of ownership or control, an issue that quickly arises is that of embeddedness: the extent to which women's labor can or should be deemed an extension of their conjugal role, or the action of autonomous individuals. The picture that emerges from the literature is a tapestry of different types of work, on different crops, in different locations, each of which is associated with greater or lesser degrees of individualized control. Berry finds, for example, that when a woman works on her husband's cocoa farm, she is not considered to be creating individualized property. In the case of a dispute, she makes her claim against her husband within the general framework of conjugal rights and responsibilities (1988a:150–151). For other crops, grown in other spaces, her rights are more distinct.

To assess the degree to which women's labor is subsumed in their conjugal role, it is necessary to examine more carefully the characteristics of the divisions of labor and rhythms of work in which women and men are involved. Whitehead (1985:42) makes a distinction between two forms of the division of labor: a sex-sequential labor process requiring inputs from each sex at different times to produce a single product, and a sex-segregated process, in which the members of the one or other sex perform all the operations necessary to produce a given product. Further, she suggests that women's claims on the property that their labor creates may be weaker in a sex-sequential process, since the basis upon which they are to be rewarded for their work is often unclear, allowing their contribution to be submerged in the definition of their conjugal role. As my Southeast Asian examples will show later, the implications of embedding labor in a sequential labor process play out differently where the essential separateness of persons is retained within the conjugal relationship—an observation that offers an opportunity to deepen Whitehead's original insights.

Jane Guyer (1991) takes the analysis of the relationship between women's and men's labor to a further level of specificity. She rejects the static formulae that catalog divisions of labor by task (men clear, women plant), or by crop (women grow groundnuts, men grow cocoa). She proposes instead a methodological focus on changing rhythms and cycles of work, observed over long periods and assessed in relation to the range of factors—from new crops and technologies to state programs—that impinge upon them. For the Beti, she shows that gendered divisions of labor and reward were extrapolated and transferred quite smoothly over a period of a decade or so when one cash crop replaced another. More serious changes, which emerged over a longer time frame, centered on the loss of space for intercrops and the abandonment of the long- and short-term land-use cycles within which women's and men's productive activities previously had been entwined. For women, the loss of these spaces and elements in the production cycle meant the "disappearance of the material embodiment of what were contingent and derivative claims" (1991:269). They responded by intensifying work on their own crops, creating the women's sphere that colonial observers mistook to be a natural (ahistorical) feature of African agrarian systems. As Beti women's and men's agricultural and nonagricultural activities became progressively separated both in space and in their associated rhythms and cycles, men, who had largely lost the capacity to requisition women's labor, made new claims on their income. Struggles within the conjugal relationship then centered on the interpretation of the old cultural formula "he clears, he eats" in the context of new regimes of production.

Guyer's analysis suggests that women's agency and the structural conditions that impinge on it can be discerned in gendered rhythms and cycles of work in Beti fields. There women articulate the nature and limits of their personhood, negotiate the ambiguities of conjugal contracts, assay the material and cultural weight of particular divisions of labor, and attempt a variety of discursive and practical strategies for making and defending claims. The conceptual repertoire developed by Guyer and the other feminist scholars whose work has been discussed in this section offers a definite advance over the dichotomies encountered in some classic formulations. As I will demonstrate, this repertoire has the potential to yield new insights when it travels.

Personhood, property, and conjugal relations in Southeast Asia

In my discussion of Southeast Asia, I draw on only a small selection of the vast ethnographic literature potentially relevant to my themes. One limitation that I impose at the outset of my inquiry is an exclusive focus on the conjugal relationship rather than other relationships (such as sibling, parent, or child) in which the individuation of women and men and their differential capacity to own or control property may also be at issue.[3] My observations focus on the area broadly known as the Malay world; I include Malaysia, Singapore, and most of Indonesia but exclude eastern Indonesia (from the Lesser Sundas) and parts of Sumatra that have a significantly different social structure (Errington 1990:39). I refer both to the lowland and urban areas in which Islam predominates, and to the interior upland populations that have been exposed to Islamic ideas to varying degrees (Tsing 1993). My analysis will indicate that, at least in relation to the topics investigated here, uplands and lowlands share elements of a common cultural repertoire. The case studies, which are drawn from the contrasting contexts of urban Singapore and upland Sulawesi, will later illustrate the ways in which cultural ideas drawn from this regional repertoire are reworked and redeployed under specific material conditions.

Personhood and property

The observation that gender relations in Southeast Asia are relatively egalitarian has become commonplace in the ethnographic literature on the region. Most often the status of women in China and India serves as the point of comparison against which the situation of women in Southeast Asia is judged (Ong 1989).[4] Distinctive cultural features held to account for the strong position of women in the region include the bilateral kinship system, an emphasis on complementarity rather than opposition in gender relations, and hierarchy based primarily on age and rank. Even when men pay bridewealth, residence patterns tend to be matrifocal, and networks among related women often form the basis of neighborhoods. In the economic sphere it is noted that women usually control the household purse and participate in farming, trade, and other professions. They tend to inherit equally with their brothers despite the male bias of Islam and are entitled to a half share of marital property upon divorce.[5]

Despite the advantages they enjoy over other women in different areas and their general (or legal) equality with men in many contexts, Southeast Asian women lag behind men in power and prestige. Feminist research on the region—such as that collected in volumes edited by Atkinson and Errington (1990), Karim (1995), and Ong and Peletz (1995)—has begun to investigate how relations that appear equal or undifferentiated on one level are differentiated in practice. This is a specific area of inquiry that can be enriched by a focus on gendered property relations drawing on the insights of feminist theorizing in Africa.

In Southeast Asia men's advantage does not appear to lie in concepts of personhood as such: Errington writes that "male and female are viewed as basically the same sorts of beings, that is, ones whose souls and functioning are very similar or are parallel" (1990:39). Instead, "differential male-female access to power tends to be located, in local theory, not at the level of the person's gendered characteristics or anatomy analogue, but in practices . . . [W]omen and men are basically the same, but because of the activities women engage in or fail to do, they tend not to become prominent and powerful" (1990:40; Atkinson 1990; Tsing 1990:124).

The research that I cited earlier is not very clear about whether women in the studied African contexts are considered to be intrinsically different from, or inferior to, men as persons. Rather, this research emphasizes, in common with feminist research in Southeast Asia, that men's greater prestige relates to the meanings attached to the particular practices in which they engage. Errington (1990:7) argues that the busyness of Southeast Asian women in economic matters and their concern with money are signs not of power (a Western reading) but of weakness. Real power is derived from the spiritual domain, and there men have the advantage.[6] This is true not only in the Muslim lowlands, but also in the highland interiors: according to Atkinson (1990) Wana women in upland Sulawesi are not forbidden to seek or acquire spiritual power and act as shamans but, because they are tied to an annual cycle of field work, they travel only rarely into the forests and distant places where spiritual power can be found. Similarly, in Kalimantan, Tsing (1984, 1990) observes that Meratus men's tasks in the gender division of labor (land preparation, forest-based hunting and gathering, and travel to trade in distant markets) are conferred special prestige. Moreover, under conditions that do not support institutionalized hierarchy and formal leadership, it is the prestige and experience gained from these activities that permit men to assert themselves in political forums in which both men and women are present and many speak, but only some men are heeded.[7]

The Southeast Asian ethnographic literature is quite explicit on the strength of culturally embedded ideas about the uniqueness of individuals.[8] These ideas derive from, or at least relate to, the Islamic notion of *nasib*, the unique fate that endows each person with special capacities. Nasib defines the individual's personality and preferences, the destined partner in love and marriage *(jodoh)*, the moment and manner of death, and, most relevant to my discussion here, the individual's economic fate, or *rezeki*. A similar concept *(rajaki)* is encountered in the Meratus mountains, an area marginal to (but not unaware of) Islam:

> Rajaki is . . . the "luck" one can strive to increase through knowledge and ritual and one can look for as game in the forest, honey in the trees, a good deal in the market place, or a winning number in the lottery. The demands of following one's rajaki separate individuals each trying to find their own living. At a basic level, the individual is the subject of rajaki; or, one could equally say, the discourse on rajaki creates individual subjects with separate needs and desires . . . [T]he formation and birth of a child highlights the individual nature of rajaki . . . [as] the child in the womb creates its own subjective needs, desires, and life course in an individual confrontation with God.
>
> [Tsing 1984:486]

As Tsing notes, in the Meratus context, parents and other kin have little in the way of accumulated resources to pass on to a child: it is the practice of living and finding one's rajaki that differentiates individuals. An individual's rajaki can be pursued in cooperation with others: with the swidden (hill rice) farming unit (usually formed around a conjugal couple), and in the larger group that seeks collective good fortune through the performance of community rituals. At the same time, "talk of rajaki explains why people, even kin, may go their separate ways" (1984:491).

Southeast Asian concepts of the uniqueness of individuals and the expectation that individuals will express personal desires and seek ways to realize their own fortunes are central to local understandings of personhood in the context of conjugal relations. As my ethnographic analysis will show, a significant dimension of gender inequality can be exposed by tracing the ways in which the spiritual uniqueness of persons plays out differently for women and men in the operational contexts of work and exchange. These are the contexts in which questions of individuation and embeddedness arise in particular and sometimes unexpected forms.

Divisions of labor and the conjugal contract

The enduring separateness of persons is integral to the concept of marriage in Southeast Asia. Banks (1983), writing about rural north Malaysia, describes the conjugal bond as a contract based on the voluntary agreement *(muafakat)* of two adults to live and work together. Although there are certain legal and customary obligations associated with marriage, the distinct interests of the partners imply that a relationship can be built only out of a willingness to cooperate and to be generous. This idea is expressed as *kesayangan*, a spiritual kinship that "links the essences of individuality in persons as whole beings" (Banks 1983:48) and is "present in all close social relationships containing a voluntary moral component" (1983:1 28). Similarly, Jay (1969:66, 124) reports that in rural Java the ideal in marriage, as in all social relationships, is *rukun*, understood as the ongoing process of harmonizing distinct and separate interests.

Founded in a voluntary contract, and based on distinct but complementary interests, the marriage relationship is, at its core, an economic partnership. As Atkinson observes for the mountain Wana, "the conjugal relationship is about work" and "both spouses are expected to be hardworking contributors to their productive unit" (1990:68). The complementary interests of men and women are clearly highlighted in the customary sequential division of labor that prevails among rice farmers both in upland swidden conditions and on the plains.[9] Yet in urban areas such as Singapore (Li 1989); in areas with mixed economies and substantial outmigration, for example Negeri Sembilan (Stivens 1988); or among classes—both landed and landless—who do not labor directly on their own fields (Hart 1991, 1992; Stoler 1977), the exchanges between husband and wife and the nature of their economic partnership take on more complex forms.

In Southeast Asia the terms of the conjugal contract are seldom made explicit. The cultural emphasis is on the ideal of complementarity between marriage partners, and the hope that harmony and kinship sentiment will prevail over narrow self-interest. This is the idea expressed by both women and men when they emphasize that they may work and accumulate property separately from their spouse, but they nevertheless eat and enjoy good fortune together. Subject as they are to the delicacies of conjugal commitment, the terms of exchange that link distinct labor practices and the expectations of husband and wife in relation to the enjoyment of benefits and ownership of assets are embedded in everyday habits and seldom discussed directly. For this reason, I have found it useful to supplement my attempts to understand the cultural framing of conjugal exchanges in "normal" times with an examination of divorce proceedings and property disputes. These occasions offer an additional source of insight into the ways conjugal relations and individuation within marriage can be accounted for in more explicit normative terms.[10]

Throughout the Malay region the customary principle of *harta sepencarian* (literally, property resulting from people working together) or *harta syarikat* (joint or collective property) is recognized as the legal basis for property division upon divorce. This principle has long been interpreted to mean that husband and wife are entitled to equal shares of the product of their joint labors (Maxwell 1884:125). Local understandings and legal precedents have contributed to evolving interpretations of the meaning of this "jointness."

The distinction between sex-sequential and sex-segregated labor processes raised in the Africanist literature offers a way of making sense of the understandings at work in property settlements in Southeast Asia. Where husband and wife work on a single crop, most commonly rice, their partnership in production is recognized in the equal division of the property acquired as a result of their joint efforts. For example, if they produce bountiful harvests and are therefore able to purchase additional land, the land is divided equally upon divorce. But where husband and wife labor separately in a sex-segregated process, the property that each produces is not considered to be jointly owned. In rural north Malaysia, according to Banks, the determination of rights to property acquired during a marriage involves a precise breakdown and calculation of the capital and labor that each partner has contributed to the acquisition of each specific material asset, such as a house or an addition to a house (1976:581). In a legal case in Perak in the 1900s, the court found that the wife had *no* claim at all on her husband's money earnings while she was at home (Ahmad 1978:271). Likewise, in the Gayo highlands of Indonesia the courts have found that a woman has no claim to the commercial tree plantations of her husband unless she has labored *directly* on the crop (Bowen 1988).[11]

The assumption that underlies these property settlements is the enduring separateness of individuals and their labor within marriage. What prevails is, in effect, a labor theory of property according to which individuals have clear entitlement only to property for which they have labored directly.[12] Such a view places an obvious tension at the heart of the conjugal contract, which is built upon the necessity and, indeed, the willingness of individuals to enter into relationships of cooperation and exchange. It suggests that men and women stand to gain or lose differentially from specific forms of the division of labor as these are enacted through varying cycles and rhythms of work. Implicit in daily practices are struggles over the meaning and value of the different labor processes in which women and men engage and the significance of the exchanges between them.[13]

Since there are multiple possibilities present within any cultural repertoire, it is important to explore the specific conditions within which ideas drawn from this repertoire become shaped and structured and to examine the day-to-day processes and negotiations through which cultural production, reproduction, and reinvention are accomplished. The strategies that men and women pursue to preserve and enhance their autonomy within marriage and to secure partnerships that meet their aspirations are explored in the two case studies that follow. In the first case, urban Singapore, I focus on transactions that relate to money earned by men and trace out the implications for gendered asset formation. In the second, rural Sulawesi, I focus on deployments of labor in the direct creation of property in the form of land and trees. In both cases, women must negotiate the meanings of things, practices, and relationships within a political-economic context that, in various and sometimes subtle ways, favors men and renders men's claims more powerful.

Gifts and personhood in a commoditized context: Malay households in Singapore

For the Malay component (about 15%) of Singapore's population,[14] property takes the form of goods and cash acquired primarily through the urban wage economy. For Malay women who work without pay in the home, asset formation depends on the claims they can make on money earned by their husbands. To acquire property and develop personal security and autonomy, they must address the individualizing potential of the wage form and negotiate the meaning of the goods and services exchanged within the domestic unit. Their strategies and practices reveal significant dimensions of the conjugal contract as a contested field.

Living in a world of goods, Singaporeans are acutely aware that every item purchased or service rendered has a market price and is actually or potentially a commodity. Women know the market price of the domestic services they perform without pay in the home, and the opportunity cost of their time in terms of wages foregone. Labor is short and large banners on factory walls close to residential areas remind young women daily of the money they could earn if they walked in through the gate. For older women, domestic services (cleaning houses, canteen cooking, running a food stall) are always saleable.[15]

While men and women are equally aware of the commoditized value of their labor, it is performed in separate spheres and differently rewarded. Men typically receive the reward for their labor as wages, an obvious and direct outcome of their individual effort. Women working in the home can claim their "pay" only indirectly, as a share of their husband's income. In this situation, a potential contradiction exists between women's sense of the value of their labor and the nonrecognition of its commodity aspect when this labor is performed in the home.

In describing exchanges between husband and wife, Singapore Malays give major emphasis to the notion of the gift (Li 1989:6–11, 18–33). This is a usage that might be explained by regarding the household as a residual sphere of natural economy uncontaminated by the market. It could also be understood as a cultural construction that has arisen in the context of, and in response to, the market. As Parry (1986:456) points out, it is market economic relations that engender an emphasis on the pure gift as their conceptual counterpart. More specifically, as Bourdieu observes, the moral burden carried by gifts makes them powerful vehicles for "getting and keeping a lasting hold over someone" (1977:191). It should therefore not surprise us that the notion of the gift becomes especially prominent in a context where the individualizing potential of the urban wage economy makes it possible for people to go their own way. It is through the exchange of gifts, or exchanges phrased as gifts, that individuals seek to build and sustain long-term relationships.[16] Moreover, gifts enable people to establish such connections while *simultaneously* asserting and enhancing their own autonomy and personhood.

Malay women who forego the direct autonomy that stems from earning wages seek to assert their identity as autonomous persons by describing the domestic labor services performed without pay for their families as gifts. They stress the voluntary nature of their commitment to nurture husband and children.[17] Men, too, when they hand over part or all of their wages to their wives, both acknowledge their obligations as husband and father and assert that they are making a gift, since they never ask for any accounting, and wives are free to save or spend any surplus as they please. Ironically but not surprisingly, many women feel rather burdened by these gifts and prefer the autonomy that arises from earning and spending their own cash. Malays often emphasize the unrepayable nature of gifts, capturing this notion in the phrase "debts of kindness go with you to the grave" *(hutang budi dibawa mati)*. Even in conjugal relationships that are built on exchange, it is less burdensome to be the donor than the recipient.

The claim that cash or unpaid labor are gifts depends on an assertion of individualized claims to labor, even within marriage, an idea drawn from the broader Malay cultural repertoire discussed earlier. The customary and legal obligations of men and women to their spouses are viewed as secondary to the voluntary acts of will that initiate and sustain marriage partnerships. Yet in claiming their unpaid labor services as gifts, women in Singapore are on thin ground. Since domestic services performed inside the home do not command a "real" wage, there is a sense in which their commodity value, and therefore also their gift dimension, is denied and devalued. Taking an alternative approach, some women press for a recognition of the direct commodity value of their labor, pointing out that the money men give them is not actually a gift but the pay they have earned for services performed. Yet they voice this claim only in a teasing or joking manner, or when pressed by anger and disappointment. Lacking the proof of worth that comes from actually receiving a wage, they know they cannot go too far in seeking to redefine the meaning and value attached to their labor. More than this, women have an emotional and financial stake in a conjugal relationship built on personal commitment and generosity. Men who have paid their debts might feel free to depart and purchase domestic services elsewhere. To emphasize the gift dimension in the relationship is to invest in the long term.[18]

The sense that they are able to construct enduring conjugal bonds through their voluntary commitments as autonomous adults, and specifically through a process of the exchange of goods and services as gifts, is a source of pride and moral worth for Singapore Malays. When gifts are used to build and maintain social relationships, the calculations embedded in them are left implicit and sometimes denied. The underlying calculations made by each partner become more explicit, however, when divorce is imminent. At that point, the couple has no more interest in framing their relationship in terms that build and sustain social ties. It is in the context of divorce, too, that the notional equivalence of women's unpaid domestic services and men's wages exchanged within the conjugal contract breaks apart. When the idiom of gifts loses its force and attraction, the vulnerability of women's position is fully exposed.

I noted earlier that not only the initial capital but also the labor power and labor product of husband and wife remain their separate property during the marriage. It is the product of their joint labor (most obviously through a sex-sequential labor process) that is divided between them upon divorce. In Singapore, the work of the homemaker wife and the wage-earning husband are conducted in separate spheres, take different forms (waged, unwaged), and carry different meanings (real work, not work). Upon divorce, it transpires that a wife's contribution of domestic services is not translatable into a claim on the property her husband has accumulated through his labor. On this point, the rulings of the Muslim family court are consistent with the views expressed by Malay men and women about their respective entitlements.

In ten cases of property settlement observed in Singapore in 1982, the wives did not make any claims on the goods the men had bought with their earnings. As one woman stated, "it is his money, it is his body that works, they are his things not mine" (Li 1989:33–40). Women's sense of the devalued status of their labor contribution was expressed especially with regard to housing bought out of the husband's wage. Wives commented that they were mere lodgers on their husband's property. They felt they had no right to the house as the husband had paid for it, while they simply stayed home and did not work. Government housing rules actually decree that the marital home is the joint property of husband and wife, regardless of who earned the money, and that each is entitled to a half share. When this information was conveyed to divorcing men by court officials, some were dismayed, asking how a wife could possibly have a share if she had not paid one cent toward the purchase. Women claimed only the goods they had purchased out of their own earnings or savings from market money, or those they had received directly as gifts.[19]

Besides paying for their house or flat, wage workers in Singapore must contribute a sizeable portion of their income (up to 25%) to the Central Provident Fund. They can access this fund for housing and medical expenses, but most of it is released in a lump sum at the official retirement age of 55. Non-wage-earning wives have no fund of their own; many Malay husbands who receive this money give their wives a share, phrased as a gift. But a woman who is divorced before her husband has access to the fund stands to lose out entirely on the long-term accumulated product of her husband's labor. In private interviews Malay women often mentioned this inequality, sometimes with bitterness, but it was not raised in any of the divorce proceedings I witnessed. The official nature of the fund, its many rules and restrictions, the inaccessibility of the cash before the appointed time, and its direct association with the labor of the individual whose wages had been channelled there combined to separate the fund, both in time and in space, from women's everyday labor in the home. Women could grumble, but they did not make claims.

The power and the property claims associated with direct access to wages derive from a reworking of the labor theory of property in a context where the rhythms and cycles of men's and women's work have become quite distinct. Women's everyday assertions of autonomy and personhood built from their giftlike provision of domestic services indicate a domain of contested meanings. Except where divorce is imminent, the contestation is usually muted and indirect.

For many women, it makes more sense to emphasize the "eating together" component of the conjugal relationship, subsuming and downplaying the issues that arise from the fact that husband and wife actually "work separately" and are differentially rewarded for their efforts.

Reworking tradition, staking claims: labor and property in upland Sulawesi

In this section I discuss the processes through which men and women create and lay claim to property in the form of land and trees. My discussion concerns the Lauje people, indigenous shifting cultivators numbering about 30,000 who inhabit the narrow coastal plain and hilly interior of the Bay of Tomini, on the northern peninsula of Sulawesi, Indonesia.[20] Lauje hill farmers grow food crops (maize, tubers, and hill rice) on swidden fields as well as cash crops for regional and international markets. Most of those on the coast and in the first range of hills are Muslim, while Christianity is slowly spreading among the Lauje living closer to the center of the peninsula.

The Lauje, like the Malays described earlier, espouse a "labor theory of property." The idea that individuals are the owners of their labor power is quite highly developed and manifest in various ways.[21] At its simplest, a man walking though the forest who spots a valuable resin or sago tree only has to work for a few minutes clearing the brush around it to claim it as his own. He can bring a legal case against anyone who disturbs his tree. He can also give or sell the tree to another person if he wishes or exchange it for a tree in a more convenient location. Any of these transactions would put an end to his claim: the property is fully his and also fully separable from him, as alienable as any commodity. If he dies and his tree is claimed by his descendants, however, their rights are much less complete: not only are there likely to be many of them, each claiming a share, but, since none of them invested labor in creating the property, they are all, in a sense, borrowing it from him. His labor invested in the product marks it as his, even after his death.[22]

Land, the key to the livelihood of swidden farmers and, in many cultural contexts, an item in which a collectivity such as a kin group has definite rights, can also be alienated by the individual who first clears a plot of primary forest. Stories from the 1920s tell of men who had cleared land and exchanged it for two axes or a new pair of pants, purchased on the coast. Some men who liked to live on the forest frontier routinely cleared land and traded it to others who moved in behind them, forming more densely settled hamlets and farming the secondary forest. Even in the absence of an "open" market in land, the huge effort required to clear a plot, combined with the possibility of trade and exchange, always marked it as an alienable commodity. Nowadays, a man can sell land he has cleared for cash or gamble it away, but the cultural basis for individualized ownership has not changed.

The strongest claims to property can be made when a person has labored alone. Most labor processes, however, involve more than one person and claims over the product are more complex. They depend on the meanings attributed to different labor practices and the negotiated terms upon which these practices are combined. As in the Singapore example, a woman's work in one sphere does not provide a claim on a man's labor product when he is engaged in a separate, apparently autonomous sphere. A man who invests a season's labor in clearing primary forest depends for his subsistence during that period on the food that his wife produces on the second-year swiddens. Yet women do not claim to own a share of the new land, which is associated only with men's direct labor. Similarly, if men go off to earn wages extracting ebony or rattan in the headwaters, women do not claim a share of their income. At most, they grumble when men bring nothing home, not even a new shirt for one of the children.

The Lauje recognize and comment on the need, desire, and, indeed, the right of men, women, and children to engage in their own productive ventures and to have control over the product of their labor. On this basis, they are able to experience the pride that comes from the ability to make gifts and thereby to enter as autonomous persons into relationships with others. In the past, a man would divide up a plot of land he had cleared and give a section to his wife and each child of working age to farm.[23] On this land, each would grow the normal food crops, hill rice and maize, and each would store the harvest separately, experiencing directly the relationship between labor and reward. Some of the food would be brought out by each person to add to the family cooking pot, "eating together" being central to the definition of the household unit. But the owner of the crop, the one who had undertaken the tedious work of weeding and tending it, was entitled to alienate at least some of the product through gifts or trade. Food has always been in short supply in the dry foothills and on the coast, and some hill farmers continue to barter their produce for salt fish, salt, clothing, knife blades, and other goods with Lauje who walk up into the hills for this purpose.

Few people today bother to divide up the food gardens, but ownership of cash crops such as onions and garlic lies quite clearly with the individual women, men, and children who plant them. Labor arrangements include share agreements, where husband and wife (or other kin or neighbors) work together on a crop and share the proceeds equally between them. Commonly, husband and wife maintain separate fields but routinely exchange labor: "Today we will plant my field, tomorrow yours." Where one party invests significant amounts of labor in the field of another without reciprocation in like form, typically, when a woman weeds and tends her husband's shallot crop during his absence, she expects her labor to be recognized with a share of the proceeds.

Husband and wife are deemed responsible for household provisioning, each drawing on separate (often secret) stores of cash when food needs to be purchased. Many women are fairly successful in retaining control over their personal income, at least in the short term. Over the longer term, however, the precarious nature of hillside farming under degraded swidden conditions, the uncertainties caused by weather, pests, and market prices, and the ever-present threat of illness and premature death have made the issue of accumulation somewhat irrelevant. Ideally, it is generosity and noncalculation which characterize the conjugal bond. Although husband and wife may work separately, they eat together, both literally and in the broader sense of being generous when their fortune is good.

Over the past few years, the introduction into the Lauje hills of tree crops such as cocoa and cloves has given the issue of property ownership an entirely new significance. Ancestral land is becoming privatized, and tree groves are squeezing out former land uses. The question of how ownership in trees and land will be established, and by whom, has become a crucial issue for negotiation within the conjugal relationship as well as among kin, neighbors, and others interested in gaining a stake in the economic potential of the Lauje hills.[24] At the center of struggles over ownership is the definition of the value and meaning of specific types of labor investment.

An individual who wishes to plant trees must first establish exclusive rights over land. Most of the pioneers who cleared the land in the Lauje hills are long gone, and their numerous descendants are engaged in a competitive process of staking individual claims to the pool of land they previously "borrowed" from their ancestors for use as swiddens. Consistent with the labor theory of property explored earlier, cousins, siblings, and other coinheritors recognize that a man who transforms ancestral land by investing a significant amount of labor in it (clearing the plot and establishing trees) has, in effect, created a new piece of property. His labor investment renders moot claims based solely on ancestral clearing generations ago. He can only be dislodged by someone willing to compensate him for that labor investment.

While everyone agrees that all descendants have equal rights to inherited land regardless of sex, men and women differ significantly in their capacity to operationalize their claims. First, men are more prominent in the public sphere of interhousehold relations, especially where it involves land disputes, and claim to have more knowledge about the boundaries of the plots cleared by men of previous generations. Second, men who have identified a patch of ancestral land can consolidate their claims by investing their labor immediately and directly. A woman must depend on her husband to do the crucial "initiating" and property-establishing work of clearing and tree planting for her. A woman whose husband is absent, incapacitated, or simply lazy cannot lay effective claim to her own inheritance and loses out to her brothers and cousins, each scrambling for a larger share. Finally, since the husband's labor is needed to clear the plot, his labor entitles him to a share in the trees. Unless they have the cash to pay someone to clear land for them, women generally end up sharing their inherited land and its newly planted trees with their husbands.

Women have engaged in two distinct strategies in their attempt to secure a stake in the land and trees on which their future and fortune depend. Both strategies involve the labor theory of property and the redeployment of familiar ideas and arguments to make new claims. One strategy involves women investing labor in a sequential process. They make themselves busy weeding and tending the young cocoa seedlings their husbands have planted, thereby asserting that the trees are jointly owned. They point to the example of neighbors or outsiders who have cleared and planted large areas, sometimes with the purpose of monopolizing land, only to end up with nothing because the overgrown seedlings died. Men's work may establish tree gardens, but women's consistent care is needed for the trees to survive and thrive and the claim over the land to be consolidated.

Although there is general agreement that successful tree gardens are established through the joint labor of husband and wife, women feel uncertain about their effective power to claim a half share in the case of divorce.[25] They have very little experience with the division of permanent, immovable property. Previously, women and men could separate and take with them their tools, personal clothing, cash savings (usually secret), and perhaps a share of the rice harvest, but there was no long-term accumulation that could be regarded as their joint product. Women also express a fear that their husbands could sell the trees to meet gambling debts, leaving their rights out of account. While men and women agree that they should consult before selling joint property, practice is another matter. Once the trees are sold, women have no means of redress.

To address their vulnerability in relation to jointly owned property, some women have adopted the alternative strategy of planting trees independently. They avoid seeking any assistance from their husbands which could jeopardize their individual ownership rights. This strategy can be used only if women have cash to pay other men to work for them. To soften their stance, women insist that no break in conjugal harmony is intended: "We work separately, but if anything comes of our efforts, we eat together."[26] By using the commoditized labor of other men, they avoid long-term obligations; debts of kindness, if help is offered as a gift; or entailments on the property itself, if the person who assists later claims a share in return for his labor. The paid worker has worked but has not invested personhood nor acquired any rights beyond the right to the agreed wage at prevailing local rates.[27] This mechanism to secure individualized property rights is double-edged. Men can also pay workers to establish and maintain cocoa gardens, never calling on their wives' labor, and therefore making it difficult for women to assert claims to these trees as joint conjugal property. Both men and women, if they have capital, can buy tree gardens already formed. In this case, their partner has no claim.

The mechanisms through which women and men make their claims to individual property are identical, but their capacity to acquire such property is not the same. Women are disadvantaged in their access to sources of capital. They are excluded from the wage opportunities in the wider regional economy to which men turn when in need of cash. Women may perform wage labor in the neighborhood but the pay is low and sometimes received in-kind: women carry home a bundle of corn to feed the family for a few days or a new sarong needed immediately as a blanket for a child. The rhythms of this type of wage work do not permit accumulation and investment. It is shallot production that provides women with lump sums of cash, but crop disease and the space and attention given over to tree crops have greatly reduced shallot incomes in the past five years.

Women's capacity to own trees independently is also hindered by the male bias of government interventions in the hills. On the primary data collection forms for the 1990 national census, all Lauje women were classified as housewives, despite the fact that the census takers (all male) were themselves Lauje and that women can be seen at work in their gardens everywhere in the hills. Now that land is being registered and taxed for the first time, women's independent claims to land and trees and their claims to own property jointly with their husbands have been entirely ignored. All the land and trees have been registered in the name of the husband as "head of household," informalizing women's claims and rendering their labor contributions invisible to the official sphere. State readings of women's roles have led to their exclusion from the farmers' groups through which free inputs (seedlings, tools, fertilizers) are distributed, thereby disadvantaging women in their struggle to hold onto resources and establish new assets during this period of agrarian transition.[28]

State laws, programs, and ideologies have provided men with a new set of idioms for asserting and legitimating claims to contested resources. So far, however, the potential damage to women's interests caused by state interventions has been limited by three factors. First, there is the practice of settling property disputes and divorce cases "on the spot," under the guidance of hamlet leaders. The Lauje hills are sufficiently remote that, at least for the moment, local understandings of the property rights deriving from labor investment prevail and the dictates of state ideology or Islamic doctrines are seldom invoked. Second, in response to the "gift" of seedlings from the state, some men have chosen to share their good fortune with their wives and other kin, thus mitigating the impact of state biases. Finally, with or without official recognition or access to state largesse, women are continuing to plant trees alone or together with their husbands and to assert property claims. While there is little public discussion of the meaning and value of the new labor processes (discussion of the kind that occurs in the context of divorce settlements), the meanings of things and relationships are being negotiated through the apparently mundane and unmarked sphere of everyday labor: being present in the field and working can be, in itself, a political act.

Conclusion

Building on the insights of feminist scholars working in Africa, this article has investigated the relationship between personhood, property, and the conjugal contract in Southeast Asia. It has attempted to expose some patterns evident in the region as a whole, as well as the unique practices and understandings which emerge when conjugal relations are negotiated through everyday work and exchange under specific material conditions. Here I draw conclusions focused around the issue of power. I also tease out some of the ways in which travel to Southeast Asia has enriched the body of theory with which

I began, pointing to the potential of comparative ethnography to do theoretical work.

In much feminist research, especially of the "women and development" genre, the expansion of market relations and the intervention of the state in local affairs are seen to occasion a one-way demise of women's economic position and a reduction in their autonomy.[29] Running counter to the image of women as victims, a growing number of empirical studies document the regional and class differences between women which lead some to gain from their engagements with market and state at the same time as others lose out.[30] To move beyond generalities and investigate the issue of women's position more closely, the Africanist literature to which I turned suggests the need to take into account a wide set of relationships and to pay attention to changes emerging over different time periods. It also argues for a more nuanced reading of power and closer attention to the range of strategies and practices through which agency is exercised.

My Southeast Asian examples illustrate the limits of state power and the contradictory effects of state interventions. State legal systems that acknowledge women's right to retain personal property and support their claims to a share in conjugal assets strengthen women's position even as other state programs and assumptions undermine it. In Singapore, the state's ruling that the marital home is joint property regardless of individual cash contributions has protected women's interests where the Malay labor theory of property left them vulnerable, but women have lost out on long-term accumulation through the official savings scheme. In the remoteness of the Lauje hills, state presence of any kind is relatively weak. Women have therefore not been protected by official legal recognition of women's property rights. Neither, however, have they been seriously undermined by the male bias of the few programs that do reach the hills. In both Singapore and Sulawesi official rules about property tend to confirm what women already know: that their strongest claims to property derive from engaging directly in production. Only labor rewarded by wages, or labor that directly produces property, permits women to negotiate conjugal exchanges from a position of strength.

My examples also demonstrate that women have much to gain, as well as the potential to lose, from an increase in market relations. Marx anticipated a unilinear trend, arguing that more commoditization would lead to the detriment of particular classes. Had he considered questions of gender, he might have gone on to investigate how, commoditization notwithstanding, individuals do in fact form enduring bonds. This question arises with particular clarity in Southeast Asia, suggesting the need to press beyond the formulations of Marx and Whitehead and develop this body of theory in a new direction. My case studies reveal some rather complex ways in which individuals use commoditized forms of exchange to link and distance themselves. In Singapore, Malay women highlight the commoditized value of their labor in order to emphasize the significance of their gift and to tie wage-earning men and children into a sphere of moral obligation. They seek not to detach themselves from conjugal exchanges but to embed themselves and their labor more deeply. But their claims are undermined by the location of their domestic labor in a sphere that is unrecognized and unrewarded by the market. In Indonesia, women make use of commoditized labor to secure their independent property claims but see their own claims against their husbands made vulnerable by the same argument.

Both men and women, anxious to secure their individual rights over property, have proceeded at times by seeking to embed their labor in a joint labor process and, at other times, by working separately. Where Whitehead found that women's property claims were strongest in a sex-segmented labor process, my studies show that both sex-sequential and sex-segregated labor processes can be vehicles for the establishment of women's property rights. In the two cases I explore, the key issue is not the form of the division of labor itself, but the extent to which the labor investment is clearly and directly connected to the creation of the property in question. Women (and men) become especially vulnerable when the connection between their work and that of their spouse is obscured. They must then make use of various practical and discursive strategies to render the connections more obvious.

My examples also point to the diverse modes and channels through which power is exercised. The power of the bureaucracy to define rights and rewards is particularly significant in urban Singapore, although less so in upland Indonesia. Malay women's attempts to advance their claims in Singapore take place largely in the arena of everyday discourse: talking about labor and relationships in ways that emphasize their gift dimension. But, confronted with market definitions of value and forced to lay out the basis of conjugal exchanges in the official setting of the family court, the language with which they frame their everyday claims is abandoned. Out of context, it has no power.

In upland Indonesia, practical strategies seem to be at least as significant as discursive ones. Men and women assert the power of their own labor by quietly deploying it in the situations where it counts—in the fields where the grounds for current and future claims to productive resources are being laid. They comment only that, while working (and owning) separately, they eat together, thereby leaving unspoken the ongoing adjustments and negotiations that arise as they reposition themselves and reconfigure their relationships with each other.

In many contexts, including those discussed here, there is no legitimate field of discourse or "counterpoint on gender" (Risseeuw 1988) in which the changing nature of rights and privileges is publicly discussed. Therefore, major changes in gender relations may occur piecemeal, without provoking crises or laying bare the workings and distribution of power. Resistance is, in such situations, too strong a word. There is, however, plenty of evidence that both women and men are active in shaping the patterns of social and economic life which emerge from their quiet, quotidian engagements.

Notes

Acknowledgments. I would like to thank the editor of *American Ethnologist* and several external reviewers for the constructive criticism that contributed significantly to the final form of this article.

1 Prominent contributions to this genre of work on Africa include Berry 1988a, 1988b; Carney and Watts 1991; Geisler 1993; Guyer 1986, 1991; Leach 1991a, 1991b; Stamp 1989; Watts 1989; and Whitehead 1981, 1984, 1985, 1990.

2 Note also that, despite decades of ethnographic research on concepts of the person, scholars attempting to characterize modernity continue to rely on a contrast between the individualistic, detached modern person and a rather unspecified, group-embedded other. See, for example, Friedmann 1994: 212–227. Carrier's critique (1992) is relevant here.

3 Djamour (1959), Swift (1963), and Rudie (1995) examine the Malay conjugal relationship and some of its economic dimensions. Peletz (1988) and McKinley (1975) pay more attention to siblingship as an axis of gendered property relations. Personhood and property in the conjugal context have been intensively scrutinized in Melanesia (e.g., Strathern 1988). Space prevents me from engaging with this literature here. The reasons for the difference in emphasis in gender studies on the various continents could, perhaps, be the subject of another paper (cf. Appadurai 1986b).

4 For a rich discussion of the problems women face in exercising control over property in south Asia, see Agarwal 1994.

5 For generalizations along these lines and specific citations, see Errington 1990; Karim 1995; Ong 1989; and Winzeler 1974.

6 Arguing for a more critical stance toward a gendered prestige system and men's much-stated spiritual potency, Brenner (1995) notes that Javanese men are widely (but informally) acknowledged to be incapable of controlling their selfish desires; women must therefore take over as guardians and promoters of family spiritual welfare, prestige, and financial stability. Peletz (1995) notes similar views of men's disabilities in Negeri Sembilan, Malaysia, but is concerned to locate the reasons for (some) men's failings as husbands in the history of the regional economy and in terms of class position.

7 Karim argues that Western feminists are still overly focused on women's exclusion from the public domain, and ignore or trivialize the invisible, informal, nonbureaucratized, and indirect forms through which women's power is exercised. These include silence, withdrawal, sorcery, avoidance of open confrontation, and "hostile harmony" (1995:12, 20). Karim argues that these forms of power are as effective as the power exercised by men in public. For an example of the indirect strategies used by women to exert control over property in Minangkabau, see Krier 1995; note, however, that in the case Krier describes, women were forced through weakness to talk too much and thus appear "out of control," while men, their power established, were silent.

8 See, for example, Banks 1983:68–71, Djamour 1959, Jay 1969, Li 1989, and Peletz 1988.

9 For a swidden example, see Tsing 1984, 1990; for the lowlands, see Jay 1969, Peletz 1988, Swift 1963, and Wilder 1982. Note also the changes that have occurred in gendered divisions of labor and reward in the context of the Green Revolution (Hart 1991, 1992; Stoler 1977; Wong 1987).

10 Divorce rates have characteristically been high in much of Southeast Asia: approximately 50 percent at least up to the 1960s. For more statistics and a discussion of the problems of measuring divorce and its impact, see Li 1989:34.

11 Bowen (1988) points out that Indonesia's national legal system assumes a gender-neutral bilateralism to be the prevailing national custom; as a result, courts may allocate women a share of inherited and marital property even in areas such as the Gayo highlands, where property was controlled by patrilineal clans. Interpretations of "tradition" in Gayo have had to work within and around national agendas.

12 The notion is not unique to Southeast Asia. For African examples, see Berry 1988a:145 and Caplan 1984. Caplan describes concepts of personhood, gendered divisions of labor, and systems of kinship and property in Muslim East Africa remarkably similar to those

encountered in Southeast Asia, confirming the need for a much more broadly based comparative study than that attempted here.

13 Geertz (1961:49–50, 125), discussing market traders in Java, suggests that all of the partners' labor and property is joint, but does not explain how, in practical terms, men and women identify and lay claim to each other's earnings, especially where their activities take place in the urban informal sector. See my discussion in Li 1989:31–33. Wazir Karim (1995) notes that, for Malay women, separate and secret savings are a normal strategy for maintaining autonomy within marriage. On the significance of autonomy for Malay women and the strategies that differently positioned women pursue, see also Rudie 1995.

14 Malays live side by side with the majority Chinese population (77%), mostly in high-rise flats, and are fully integrated into the urban economy (Li 1989:93–98). For more information on Singapore, the methodology and results of my field study, and a full discussion of the issues raised briefly in this section, see Li 1989. See also Li 1997b.

15 If they work outside the home, women are expected to meet the costs of their domestic-labor substitutes (child care, cooked food) out of their own incomes. Women therefore make sensitive calculations about the point at which it is worth their while to engage in paid work. As opportunities and pay have improved, so has their rate of labor-force participation. In the age group 25–29 (those with maximum reproductive responsibilities), 7 percent were in the labor force in 1957, and 55 percent in 1980 (Li 1989:104).

16 There are cultural precedents for the emphasis on gifts in the rural Malay world. In Java, Jay observed that substance is "the coin of social relations" and noted the significance of "making a gratuitous gift within the framework of kin expectations" (1969:44). In Malaysia, Banks (1983:138, 157) observed that inheritance is viewed as a gift, and Peletz (1988) comments on the ambivalence in relations between kin, especially siblings, where obligations created through gifts and other forms of assistance become onerous burdens.

17 Similarly, young adults phrase the portion of their wages that they give to their mother as gifts from the heart; they thereby deny that their parents have claims over their labor and emphasize their choice, as autonomous young adults, to build and maintain a relationship with their parents based on concern and generosity. See Li 1989:41–73.

18 Some women are more sanguine in demanding direct compensation from working children: "I wash your clothes and cook your food, now I want my pay!" In claiming their pay, however, mothers potentially undermine the gift dimension of these transactions and the pride that comes from having children who are loving and generous. In contrast to McKinley (1975) and Carsten (1989), I argue that, far from being a pure realm unsullied by market considerations, the household *is* a sphere of economic transactions, although the calculations underlying these transactions are seldom made explicit. Economic transactions are not the antithesis of emotional bonds, but part of the process through which such bonds are created and sustained.

19 Classic studies of Malay marriage and divorce, such as Djamour 1959 and Swift 1963, emphasize the significance of jewelry as a form of long-term saving for women. They also note that husbands make gifts of jewelry with the dual purpose of building the relationship and endowing their wives with assets that will enable them to withstand contingencies such as widowhood and divorce.

20 For general descriptions of the area, see Nourse 1989 and Li 1991.

21 Although Nourse's thesis (1989) is focused mainly on Lauje beliefs and practices surrounding birth, curing, and communication with the spirit world and does not directly address the relationship between personhood and property, some of her observations indicate a sense of the spiritual uniqueness of individuals akin to that encountered in other literature on the Malay world explored earlier (see especially Nourse 1989: 314–320).

22 Similarly, Singapore Malays who inherit property feel an ongoing burden of debt toward the person whose actions first created it (see Li 1989:70–71).

23 See Li 1996b for a discussion of Lauje children's rights to the product of their labor and the absence of "the household" as a unit of ownership or production.

24 For a discussion of the class dimensions of this transition and the role of the state, see Li 1996a and 1997a.

25 There is no documentation on either marriage or divorce in the Lauje hills, and most of these transactions are never formally registered. My impression is that divorce in the early years of marriage is quite common, but that well-established relationships rarely falter. Regardless of the rate, women frequently mentioned the topic to me. From this I conclude that the possibility or threat of divorce is significant to them, and they shape their strategies accordingly. Divorce cases are settled by a group of informally acknowledged and officially designated hamlet leaders, all male. Islamic ideas may weigh into these proceedings, but do not necessarily dominate them. It is rare for an Islamic official to be present. Some of the observations made by Tsing (1990) on women's disadvantages in informal dispute settlement processes may be relevant here.

26 Young unmarried women and men also use the strategy of working alone: a father described watching his teenage daughter struggle to transport tree seedlings on her own. He observed that she did not want to ask for his help, in case he later claimed a return on his labor.

27 Note that, rather than opposing commodities and commoditized relationships as some of the eco-feminist literature would imply (e.g., Mies and Shiva 1993), Lauje women prefer to make use of the wage form since it permits an effective separation of property from labor.

28 Benda-Beckmann (1988) observes that development programs and projects often have quasi-legal implications in the way they sequester and redistribute resources according to their own sets of rules. In one Indonesian case, she found the results to be markedly "gender-skewed" to the detriment of women. For a parallel argument in relation to Africa, see Whitehead 1990:62.

29 See, for example, Stamp 1989 for a review of evidence that involvement in commodity production has impoverished African women in relation to men. See Mohanty 1991 and Moore 1988 for critiques of the portrayal of women in the developing world as perennial victims.

30 See, for example, Agarwal 1991; Berry 1988a, 1988b; Hart 1991, 1992; Leach 1991a, 1991b; Moore 1988; Risseeuw 1988; Stivens 1988; Stoler 1977; Whitehead 1990; and Wong 1987.

References

Agarwal, Bina. 1991 Engendering the Environment Debate: Lessons from the Indian Subcontinent. CASID Distinguished Speaker Series, 8. East Lansing, MI: Center for Advanced Study of International Development.

——. 1994 Gender and Command over Property: A Critical Gap in Economic Analysis and Policy in South Asia. World Development 22(10):1455–1478.

Ahmad, Ibrahim. 1978 Family Law in Malaysia and Singapore. Singapore: Malayan Law Journal.

Appadurai, Arjun. 1986a The Social Life of Things. Cambridge: Cambridge University Press.

——. 1986b Theory in Anthropology: Center and Periphery. Comparative Studies in Society and History 28(2):356–361.

Atkinson, Jane Monnig. 1990 How Gender Makes a Difference in Wana Society. *In* Power and Difference: Gender in Island Southeast Asia. Jane Monnig Atkinson and Shelly Errington, eds. Pp. 59–94. Stanford, CA: Stanford University Press.

Atkinson, Jane Monnig, and Shelly Errington, eds. 1990 Power and Difference: Gender in Island Southeast Asia. Stanford, CA: Stanford University Press.

Banks, David J. 1976 Islam and Inheritance in Malaya. American Ethnologist 3:573–586.

——. 1983 Malay Kinship. Philadelphia: Institute for the Study of Human Resources.

Benda-Beckmann, Keebet von. 1988 Development, Law, and Gender-Skewing: An Examination of the Impact of Development on the Socio-Legal Position of Indonesian Women, with Special Reference to Minangkabau. Paper prepared for the seminar "The Social-Legal Position of Women in Changing Society, IUAES Commission on Folk Law, and Legal Pluralism," Zagreb, June.

Berry, Sara. 1988a Property Rights and Rural Resource Management: The Case of Tree Crops in West Africa. *In* Production and Autonomy. John W. Bennet and John R. Bowen, eds. Pp. 143–161. Boston: Society for Economic Anthropology and University Press of America.

——. 1988b Concentration without Privatization? Some Consequences of Changing Patterns of Rural Land Control in Africa. *In* Land and Society in Contemporary Africa. R. E. Downs and S. P. Reyna, eds. Pp. 53–75. Hanover, NH: University Press of New England.

Bourdieu, Pierre. 1977 Outline of a Theory of Practice. Cambridge: Cambridge University Press.

Bowen, John R. 1988 The Transformation of an Indonesian Property System: *Adat,* Islam, and Social Change in the Gayo Highlands. American Ethnologist 15:274–292.

Brenner, Suzanne. 1995 Why Women Rule the Roost: Rethinking Javanese Ideologies of Gender and Self-Control. *In* Bewitching Women, Pious Men: Gender and Body Politics in Southeast Asia. Aihwa Ong and Michael Peletz, eds. Pp. 19–50. Berkeley: University of California Press.

Caplan, Patricia. 1984 Cognatic Descent, Islamic Law and Women's Property on the East African Coast. *In* Women and Property—Woman as Property. Renee Hirschon, ed. Pp. 23–43. London: Croom Helm.

Carney, Judith, and Michael Watts. 1991 Disciplining Women? Rice, Mechanization, and the Evolution of Mandinka Gender Relations in Senegambia. Signs 16(4): 651–681.

Carrier, James. 1992 Occidentalism: The World Turned Upside Down. American Ethnologist 19:195–212.

Carsten, Janet. 1989 Cooking Money: Gender and the Symbolic Transformation of Means of Exchange in a Malay Fishing Community. *In* Money and the Morality of Exchange. Jonathan Parry and Maurice Bloch, eds. Pp. 117–141. Cambridge: Cambridge University Press.

Djamour, Judith. 1959 Malay Kinship and Marriage in Singapore. London: Athlone Press.

Errington, Shelly. 1990 Recasting Sex, Gender and Power: A Theoretical and Regional Overview. *In* Power and Difference: Gender in Island Southeast Asia. Jane Monnig Atkinson and Shelly Errington, eds. Pp. 1–58. Stanford, CA: Stanford University Press.

Friedman, Jonathan. 1994 Cultural Identity and Global Process. London: Sage Publications.

Geertz, Hildred. 1961 The Javanese Family. New York: Free Press of Glencoe.

Geisler, Gisela. 1993 Silences Speak Louder than Claims: Gender, Household, and Agricultural Development in Southern Africa. World Development 21(12): 1965–1980.

Gregory, C. A. 1982 Gifts and Commodities. London: Academic Press.

Guyer, Jane I. 1986 Intra-household Processes and Farming Systems Research: Perspectives from Anthropology. *In* Understanding Africa's Rural Households and Farming Systems. Joyce Lewinger Moock, ed. Pp. 92–104. Boulder, CO: Westview Press.

——. 1991 Female Farming in Anthropology and African History. *In* Gender at the Crossroads of Knowledge. Micaela di Leonardo, ed. Pp. 257–277. Berkeley: University of California Press.

Hart, Gillian. 1991 Engendering Everyday Resistance: Gender, Patronage and Production Politics in Rural Malaysia. Journal of Peasant Studies 19(1):93–121.

——. 1992 Household Production Reconsidered: Gender, Labour Conflict, and Technological Change in Malaysia's Muda Region. World Development 20(6): 809–823.

Jay, Robert. 1969 Javanese Villagers: Social Life in Rural Modjokuto. Cambridge: MIT Press.

Karim, Wazir Jahan, ed. 1995 "Male" and "Female" in Developing Southeast Asia. Oxford: Berg.

Krier, Jennifer. 1995 Narrating Herself: Power and Gender in a Minangkabau Woman's Tale of Conflict. *In* Bewitching Women, Pious Men: Gender and Body Politics in Southeast Asia. Aihwa Ong and Michael Peletz, eds. Pp. 51–75. Berkeley: University of California Press.

Leach, Melissa. 1991a Locating Gendered Experience: An Anthropologist's View from a Sierra-Leonean Village. IDS Bulletin 22(1):44–50.

——. 1991b Engendered Environments: Understanding Natural Resource Management in the West African Forest Zone. IDS Bulletin 22(4):17–24.

Li, Tania Murray. 1989 Malays in Singapore: Culture, Economy, and Ideology. New York and Singapore: Oxford University Press.

——. 1991 Culture, Ecology, and Livelihood in the Tinombo Region of Central Sulawesi. Halifax, NS: School for Resource and Environmental Studies.

——. 1996a Images of Community: Discourse and Strategy in Property Relations. Development and Change 27(3):501–527.

——. 1996b Household Formation, Private Property and the State. Sojourn 11 (2):259–288.

——. 1997a Producing Agrarian Transformation at the Indonesian Periphery. *In* Economic Analysis beyond the Local System. Richard E. Blanton, Peter N. Peregrine, Deborah Winslow, and Thomas D. Hall, eds. Pp. 125–146. Lanham, MD: University Press of America.

——. 1997b Constituting Capitalist Culture: The Singapore Malay Problem and Entrepreneurship Reconsidered. *In* Market Cultures: Society and Morality in the New Asian Capitalisms. Robert Hefner, ed. Pp. 147–172. Boulder, CO: Westview Press.

Maxwell, Hon., W. E. 1884 The Law and Custom of the Malays with Reference to the Tenure of Land. Journal of the Straits Branch of the Royal Asiatic Society 13:75–220.

McKinley, Robert. 1975 Knife Cutting Water: Child Transfers and Siblingship among Urban Malays. Ph.D. dissertation, University of Michigan.

Mies, Maria, and Vandana Shiva. 1993 Ecofeminism. London: Zed Books.

Mohanty, Chandra T. 1991 Introduction: Cartographies of Struggle. *In* Third World Women and the Politics of Feminism. Chandra T. Mohanty, Ann Russo, and Lourdes Torres, eds. Pp. 1–50. Bloomington: Indiana University Press.

Moore, Henrietta L. 1988 Feminism and Anthropology. Minneapolis: University of Minnesota Press.

Nourse, Jennifer. 1989 We Are the Womb of the World: Birth Spirits and the Lauje of Central Sulawesi. Ph.D dissertation, University of Virginia.

Ong, Aihwa. 1989 Center, Periphery, and Hierarchy: Gender in Southeast Asia. *In* Gender and Anthropology: Critical Reviews for Research and Teaching. Sandra Morgen, ed. Pp. 294–312. Washington, DC: American Anthropological Association.

Ong, Aihwa, and Michael Peletz, eds. 1995 Bewitching Women, Pious Men: Gender and Body Politics in Southeast Asia. Berkeley: University of California Press.

Parry, Jonathan. 1986 The Gift, the Indian Gift and the "Indian Gift." Man (n.s.) 21:453–473.

Parry, Jonathan, and Maurice Bloch. 1989 Introduction: Money and the Morality of Exchange. *In* Money and the Morality of Exchange. Jonathan Parry and Maurice Bloch, eds. Pp. 1–32. Cambridge: Cambridge University Press.

Peletz, Michael. 1988 A Share of the Harvest. Berkeley: University of California Press.

——. 1995 Neither Reasonable nor Responsible: Contrasting Representations of Masculinity in a Malay Society. *In* Bewitching Women, Pious Men: Gender and Body Politics in Southeast Asia. Aihwa Ong and Michael Peletz, eds. Pp. 76–123. Berkeley: University of California Press.

Risseeuw, Carla. 1988 The Fish Don't Talk about the Water: Gender Transformation, Power, and Resistance among Women in Sri Lanka. Leiden, The Netherlands: E. J. Brill.

Rudie, Ingrid. 1995 The Significance of "Eating": Cooperation, Support, and Reputation in Kelantan Malay Households. *In* "Male" and "Female" in Developing Southeast Asia. Wazir Jahan Karim, ed. Pp. 227–246. Oxford: Berg.

Said, Edward. 1983 The World, the Text and the Critic. Cambridge, MA: Harvard University Press.

Stamp, Patricia. 1989 Technology, Gender, and Power in Africa. Ottawa: International Development Research Centre.

Stivens, Maila. 1988 Sexual Politics in Rembau: Female Autonomy, Matriliny, and Agrarian Change in Negeri Sembilan, Malaysia. *In* Development and Displacement: Women in Southeast Asia. Glen Chandler, Norma Sullivan, and Jan Branson, eds. Pp. 79–114. Clayton, Australia: Monash University Centre of Southeast Asian Studies.

Stoler, Ann. 1977 Class Structure and Female Autonomy in Rural Java. *In* Women and National Development: The Complexities of Change. Wellesley Editorial Committee, eds. Pp. 74–89. Chicago: University of Chicago Press.

Strathern, Marilyn. 1988 The Gender of the Gift: Problems with Women and Problems with Society in Melanesia. Berkeley: University of California Press.

Swift, Michael. 1963 Men and Women in Malay Society. *In* Women in the New Asia: The Changing Social Roles of Men and Women in South and Southeast Asia. Pp. 268–286. Paris: UNESCO.

Tsing, Anna Lowenhaupt. 1984 Politics and Culture in the Meratus Mountains. Ph.D. dissertation, Stanford University.

——. 1990 Gender and Performance in Meratus Dispute Settlement. *In* Power and Difference: Gender in Island Southeast Asia. Jane Monnig Atkinson and Shelly Errington, eds. Pp. 95–126. Stanford, CA: Stanford University Press.

——. 1993 In the Realm of the Diamond Queen. Princeton, NJ: Princeton University Press.

Watts, Michael. 1989 The Agrarian Question in Africa: Debating the Crisis. Progress in Human Geography 13(1):1–14.

Whitehead, Ann. 1981 "I'm Hungry Mum": The Politics of Domestic Budgeting. *In* Of Marriage and the Market: Women's Subordination in International Perspective. Kate Young, Carol Wolkowitz, and Roslyn McCullagh, eds. Pp. 88–111. London: CSE Books.

——. 1984 Women and Men: Kinship and Property: Some General Issues. In Women and Property—Women as Property. Renée Hirschon, ed. Pp. 176–192. London: Croom Helm.

——. 1985 Effects of Technological Change on Rural Women: A Review of Analysis and Concepts. *In* Technology and Rural Women: Conceptual and Empirical Issues. Iftikhar Ahmed, ed. Pp. 27–64. London: George Allen and Unwin.

——. 1990 Food Crisis and Gender Conflict in the African Countryside. *In* The Food Question: Profits versus People. Henry Bernstein, Ben Crow, Maureen MacKintosh, and Charlotte Martin, eds. Pp. 54–68. New York: Monthly Review Press.

Wilder, William. 1982 Communication, Social Structure, and Development in Rural Malaysia. London: Athlone Press.

Winzeler, Robert. 1974 Sex Role Equality, Wet Rice Cultivation, and the State in Southeast Asia. American Anthropologist 76:563–567.

Wong, Diana. 1987 Peasants in the Making: Malaysia's Green Revolution. Singapore: Institute of Southeast Asian Studies.

11 Migrant pathways to resource access in Lampung's political forest

Gender, citizenship and creative conjugality

Rebecca Elmhirst

Introduction

Within debates over the political ecologies of enclosure, resource access and control in South East Asia, the phrase 'political forest' has been coined to capture a particular constellation of territorialized power, expressed in ideas, practices and institutions that place spatial limits of peoples' ability to access and utilize lands, providing recognition and legitimacy to some, whilst excluding and criminalizing others (Peluso and Vandergeest, 2001). In Indonesia, analyses have centred on conflicting and (almost always) incomplete projects designed to facilitate resource extraction, conservation and/or development and welfare by rationalizing the use of land, moving populations from one place to another, dividing farm from forest, and in doing so "realigning landscapes and livelihoods" (Li, 2007, p. 1, see also Dove, 1999; Li, 2000; McCarthy, 2004; McCarthy and Cramb, 2009; Peluso, 1992; Tsing, 2004). Such work has revealed the multiplicity of ways that people, including rural migrants, have sought to derive benefits from resources, and specifically the investments they make in various mechanisms, processes and social relationships to develop and maintain resource access, including, but not restricted to socially recognised resource rights (Ribot and Peluso, 2003, p. 155, McCarthy, 2004; Li, 2000; Peluso, 1996; Witasari et al., 2006).

Whilst class and 'race' loom large in these analyses, far less attention has been paid to the gender dynamics that inhere in the regularization of land and livelihood, the ordering of unruly upland spaces and the negotiation of resource access in Indonesia's 'political forest'. In this paper, I draw on recent feminist and queer theorizing about subjectivity, personhood and gendered citizenship in Indonesia (e.g. Blackwood, 2000; Boellstorff, 2005; Robinson, 2009), to explore the ways in which the 'political forest' might be interpreted as a *gendered* project, specifically through the centrality of conjugal partnership and the hetero-normative family in processes and practices of colonial and post-colonial statemaking, and in the practices and engagements of migrants as they seek access to resources of various kinds. The analytical strategy outlined in this paper contributes to a recent wave of feminist political ecology work that emphasizes subject formation in struggles around livelihoods and natural resource access (e.g. Harris, 2006; Nightingale, 2006, 2011; Sundberg, 2004), where "gender" is seen as an analytical concept for considering the workings of power, rather than as a descriptive term or categorization (Butler, 2004; Cornwall, 2007).

Analysis centres on the experiences of migrant farmers in the Indonesian province of Lampung, located on Sumatra's southernmost tip. As happens elsewhere in the global South, in Lampung there is a profound ambivalence towards unregulated rural migrants, who are on the one hand, positioned as a source of labour for commercial enterprises, and at the same time, excluded by forms of enclosure (e.g. for conservation or commercial agriculture) and by the re-assertion of resource claims associated with place-based customary identities (Koczberski and Curry, 2004; Dressler, 2006). Lampung has long been relatively accessible for land-poor migrants from the densely populated islands of Java, Bali and Madura, and this was given impetus firstly by the Dutch colonial authorities' resettlement of landless Javanese in the early 20th century (Kingston, 1987; Utomo, 1967), and more recently, through the migration of poor people from Java following in the wake of the Indonesian government's official transmigration resettlement programme (Pain et al., 1989; Charras and Pain, 1993). In recent years, the province's agro-ecology has been transformed as the last tracts of rainforest have been removed and replaced by marginal upland food cropping, oil palm estates and, in the mountains, coffee and other smallholder tree crops. Resource governance has also evolved, prompted by a growing desire on the part of the authorities to contain spontaneous movement and implement watershed management programmes. In the early 1980s, a central plank of this new resource governance regime was the establishment of the "local transmigration programme" (*Translok*) through which so-called "forest squatters" were forcibly removed from protected areas and resettled in areas designated for food crop agriculture.

Discussion draws on ethnographic research comprising repeat household surveys (covering household demography, land access and sources of livelihood), observation and interviews with migrants (from a sample of 40 households) over a total of 20 months of intensive fieldwork (in 1994–1995, 1998 and 2005) in the *Translok* resettlement site of Negara Jaya to which households had been resettled from protected forests elsewhere in the province. Interviews were also conducted with government officials (at district, provincial and national level), community leaders and NGO personnel, and official documents and associated "grey literature" were analyzed. Interviews were undertaken with both male and female migrants, and where possible, these were held in relatively private settings. Amongst the key questions being explored during the research were issues around individual and community livelihood histories and the negotiation of access to resources (including land), family and kinship dynamics, and gender norms where the latter were expressed in policy discourses, migrant narratives and everyday livelihood practices that evolved over the 10 years of the research.

The paper argues that through the everyday practices involved in traversing a landscape of access and exclusion in Lampung, a particular normative framing of gender relations materializes: one that hinges on conjugal partnership. In this context, marriage (the forging of a conjugal partnership) is an important element in the "means, processes and relations by which actors are enabled to gain, control and maintain access to resources" (Ribot and Peluso, 2003, pp. 159–60) within Lampung's political forest, and thus the negotiation of resource access is simultaneously a process of regulation, (self) discipline and subject-making that cements gender categories and inculcates gendered (and hetero-normative) ideologies of the "ideal citizen". Effectively the reproduction of the political forest hinges on ideologies of family and spatialized practices of reproduction (domesticated conjugal partnerships) into which people discipline themselves as gendered beings (Boellstorff, 2005) – a gendered framing of what Agrawal has described as "intimate government" (Agrawal, 2005, p. 193).

The paper is organized as follows. It begins by exploring the potential interconnections between theorizations of governmentality and resource access, and feminist theorizations of gender, hetero-normativity and citizenship in Indonesia, highlighting the ways in which both illuminate, albeit in different ways, processes of regulation, discipline and subject-making in a context where everyday practices fold into and unfold from a moral hierarchy that mediates contradictions between individual and community in culturally specific ways that are rooted in and reinforced by a complex combination of nation, community and religious traditions of personhood and respectability. These conceptual tools are then put to work to develop a feminist political ecology analysis of Lampung's political forest, focusing on three aspects of the control of people and the control of space in this context: (i) independent settlement by Javanese migrants in Lampung's uplands; (ii) local transmigration resettlement; and (iii) reform-era uncertainty, where livelihoods are increasingly multi-local. Whilst to a degree these three aspects reflect changing environmental governance and resource access regimes within Lampung, they have been selected for what they reveal about the linkages between norms and practices of conjugality, and everyday practices involved in negotiating, reproducing and reiterating the political forest.

Political forests, resource access and gendered subjects

Within the field of political ecology in Asia generally, Peluso and Vandergeest's (2001) concept of 'political forest' is important in revealing how resource access and exclusions are produced through the expansion and reworking of colonial and post-colonial authority over people and space. Drawing on a Foucaultian notion of governmentality, work in this vein has considered how various techniques of rule bring into being and make visible particular subject positions that then may be granted or denied various kinds of access to resources

(Li, 2000, 2007; Sivaramakrishnan, 1999; Sowerwine, 2004; Vandergeest and Peluso, 2006a,b; Yeh, 2009; see also Rutherford, 2007 for an overview). Resource access is shaped by the power to produce categories of knowledge; through the systematic "scientific" classification of uplands as backward, empty, or lacking development, and through racialized ideas of legitimate belonging, entitlement and territorial attachment (Ribot and Peluso, 2003; Li, 2000; Sowerwine, 2004; see also Moore, 2005). In Indonesia, for example, the codification of customary practices of "native" people outside of Java as Customary Law or *adat* produced a form of racialized territorialization (Vandergeest and Peluso, 2006a,b), and continues to resonate to this day in individualized forms of identification (as "Dayak", "Minangkabau", "orang Lampung"), claims for recognition, and the construction of new administrative arrangements for resource entitlement with their accompanying forms of exclusion, often directed at rural migrants and others unable to articulate their identity in line with prevailing notions of belonging (Li, 2000).

One branch of this work has adopted a more agency-inflected analysis of governmentality, focusing on subject formation as a productive process, "where individuals are the vehicles of power, not its points of application" (Foucault, 1980, p. 98). Subjectivities are rendered through projects of "self-discipline", techniques of the self involving management, self-surveillance and regulation of behaviour (Rutherford, 2007), where "images of personhood and pedagogies of conduct and comportment . . . provide the means for understanding and acting on the self" (Rose, 1999, p. xx). The process of "assembling subjects" is constituted through moral–ethical regimes: "the everyday, practical procedures, systems and regimes of injunction, prohibition, judgement through which human beings come to understand, and act upon their daily conduct" (Rose, 1999, p. xx; see also Rankin, 2001). Following this thread, Agrawal's (2005) study of governmentality, subject formation and community-based conservation in India connects closely with the concerns of political ecology. In this work, he describes how the environmental knowledges underpinning devolved environmental management were internalized, producing environmental subjects as villagers' habits, aspirations, beliefs and ethical orientations towards "nature" changed in ways that hinged around the negotiation of resource access.

For the most part, analyses of Indonesia's political forest that highlight the links between the governance of space (territorialization) and the production of governable subjects have, in numerous ways, shown how racialized identities are forged and mobilized to establish authority and differentiated levels of entitlement in the process of the state's territorialization of resources (Li, 2007; Peluso, 2009; Tsing, 2004). These kinds of insights invite consideration of other forms of social identity that are entwined with producing the political forest, and in particular, notions of gendered citizenship that are linked to the Indonesian state's assertion of "the family" as the principal vehicle for "development". Inspiration for this type of analysis may be drawn from feminist and queer perspectives on nation-building in Indonesia, which both consider the centrality of gender identities and gender relations to the politics of rule and to processes and practices of colonial and post-colonial state-making. This kind of perspective builds on and extends insights from feminist political ecology (and gender analyses of resource access more generally) where the family and conjugal partnership are viewed as a critical site for politics within political ecology, and where resource access is reworked and negotiated (Agarwal, 1994, 2003; Carney, 2004; Gezon, 2002; Gururani, 2002; Razavi, 2003; Resurreccion and Elmhirst, 2008; Schroeder, 1999; Thomas-Slayer and Rocheleau, 1995; Whitehead and Tsikata, 2003). Linking feminist and queer perspectives on "the family" and gendered citizenship with an analytic of "the political forest" provides a potentially fruitful avenue for analysis where, following the distinction between property and access made by Ribot and Peluso (2003), the ability to derive benefits from resources is contingent on social relationships that constrain or enable the realization of such benefits (see also Sikor and Lund, 2009).

Sharing both Peluso's and Li's concern with the ideological power vested within a central ruling authority which attempts to elicit the compliance of citizens, feminist and queer perspectives on nation-building in Indonesia have shown how the hetero-normative family constitutes a site where state values are instilled and reproduced, and how its constitution has been integral to 20th century state formation processes (Blackwood, 2000; Boellstorff, 2005; Robinson, 2009). As Boellstorff writes, "in modern societies, forms of kinship and forms of governmentality shape each other" (Boellstorff, 2005, p. 117). Such work seeks to illuminate the processes and practices through which conjugality and gendered personhood are cast and recast, and through which ideas about government through gender might be contextualized. What stands out in this literature is a sense in which a key element in Indonesian state ideology is the "family principle", placing the family as the fundamental unit of the nation. "It is this 'public domesticity' that the state equates with citizen subjectivity and summons into being through a range of development practices" (Boellstorff, 2005, p. 117; Suryakusuma, 1996). Scholars have shown how this principle has emerged historically from a constellation of processes that have the effect of naturalizing particular framings of gender difference, family forms and the organization of modes of social reproduction (Cowen, 2004) into which people attempt to discipline themselves as gendered subjects. Elements that make up this constellation include indigenous notions of masculinity and femininity, gender equality and complementarity, prestige and stigma (Atkinson and Errington, 1990); Islam and its resurgent modernizing forms (van Doorn-Harder, 2006); colonialism (Stoler, 1992; Gouda, 1995) and nationalism (Djajadiningrat-Nieuwenhuis, 1987; Wieringa, 2002); regimes of capitalist accumulation and global cultural flows (Ong and Peletz, 1995; Koning et al., 2000; Blackwood, 2000; Robinson, 2009), and the everyday practices that are embedded in and give shape to all of these (Newberry, 2006).

Three themes are particularly relevant to the argument being developed in this paper. First, whilst historically (and at present) many different family forms and gendered subject positions exist in Indonesia, marriage or conjugal partnership is an almost universal fact of life (Blackwood, 2005; Boellstorff, 2005). Not marrying makes no sense within the moral and cultural logics that make up Indonesia. Marriage is associated with reaching social maturity: as one interviewee in this study put it, marriage is necessary to "menjadi orang", (literally, "become a person"). For both men and women, access to various social resources becomes possible on reaching social maturity, and therefore is contingent on forging a conjugal partnership: a fact which is important for women *and men* as they negotiate the insecurities of life within Lampung's political forest. Moreover, across Indonesia, but with particular emphasis amongst Javanese people, there is a strong connection made between marriage, or principles of "family", and ideas of harmony (*rukun*) over self-interest, or as Karim puts it: "the need to maintain social relationships through rules of complementarity and similarity rather than hierarchy and opposition, and the need to reduce imbalances of power through mutual respectability and cooperation, rather than oppression and force" (Karim, 1995, p. 16). Following this line of thought, other scholars emphasize how a moral hierarchy of "family", particularly in Java and amongst Muslims, "stresses the well-being of the collective through the enactment of social roles and obligation" (Adamson, 2007, p. 6). In other words, conjugal partnership – "family" – is a means by which the unsettling desires of the individual can be domesticated, so that community order and stability can be maintained (Brenner, 1998). The significance of this connection between conjugal partnership and ideas of order and stability is also revealed by the anxieties provoked by large groups of unmarried male migrant workers, for example, plantation workforces of the colonial and post-colonial agricultural frontier described by Breman (1989) (see also Stoler, 1985), and, as this paper goes onto show, by the arrival of unmarried male agricultural migrants in Lampung, a story which has been integral to the forging of Lampung's political forest.

Secondly, the power of this notion of conjugality as signalling social maturity, responsibility and order has been heightened through nationalist discourses, which contrasted the Indonesian "family principle" with Western ideas about "individualism" (Djajadiningrat-Nieuwenhuis, 1987). More recently, and corresponding with the time frame for this study, "the family" was explicitly constructed as the place for nation-building, through former president Suharto's New Order government, in power from 1965 until 1998. The idea of family enshrined here was a hetero-normative nuclear family: father, mother, two children (Suryakusuma, 1996). By disciplining themselves into such families, men and women create the building blocks of the nation, and "through this choice, they also make themselves into proper authentic citizens who will be recognised by the nation", an idea that remains powerful even following Suharto's downfall (Boellstorff, 2005, p. 199). The clearest expressions of the New Order's biopolitics are to be found in its development programmes, where the "will to improve" (Li, 2007) takes its form in interventions into the most intimate corners of Indonesian lives. For example, the Pembinaan Kesejateraan Keluarga (PKK – literally, support for the prosperous family), is a state-sponsored grass roots organisation that operates at village level across Indonesia, which during the New Order period was responsible for implementing training for women around health, hygiene and family planning (Newberry, 2006). The policies and practices associated with this reflected a powerful ideology of domesticity and citizenry, and whilst numerous commentators have remarked on the disjuncture between its programmes and the realities of most women's lives (e.g. Elmhirst, 2002; Robinson, 2009), it nevertheless has provided a powerful moral framework against which people lived their lives: "to give a credible account of oneself within kampung community may mean making use of the resources of PKK" (Newberry, 2006, p. 145). Through television, radio, billboards and posters in even quite remote settings (including Lampung's uplands), the PKK, together with the New Order government's family planning programme, have been representative of the massive effort by the Indonesian state to inculcate an ideology that links ideally gendered men and women into the citizen family (Boellstorff, 2005). A similar process of nation-building through "conjugal production" is evident in the forging of Lampung's political forest, through the local state's transmigration resettlement programme, into which migrants were disciplined (and disciplined themselves), with varying degrees of success.

Whilst there is an evident gap between ideological characterizations of conjugality and idealized gender identities, and the realities of most Indonesian lives in a context marked by divorce, delayed marriage, and where family lives often straddle different localities (and may even be transnational), the durability of marriage as both an aspiration and as a norm is notable. Beyond the cultural and political dissonance associated with not being married, conjugal partnership is an important part of everyday life and everyday livelihoods. A final theme therefore, focuses on conjugality as made and remade through the everyday productive and reproductive practices of men and women in livelihoods marked by a "patchwork of formal and informal sector work, makeshift jobs, local exchange of goods and services, and reproductive tasks [that] are combined in a pattern that is shifting and fluid" (Newberry, 2006, p. 117). For women and men, conjugal partnership is a way of mitigating material risk and vulnerability, as livelihoods are made possible through domestic forms of cooperation, even in contexts where such partnerships are rapidly made and dissolved (as is the case in many parts of Indonesia, where divorce and remarriage is common). The quotidian, routinized and habitual practices of daily life (engagement in agricultural tasks, waged work, collecting water or fuelwood, childcare, rituals of kinship or community) are, for many rural Indonesians, accomplished through social relations of gender – co-operation, negotiation, shared and separate interests, emotional work – that are effected in the performance of conjugality. For example, Koning's work on the "nuclearization" of Javanese families traces such negotiations, and shows very clearly how conjugal relations are reframed to make possible "modern" livelihoods, as such livelihoods shift away from agriculture (and exchange labour) and towards multi-local forms of income generation (Koning, 1997).

Jackson's work on "creative conjugality" expands analysis of conjugal partnership away from stereotypical ascriptions of gender hierarchy, individualised interests and conflict, and towards a more nuanced account of how marriage may constitute a form of insurance against risk where performances of conjugality may yield a cooperative dividend for those who can manage to limit the sacrifices associated with marital cooperation (Jackson, 2007). In line with Kandiyoti's "patriarchal bargain" (Kandiyoti, 1998), she develops a "performative" view of conjugality, showing how women particularly may barter power and personhood for material security and protection in contexts where these are essential for survival, and where the cultural contents (and ethics) of conjugality offer a symbolic resource for women in negotiating access to resources within the conjugal partnership. As Jackson puts it: "conjugality also offers possibilities for women [and men] to manipulate discourses of respectability, manage ironic performances of compliance, and engage in cultural inversions and mimicry of the gender order" (Jackson, 2007, p. 124).

Whilst Jackson focuses on access to resources within the domestic unit, the Indonesian literature also indicates that everyday conjugal performances sustain respectability in the community (Newberry, 2006), and perform permissible forms of gendered citizenship (Boellstorff, 2005), both of which open up other forms of resource access and entitlement, from the community and from the state that are relevant not just for women (and the production of feminine gendered subjects) but also for men (and the production of masculine gendered subjects). Such a perspective can illuminate the myriad ways that disciplining selves in the construction of intimate geographies underpins placemaking and power at wider scales; in this case, Lampung's political forest. Subsequent sections of this paper take up these conceptual threads in order to explore the importance of conjugal partnerships as an element in gaining and maintaining access to resources – which includes different forms of social power and state largesse – in Lampung. This, in turn, invites consideration of how the negotiation of resource access is simultaneously a process of regulation, discipline and subject-making that cements gender categories and inculcates gendered (and hetero-normative) ideologies of the "ideal citizen".

Governing through gender in Lampung's political forest

The genealogy of Lampung's political forest has produced a highly politicized landscape of resource access that includes elements of 'fortress conservation', large-scale commercial agroforestry (oil palm and rubber), intensive rice cultivation and pockets of resource-poor and highly marginal food cropping associated with the resettlement of so-called 'forest squatters'. Particularly striking in this genealogy are the changing regimes of resource governance pursued by the colonial and post-colonial state, expressed most clearly in shifting approaches towards the control of people (migrants and local Lampungese) and the control of space. An important element in this story has been the experience of landless rural migrants in Lampung who came originally from the densely populated island of Java in search of land and livelihoods in the early part of the 20th century. Within the context of changing colonial and post-colonial resource governance regimes, landless Javanese migrants and their descendents, have, at times, been welcomed into the province as potential agents of modernization and development, or else have been seen to pose a demographic or social threat to resource stability, an unruly element requiring 'management' (de Haan and Rogaly, 2002; Curran and Agardy, 2002; Deshingkar, 2005). Migrants' negotiation of resource access must therefore be understood as reflecting continually evolving resource governance regimes, and their shifting legitimacy and recognition as bodies that are frequently regarded as 'out-of-place'. Whilst for clarity of explication the remainder of the paper is organized around various 'phases' in the genealogy of Lampung's political forest, it should be noted that this represents an overly simple periodization of resource access regimes in Lampung: each 'phase' overlaps and bleeds into the others, and there are important continuities rather than stark reversals in fortune for migrants. What each of these 'phases' reveals, albeit in different ways, is that the political forest is a gendered project: conjugal status looms large across all pathways to resource access, and this being so, the negotiation of Lampung's landscape of access and enclosure has the effect of re-citing gendered hetero-normativity.

Accessing upland resources: hetero-normativity and community order

Whilst the migration of landless Javanese to Lampung is generally traced back to the Dutch colonial authorities' *Kolonisatie* settlement projects in the early 1900s (Kingston, 1987), large scale Javanese migration into the province prior to Indonesian independence took place mostly from the late 1920s onwards, during the colonial Netherlands Indies' so-called Ethical Policy. This change of approach by the Dutch to development, labour relations and social policy, coupled with financial constraints that limited direct support of migrants, encouraged large-scale spontaneous migration of land-poor Javanese farmers to Lampung. Resource governance in Lampung during this period was made up of a combination of forest reserves (administered by the colonial authorities), official resettlement schemes (colonies of government-supported Javanese migrants working irrigated land, such as Gedong Tataan, founded in 1905), lands leased to colonial private capital (rubber and coffee plantations largely controlled from Europe), and land under the authority of the leaders of indigenous Lampungese clans (*Marga*), which, since 1928, held the status of *inlandse gemeente* (native community) (Ter Haar, 1948; Utomo, 1967). Colonial census data from this period indicates a rapid increase in the population of Javanese migrants in Lampung: from 6073 in 1911 to around 74,000 by 1941 (Pelzer, 1945).

The centrality of conjugal partnership as an element in the "means, processes and relations by which [migrants] are able to gain, control and maintain access to resources" (Ribot and Peluso, 2003, pp. 159–60) in Lampung's political forest is best understood by considering the kinds of resources for which access was sought by migrants, and through analysis of the workings of different sites of power that sanction and validate access (c.f. Sikor and Lund, 2009), which together have the effect of emphasizing domesticated conjugal partnership as a source of legitimacy and entitlement for migrants during this late colonial period in Lampung. 'Resources', in this instance, includes not only access to land (in terms of an ability to derive benefits from land), but access to state largesse: in material terms, and also in terms of recognition – in this case, legitimacy and permission to move from Java to Lampung in the first place. Conjugal partnership was cemented firstly in the governance of mobility (i.e. access to resources that permit mobility), and secondly, in and through access to land resources once in Lampung itself. Here, these are considered in turn.

The movement of migrants from Java to Lampung at this time was controlled in various ways by the colonial authorities through a regime which identified what sorts of 'bodies' were acceptable as migrants. Potential settlers were recruited from rural Java by local village leaders, who drew up lists on the basis of directives from the colonial government's Central Colonisation Committee in Lampung regency who oversaw migration to the area (Kingston, 1987). In colonial reports from the period, Netherlands Indies colonial administrator H. G. Heyting, founder of the early migrant colonies in Lampung, issued a number of statements guiding the selection of migrants: "Select the colonists with the greatest possible care. Take only strong, young people who are really farmers. Do not take persons who have been labourers on Western plantations; they have lost a great deal of initiative and capacity for working on their own and, above all, have become used to regular weekly or monthly payments of wages, whereas the farmer must wait for months before he gets a return for his labour. *Take only married people; single men are likely to desert as soon as they are confronted by a difficulty*" (from Schalkwijk, 1921 '*De kolonisatie met van Java afkomstige gezinnen*, cited in Pelzer, 1945, p. 200, emphasis added). Later, this emphasis on the qualities of potential migrants and particularly their conjugal status appears in the *Ten Commandments for Selection*, authored by C. C. J. Maassen of the Central Colonisation Committee:

1 Select real farmers, non-farmers are a burden for a colony and endanger its success.
2 Select physically strong people; only they can stand the hardships of pioneering.
3 Select young people, by taking them one reduces future population increase in Java.
4 Select families; families are the foundation of peace and order in the colonies.
5 Do not select families with many young children, the working members of the family cannot carry that burden at the start.
6 Do not select former plantation labourers; in 90% of all cases they are the cause of discontent in the colonies.
7 Do not allow so called colonisation marriages: they are a source of unrest in colonies.

8 Do not accept expectant mothers; the pioneer settler needs the full help of the wife during the first year.
9 Do not accept bachelors: sooner or later they will become involved with somebody else's wife.
10 Allow *desas* or *kampongs* (rural and urban villages) to migrate as a whole, in such a case the first nine commandments can be ignored.

(Maassen, 1939 'Javanese Agricultural Migration').

These directives reflect the gender politics of the colonial Ethical Policy, which offered an orientalist idealization of 'the Javanese family' as a source of harmony and stability (Locher-Scholten, 2003), which was contrasted with the 'unruliness' of single male migrants. During the late 1920s, the significance of this contrast is linked to the anxieties provoked amongst the colonial authorities by a communist uprising in West Sumatra (Kahin, 1999), and associations of labour unrest in the colonial plantations of Deli in East Sumatra, populated largely by single male 'coolies' (Breman, 1989). As commentators have noted, the association between family (meaning, hetero-normative conjugal partnership) harmony and stability at a number of different political scales continues as a powerful discourse in contemporary Indonesia (Boellstorff, 2005; Robinson, 2009).

The possibility of fully ascertaining the ways in which migrants disciplined themselves as gendered subjects, through domesticated conjugal partnership, is limited. However, writings by administrators and academics from this period make frequent reference to the common occurrence of unmarried potential migrants marrying hastily before emigrating, in order to have access to state support of various kinds. Women marrying in this way were known as voyage wives (*bini jalan*) or tow wives (*bini gandingan*). Interviews with older Javanese migrants in this study (in most cases describing their parents' stories) reveal a number of instances whereby women came to Lampung from Java as *bini jalan*, in some cases through a marriage arranged between two families, to men that were considerably older than they were, and to whom they had had little prior connection.

Access to land in Lampung's late colonial political forest was also built upon and had the effect of cementing conjugal partnership. Representations of migration and resource control during this period (and indeed up to the 1970s) hinge on the role played by these 'pioneers', who are seen as agents of 'development', entrepreneurs opening up a 'backward' agricultural frontier, with minimal cost to government (Pelzer, 1945; Utomo, 1967). For migrant farmers, access claims centred on the following interconnected points: permission to cultivate, access to sharecropping opportunities, and, importantly, where resource claims remained tenuous, access to state largesse. The negotiation of a complex, scaled and racialized landscape of access to resources (in the broadest sense) involved various kinds of positionings for migrant farmers in terms of claims for recognition and entitlement, and by extension, projects of "self-discipline" – habits, aspirations, beliefs, orientations – that were cemented and reinforced through the material demands of frontier livelihoods.

During this period, land ostensibly controlled by Lampungese elites (*Marga* leaders) was opened up to migrants, in return for receipt of an indemnity from the colonial government. Otherwise, much of the settlement occurred on unopened forest land under the control of the colonial authorities. In both instances, access to land was acquired through a particular sequence, in which a chief clearer (*kepala tebang*) would negotiate with local Lampung leaders, and thereafter, lead a group of largely male migrants to clear land for dry-land cropping. Following successful cropping, further migrants would join the group, including wives and children, who hitherto would have been engaged as labourers either on the pepper or coffee gardens of Lampungese smallholders or in the fields of already established migrants. The material demands of pioneer agriculture (the labour inputs required to clear forest for cultivation, to participate in coffee or pepper harvests) were very high, but also access to land hinged on demonstrating investment in the resource by applying family labour – itself a resource to which access was made possible through marriage. Everyday labour practices amongst migrants in clearing land and establishing access therefore cemented conjugality.

However, as Ribot and Peluso (2003) point out, access also involves maintenance, and in this instance, this was most effectively achieved through the acquisition of community recognition. Whilst many initial forays into new areas were undertaken by men alone, the establishment of 'village communities' and qualification for recognition by the colonial state (and the resources that followed) rested on the capacity of migrant leaders to attract additional migrants and more specifically, those with wives and families. For migrants, this involved village expansion, and a demonstration of community structures that aligned with what Kingston has described as the colonial government's 'Javanese paradigm' in which 'Javanese *sawah* (irrigated rice fields) and *desa* (villages) carved out of the jungle in Lampung would provide for the maintenance and reproduction of a malleable and politically inarticulate labour force as they had in Java' (Kingston, 1987, pp. 134–35). Women, therefore whilst often invisible in contemporaneous accounts as pioneer farmers per se, were critical for solidifying communities: their presence necessary for community recognition, generally granted to those areas with stable populations of established households, and for the entitlements that accompanied this. Marital status was therefore not only a signal of social maturity for individuals, but also for communities at large – for the recognition of 'community'. The significance of this connection between conjugal partnership and ideas of order and stability is also revealed as a mechanism for allaying the anxieties provoked by large groups of unmarried male migrant workers, for example, plantation workforces of the colonial and postcolonial agricultural frontier described by Breman (1989) (see also Stoler, 1985), and tellingly by Erman, who provides an account of the ways in which H. G. Heyting (founder of the early migrant colonies in Lampung) also sought to develop 'communities' of workers around the coalmines of West Sumatra during this period, ostensibly to establish stability where miners "missed their family, . . . they had no moral support for work, . . . there was a great deal of homosexuality and conflicts over women strengthened the argument for a change in management policy" (Erman, 2004, p. 371). In Lampung's political forest, especially where demonstration of closeness to the state became an important safeguard for securing resource access in the context of somewhat tenuous and uneasy relationships with local Lampung people, it is through the practice of community recognition that gender was inscribed around the performance of migrant conjugal relationships, built around the necessity for successful pioneer farmers to be in possession of a 'stabilizing' wife, with little room for those (especially women) whose subject positions elided this. As the next section shows, falling in with this moral–ethical regime (c.f. Li, 2007) was no guarantee of access to resources for Javanese migrants as the landscape of Lampung's political forest shifted in the post-colonial period, and through the height of former President Suharto's New Order. However, similar themes shape resource access. The links between the state's territorializing projects, 'intimate government' and forms of gendered citizenship that facilitate migrants' recognition and resource access are remade: cementing the importance of conjugal partnership in new ways.

Resource access through resettlement: hetero-normativity and legitimate citizenship

Following a period of relative laissez-faire towards forest clearance and the settlement of Javanese migrants in Lampung's uplands, a new landscape of resource access and control began to emerge in the late 1970s, as Indonesia's New Order government increasingly pursued policies favourable towards large-scale forest exploiters, concession holders and plantation development. At the same time, this converged with efforts to respond to a globalized discourse of environmental protection, creating a policy climate in which resource access was regulated, on paper at least, through the re-assertion of state control over 'forest land', and the zoning of particular spaces for particular functions: commercial forestry and plantation agriculture, smallholder cultivation and watershed protection. Through a series of regulations, including the Basic Forestry Law of 1967 (UUK No. 5/1967), policies focused on the removal of people from areas now designated as conservation forest lands, including national parks (*hutan suaka alam* or *taman nasional*), defined as areas prioritizing nature conservation and where agriculture is heavily restricted, and from watershed protection forest (*hutan lindung*), designated to protect downstream commercial agriculture.

The resulting landscape of resource access in Lampung was made up of a series of interlocking (and frequently overlapping) spaces representing different modes of resource access or enclosure. These included, firstly, areas of 'fortress conservation', protected by military and other representatives of the provincial state, where migrants found themselves labelled as 'forest squatters' (*perambah hutan*), and from which they were forcibly removed, their crops and houses burned if they did not comply with the authorities. A second space is also associated with enclosure, but this time, through large scale capital investment (state, private or some combination), where land has been acquired for fast-growing timber or agro-industrial plantations (sugar, oil palm, rubber), again involving the displacement of other kinds of claims on resources. There is an ongoing contestation between such development and forms of access increasingly limited to the interstices of state and large-scale private sector enclosure – the third type of space evident in Lampung, where forms of access are associated with membership of 'original' Lampung clans (*Marga*). A final 'space' evident at this time is 'state' land designated for conversion to agriculture. During the 1980s land was given over for the development of government sponsored transmigration resettlement sites, designed to receive people forcibly displaced from other spaces. In these sites, based largely on dry-land food cropping and smallholder cash crops, access to resources has been characterized by individualized land title granted to households participating in Lampung's local transmigration resettlement scheme.

Room for manoeuvre within this kaleidoscope of resource access and control has been problematic for Javanese migrants. As environmental protection, the expansion of commercial activities and the fostering of particular kinds of smallholder agricultural development were being pursued by the provincial government and other parties, views about spontaneous migrants became increasingly hardened, and their cultivation activities criminalized. As various pathways to resource access were abruptly terminated, securing a livelihood became conditional on fostering a closer alignment to the local state through participation in Lampung's local transmigration scheme. Whilst participation in the resettlement scheme was voluntary, staying put in 'protected forests' was not an option. Those who remained had their houses and crops burned by representatives of the provincial government and by the army. By 1992, around 370,000 people had been resettled in the north of the province, in land designated by the provincial government as 'empty' and in need of development.

Various aspects of local transmigration in Lampung have had the effect of reinforcing hetero-normative conjugality as a relationship through which resource access may be realized. First, the process of resettlement itself, rested on a concept of 'the household', comprising male breadwinner and female dependent, in a manner that resonated with prevailing state discourses around masculinity, femininity and citizenship enshrined in the New Order government's general development guidelines (*Repelita III*) from this period (Smyth, 1991). Within the state's tiered and scalar conceptualization of rule, the nuclear family/household is a building block of the nation, with the male household head seen as the point of contact between state and people. This concept is apparent both in the inventories of 'forest squatters' drawn up by the provincial authorities as a prelude to resettlement, the categorization of migrants and their resettlement, and in the 'units' conjured up to receive resources transferred from the state in the process of implementing resettlement. Participation in the official resettlement programme offered an opportunity to gain legitimacy and to circumvent the intolerance of migrants that is emblematic of many forest margin areas. However, participation rested on demonstrating membership of household/family units, and on articulating clearly-defined gender roles – male breadwinner, female homemaker and socialiser of children. Resettlement inventories did not include single men or women, unless they were subsumed into membership of a household unit, headed by a married couple. Settler support (i.e. provision of a house, 2 ha of land, seeds and fertilizer at the outset, and thereafter agricultural extension services and credit as the settlement became established) was directed through such units, headed by a male breadwinner, effectively inscribing conjugal partnership as a conduit for access to (state) resources, for both men and women, and once again inscribing the link between hetero-normativity, conjugal partnership and 'order'.

Secondly, and related to this point, eligibility to resources made available by the state through transmigration resettlement also rested on the household/family concept, and therefore on conjugal partnership. A notable aspect of resettlement has been the institutionalization of state-regulated private property as a mechanism for access to land. Official title to land was granted to settler households by the provincial government, with certificates of ownership issued to the male household head who was vested with control and (to some degree) disposal rights. Migrants received 2 ha of land, of which 0.25 ha was a home garden (with a simple wooden house), 1 ha had been mechanically cleared of large trees, ready for manual field preparation for food crop cultivation, and a further 0.75 ha remained under bush (*belukar*), on the understanding that the settler family would be responsible for clearing it in due course. Migrants were provided with food crop seeds, including rice, cassava and a range of vegetable crops, and seeds for growing tree crops, principally in the home garden (*pekarangan*), these included coconut, banana, pineapple, and so on. Fertilizer was provided for the duration of a year, and an extension service was expected to provide support. Thus, in effect, the practices of land control and property relations fostered by the government in the implementation of *Translok* served to entrench and reinforce the idea of women's dependence on men, reproducing an image of women as housewives, men as breadwinners.

At this time (the early 1980s), private land ownership was seen as providing an economic incentive to promote 'settled' sustainable livelihood practices and other desired forms of rural modernity among 'unruly' forest squatters (McCarthy and Cramb, 2009), a theme that emerges in the rules and regulations under which newly-resettled migrants were expected to live. *Translok* migrants were obliged to cultivate the land they had been given, maintain their houses, practise religious observance, keep the settlement orderly, follow any instructions given to them by settlement officials, and observe good relations with the local population (Kantor Wilayah Transmigrasi, 1986, 1994). A key aspect of these rules is the ways in which access to resources for migrants became conditional on their relationship to the state, largely via the household/family units that provided the conduit for state largesse. This route to resource access was all the more significant in the absence of wider kin networks through which resource access might otherwise be realised. Resettled migrants were forbidden from selling land or their houses within 10 years of arrival, and from leaving the settlement without permission from settlement authorities – as late as 1994 (12 years after the establishment of the settlement), 71% of households surveyed had acquired land through government official title.

Within particular resettlement sites, the enforcement of such rules was effected through the village head, a person nominated through the hierarchy of local government, and it is through this particular embodiment of state power that resource access continued to be mediated by the provincial authorities. In the research site, a decade after resettlement, the village head – recently retired from the military, and posted to the *Translok* settlement to tackle security issues in transmigration areas – had elaborated on this role by exerting considerable control over various mechanisms of land transfer, including inheritance, purchase or sale and land subdivision. In a number of instances, he had refused permission for people to sell land (primarily to local speculative interests rather than to incoming migrants) and refused permission for families to subdivide and bequeath land to their children, in both cases, justifying his decision in terms of promoting community stability and the long-term viability of small-holder agrarian livelihoods in the area: both of which, he suggested, were best attained through a particular family/household constellation (i.e. comprising a hetero-normative conjugal partnership).

Normative social forms also resonated with the ways in which settlement bureaucrats (the village head and others closely aligned with village governance, formally and informally) dealt with the issue of underemployed, landless sons of those originally resettled: as time passed and the children of migrants grew up, their presence was regarded as a potential threat to community stability and order. In 1994, logged over forest adjacent to the *Translok* settlement was cleared to create a new hamlet, following a working agreement between the village council, a nearby original Lampung community and a local businessman engaged in logging and oil palm development. Land was parcelled up and made available (through various share-cropping arrangements) to young men that had hitherto been living with their parents. However, not all young men were able to access land in this way: in part, access was easiest for those from families better connected within the community, but it was also the case that those that failed to match up to a prevailing view of a 'deserving' and honourable household were also disregarded, in ways that mimicked the workings of the local government's resettlement plan some 10 years earlier. Young men consolidated their claims to this land by establishing themselves as family men with a wife and a baby, displacing anxieties about underemployed single young men, positioning themselves as 'adults' within Javanese cultural logics, and as established, responsible members of the community. Those that succeeded in enhancing the legitimacy of their claims were those that were able to position themselves within prevailing ideas about citizenship, community and social order.

Like the better-known Indonesian transmigration programme, local transmigration (*Translok*) is a distilled version of the Indonesian government's development effort, and similarly promulgates public domesticity – the 'family principle' – wherein the hetero-normative nuclear family is explicitly constructed as a site for nation-building. Boellstorff (2005) and others have shown how people choose to align themselves with this principle and thereby produce themselves as proper 'authentic' citizens who will be recognized by the nation. Within the local transmigration programme in Lampung, this process has been particularly explicit: every intervention is cast through particular ideas about men and women, and by extension, hetero-normative conjugal partnership, inherent in most of the statistical and economic concepts on which policy planning is based (Dawson, 2008). Moreover, as the above section has suggested, as settlements have 'matured' and links with the local transmigration planning office loosened, other forms of authority over resource access within the settlement continue to invoke processes of 'legitimization, inclusion and exclusion' (Sikor and Lund, 2009, p. 2) that cement conjugality as an important positioning strategy for resource-poor migrants. Negotiating access to resources through local transmigration involves migrants positioning themselves as eligible clients of the state: aligning their lives and aspirations with those valorized in development interventions (through the performance of appropriately gendered citizenship), and disciplining themselves into the 'family principle', even where the material hardships of migrant life undermine this possibility. The next section focuses more explicitly on the material dimensions of the relationship between resource access, conjugality and the capacity to derive benefit from resources, a relationship that is made explicit as the landscape of resource access in Lampung's political forest has become more geographically fragmented.

Realizing the benefits of resource access: conjugality and access to labour

In recent years, and particularly since the downfall of Soeharto's presidency in the late 1990s, the contours of resource access associated with Lampung's political forest continue to exhibit processes of enclosure, but these increasingly reflect a diversity of interests beyond the state. This includes a non-local private sector and corporate stake holders involved in securing land for commercial crops (timber, oil palm, sugar cane), infrastructure projects associated with multi-lateral donors (e.g. a major hydro-power project funded by the World Bank in the west of the province), and efforts to build community-based natural resource management frameworks (CBNRM) adjacent to protected areas (Suyanto, 2004; Witasari et al., 2006). At the same time, land conflicts have become more prevalent, partly because of the assertion of corporate claims, but also as the local Lampung population has asserted its claims more forcefully, enabled by an Indonesia-wide movement to give greater recognition to 'original' (*asli*) populations (Li, 2000). Migrants have thus found themselves caught amidst these contested interests. In terms of access to resources (land and state largesse) the changing political landscape has disrupted the taken-for-granted structural advantage of transmigrants and transmigration settlements, and indirectly brought challenges to migrants' capacity to assume resource access through alignment with the local state, a shift that accords also with neoliberalization and an emphasis on self-reliance and livelihood diversification more generally. Whilst preceding sections have shown how hetero-normative conjugality is effectively reinforced through the representational strategies – positionings – of people and communities as they claim access to resources, in this section attention focuses on the ways in which material practices necessary to realize the benefits from resource claims also reiterate hetero-normative conjugal partnership, in the context of increasingly geographically fragmented livelihood options associated with the politics and agro-ecology of Lampung's political forest in the early 21st century.

In the extremely poor agro-ecological setting of the transmigration settlement, resource access is only of benefit where there is sufficient labour to realize resource claims. As late as 1995, of the households surveyed, none had brought their entire 2 ha allocation of land under cultivation, in part because of the cost of deploying labour to clear secondary forest and scrub, but also because maintaining soil fertility in fragile upland soils required large investments of labour and capital, to which most people had no access. By 2005, when previously uncultivated fields had been cleared, this had been achieved by transmigrants leasing land to large-scale private interests for the cultivation of oil palm – where these private interests had capital to invest and were able to deploy labour gangs to undertake the necessary work. Access to labour was also made more problematic by the fact of resettlement itself – where reciprocal labour arrangements between households were slow to emerge between people who were ostensibly strangers to each other. In this context, family labour took on a particular significance, where a sense of moral and ethical entitlement to assistance between married couples and other family members was stronger than that between neighbours. In other words the ability to turn resource access into material benefit, in a situation of poor access to non-family labour, hinged on conjugal partnership, effectively reinforcing its significance as a pathway to resource access, not only for women but for men too. In interviews it was common to hear people describe how tasks normally associated with men (heavy ploughing, for example) were undertaken by women. Since the settlement was opened in 1982, it has not been possible for households to meet subsistence needs through own-account agriculture. Until the late 1990s, incomes were supplemented to a large degree by wage work at a nearby sugar plantation, for which women were recruited as day labourers for year-round crop maintenance, and men for work in the harvest. Together, their wages comprised up to 70% of household income. The necessity of men and women working collectively in this way had the effect of reinforcing the unity of the household in most cases, and this was frequently reflected upon in interviews.

After 2000, as land conflicts between corporate interests and local Lampung people escalated, the symbiotic relationship between plantation work and resource-poor subsistence agriculture in the transmigration settlement was broken when escalating conflict forced the plantation to cease operations, thus curtailing a localised source of off-farm income. At the same time, a combination of infrastructure investment (associated with the re-assertion of political authority by local Lampung people) and wider confusion over province-level resource control led to transmigrants exploring more geographically dispersed forms of resource access: plantation work in other parts of the province (facilitated by labour recruiters) and work in newly cleared coffee plantations in protected uplands to the west of the province. In both cases, those pursuing these practices for income generation were men, but such practices have been made possible by the continued labour investments of women in the transmigration settlement. In remaining in the transmigration settlement, migrant women were able to maintain the legitimacy of resource access, whilst at the same time, they provided a material basis, a safety net of entitlement, for the men's more risky but potentially rewarding ventures. A second set of livelihood diversification practices is apparent in the involvement of transmigrant women in transnational labour migration. In the early 2000s, labour brokers had arrived to recruit women for domestic work in Malaysia, Hong Kong and Saudi Arabia. Women's exclusion from other more localized non-farm work made them a ready 'reserve army of labour' once recruitment networks had become sufficiently institutionalized and trusted within the community, and lines of credit to pay for their involvement had been established.

A common pattern of labour deployment therefore was for older women to remain engaged in year-round food crop production, men and older sons to spend part of their time in coffee uplands working in plantations, whilst younger women took up 2-year contracts as overseas domestic workers. Whilst such forms of labour deployment suggest an individualization of livelihoods, in all cases practices hinge on a conjugal partnership, acknowledging that in many instances this relationship is imperfect, unstable and potentially problematic. Whilst the spatial dynamics of conjugality (and reproduction) have changed, its meanings and obligations have, if anything, strengthened, reiterating and reinforcing gender categories in ways that are linked to the shifting landscape of resource access in Lampung more generally. Political instabilities and concerns over the rapid emergence and dissolution of forms of resource access have reiterated the significance of settled agriculture for transmigrants. Of the original 40 case-study families surveyed in 1994–1995, in 1998 at the height of Indonesia's political and economic crisis, more than a third had apparently left the settlement. Yet by 2005, nearly all had returned, the exception being a landless family that had depended on the now redundant sugar plantation for work, and two others who had sold their land to pay for their children's education. As part of a wider repertoire of multi-local livelihoods, the transmigration settlement was perceived as a foothold less tenuous than other extra-local sources of livelihood, particularly where recent experiences of exclusion from protected areas colour livelihood decisions. Put another way, pathways to resource access were more clear-cut in the transmigration settlement, but the benefits of access could only be realised by geographically dispersed allocation of family labour that was made possible through a multi-local dynamic of obligation, guilt, shame and fear in a process similar to that explored in the literature on transnational migration (e.g. Velayutham and Wise, 2005). This in turn reiterates hetero-normative conjugality, and disciplines gendered bodies across space.

Conclusion

This paper has sought to extend analysis of enclosure, resource access and subject formation in Southeast Asia's "political forests" through a feminist political ecology analysis that draws on recent feminist and queer theorizing about subjectivity, personhood and gendered citizenship in Indonesia. Within this literature, much is made of the importance of conjugality – the hetero-normative family – as a social relationship, a vehicle for development, into which people discipline themselves and through which "rule" is accomplished (or at least, sought) in the Indonesian context. Given the significance accorded to "the family" in accomplishing "rule", it is perhaps surprising that family and conjugal partnership have not been explored in the richly detailed studies of the processes and tactics through which both colonial government and the modern state have attempted to forge control over people and territories in Indonesia's unruly uplands. By drawing on both these literatures, this paper has sought to cast light on the ways in which "conjugal partnership" is significant in the tactics and powerplays inherent in the negotiation of resource access, and in materializing its benefits. Through an empirical examination of processes of resource acquisition in Lampung associated with unregulated frontier migration in the late 1960s and early 1970s, of government-directed resettlement in the 1980s, and of the strategies adopted by migrants in the face of uncertain resource governance regimes and transnational livelihood possibilities in the early 21st century, what is revealed is the role of "conjugal partnership" as a conduit for access to natural resources and associated state largesse. As resource access has been negotiated, "conjugal partnership" has been remade and reiterated, directly through the social engineering of the local state, but also through projects of self-discipline as migrants aligned themselves into family forms that signalled stability, maturity and qualification for resources, and also through material practices required to realise the fruits of resource access, namely labour requirements in the absence of alternative networks of obligation and support.

Such an analysis invites the following conclusions. First, whilst this analysis departs somewhat from other analyses of family, conjugality and resource access in feminist political ecology, it also signals the importance of considering the ways in which spatialized practices of reproduction and normative heterosexuality underpin the power geometries of territorialization and the production of governable subjects in Indonesia's uplands. In other words, the reproduction of the political forest hinges on ideologies of family, whilst at the same time, processes of subject-making associated with the negotiation of resource access solidify and naturalise domesticated conjugal partnership. Secondly, and related to this point, the study invites an appreciation of how the conjugal relationship is an important conduit for resource access, not just for women (as many feminist analyses of gender-differentiated resource access have highlighted) but also for men too. The significance of this observation is twofold: it cautions against an overemphasis on the individualization of livelihoods, an interpretation commonly associated with agrarian transformation and the increasing prevalence of geographically-dispersed households. At the same time, it suggests that the attention frequently paid to the analysis of community-level cooperative arrangements for achieving sustainable natural resource management should not neglect consideration of the role of cooperative arrangements that are cemented within the moral–ethical regimes of the hetero-normative family and conjugal partnership. A better approach would be to consider the multiple layerings of social cooperation through which resource access and sustainable livelihoods are accomplished. Moreover, the "naturalness" of particular forms of conjugal partnership should be interrogated alongside the more conventional analytical strategy of questioning of the "naturalness" of community.

Acknowledgements

Research for this paper was funded by the Economic and Social Research Council (R000222089) and the British Academy (Committee on South East Asia). The author wishes to thank the Indonesian Institute of Sciences (Lembaga Ilmu Penelitian Indonesia) for granting permission and the people of Negara Jaya for participating in the research. Research was facilitated by staff at Universitas Lampung and the World Agroforestry Center (ICRAF), and assistance provided at various times by Dwi Cahyono, Hermalia, Herna and Yulianti. The paper has benefited from discussions with Babette Resurreccion and Andrea Nightingale, and the helpful comments of referees.

References

Adamson, C., 2007. Gendered anxieties: Islam, women's rights, and moral hierarchy in Java. Anthropological Quarterly 80 (1), 5–38.

Agarwal, B., 1994. A Field of One's Own: Gender and Land Rights in South Asia. Cambridge University Press, Cambridge.

Agarwal, B., 2003. Gender and land rights revisited: exploring new prospects via the state, family and market. Journal of Agrarian Change 3 (1 and 2), 184–224.

Agrawal, A., 2005. Environmentality: Technologies of Government and the Making of Subjects. Duke University Press, Durham, NC.

Atkinson, J.M., Errington, S., 1990. Power and Difference: Gender in Island Southeast Asia. Stanford University Press, Stanford.

Blackwood, E., 2000. Webs of Power: Women, Kin and Community in a Sumatran Village. Rowman and Littlefield, Lanham, MD.

Blackwood, E., 2005. Gender transgression in colonial and postcolonial Indonesia. The Journal of Asian Studies 64 (4), 849–879.

Boellstorff, T., 2005. The Gay Archipelago: Sexuality and Nation in Indonesia. Princeton University Press, Princeton, NJ.

Breman, J., 1989. Taming the Coolie Beast. Plantation Society and the Colonial Order in Southeast Asia. Oxford University Press, Delhi.

Brenner, S., 1998. The Domestication of Desire: Women, Wealth and Modernity in Java. Princeton University Press, Princeton.

Butler, J., 2004. Undoing Gender. Routledge, London.

Carney, J., 2004. Gender conflict in Gambian wetlands. In: Peet, R., Watts, M. (Eds.), Liberation Ecologies. Environment, Development, Social Movements, second ed. Routledge, London, pp. 316–336.

Charras, M., Pain, M., 1993. Spontaneous agricultural settlement in Indonesia. In: Charras, M., Pain, M. (Eds.), Spontaneous Settlements in Indonesia: Agricultural Pioneers in Southern Sumatra. ORSTOM-CNRS, Bondy, and Department of Transmigration, Jakarta, pp. 13–34.

Cornwall, A., 2007. Revisiting the "gender agenda". IDS Bulletin 38 (2), 69–78.

Cowen, D., 2004. From the American lebensraum to the American living room: class, sexuality, and the scaled production of 'domestic' intimacy. Environment and Planning D: Society and Space 22 (5), 755–777.

Curran, S.R., Agardy, T., 2002. Common property systems, migration and coastal ecosystems. Ambio 31 (4), 303–305.

Dawson, G., 2008. Keeping rice in the pot: women and work in a transmigration settlement. In: Ford, M., Parker, L. (Eds.), Women and Work in Indonesia. Routledge, London, pp. 41–60.

de Haan, A., Rogaly, B., 2002. Migrant workers and their role in rural change. Journal of Development Studies 38 (5), 1–14.

Deshingkar, P., 2005. Improved livelihoods in improved watersheds: can migration be mitigated? In: Sharma, B. R., Samra, J. S., Scott, C. A., Wani, S. P. (Eds.), Watershed Management Challenges: Improving Productivity, Resources and Livelihoods. International Water Management Institute, Colombo, Sri Lanka, pp. 144–156.

Djajadiningrat-Nieuwenhuis, M., 1987. Ibuism and priyayization: path to power? In: Locher-Scholten, E., Niehof, A. (Eds.), Indonesian Women in Focus: Past and Present Notions. Foris Publications, Dordrecht, pp. 43–51.

Dove, M., 1999. Representations of the "other" by others: the ethnographic challenge posed by planters' views of peasants in Indonesia. In: Li, T. M. (Ed.), Transforming the Indonesian Uplands: Marginality, Power and Production. Harwood Academic Publishers, Amsterdam, pp. 203–229.

Dressler, W.H., 2006. Co-opting conservation: migrant resource control and access to national park management in the Philippine uplands. Development and Change 37 (2), 401–426.

Elmhirst, R., 2002. Negotiating land and livelihood: agency and identities in Indonesia's transmigration programme. In: Yeoh, B., Huang, S., Teo, P. (Eds.), Gender Politics in the Asia-Pacific Region. Routledge, London, pp. 79–98.

Erman, E., 2004. Labouring communities: women's roles in the Ombilin coal mines of Sumatra (1892–1965). In: Elmhirst, R., Saptari, R. (Eds.), Labour in Southeast Asia: Local Processes in a Globalised World. Routledge Curzon, London, pp. 359– 386.

Foucault, M., 1980. Power/Knowledge. Pantheon Books, New York.

Gezon, L., 2002. Marriage, kin, and compensation: a socio-political ecology of gender in Ankarana, Madagascar. Anthropological Quarterly 75 (4), 675–706.

Gouda, F., 1995. Dutch Culture Overseas: Colonial Practice in the Netherlands Indies, 1900–1942. Amsterdam University Press, Amsterdam.

Gururani, S., 2002. Forests of pleasure and pain: gendered practices of labor and livelihood in the forests of the Kumaon Himalayas, India. Gender, Place and Culture 9 (3), 229–243.

Harris, L.M., 2006. Irrigation, gender and social geographies of the changing waterscapes of south eastern Anatolia. Environment and Planning D: Society and Space 24, 187–213.

Jackson, C., 2007. Resolving risk? Marriage and creative conjugality. Development and Change 38 (1), 107–129.

Kahin, A., 1999. Rebellion to Integration: West Sumatra and the Indonesian Polity. Amsterdam University Press, Amsterdam.

Kandiyoti, D., 1998. Gender power and contestation: rethinking "bargaining with patriarchy". In: Jackson, C., Pearson, R. (Eds.), Feminist Visions of Development; Gender Analysis and Policy. Routledge, London, pp. 135–152.

Kantor Wilayah Transmigrasi Lampung, 1986. Statistik Translok, Propinsi Lampung. Kantor Wilayah Transmigrasi, Bandar Lampung.

Kantor Wilayah Transmigrasi Lampung, 1994. Statistik Translok, Propinsi Lampung. Kantor Wilayah Transmigrasi, Bandar Lampung.

Karim, W., 1995. Bilateralism and gender in Southeast Asia. In: Karim, W. (Ed.), "Male" and "Female" in Developing Southeast Asia. Berg Publishers, Oxford, pp. 35–74.

Kingston, J.B., 1987. The Manipulation of Tradition in Java's Shadow: Transmigration, Decentralization and the Ethical Policy in Colonial Lampung. Unpublished PhD, Columbia University.

Koczberski, G., Curry, G.N., 2004. Divided communities and contested landscapes: mobility, development and shifting identities in migrant destination sites in Papua New Guinea. Asia Pacific Viewpoint 45 (3), 357–371.

Koning, J., 1997, Generations of Change: A Javanese Village in the 1990s. University of Amsterdam, Unpublished PhD.

Koning, J., Nolten, M., Rodenburg, J., Saptari, R. (Eds.), 2000. Women and Households in Indonesia: Cultural Notions and Social Practices. Routledge-Curzon, London.

Li, T.M., 2000. Articulating indigenous identity in Indonesia: resource politics and the tribal slot. Comparative Studies in Society and History 42, 149–179.

Li, T., 2007. The Will to Improve. Governmentality, Development, and the Practice of Politics. Duke University Press, Durham, NC.

Locher-Scholten, E., 2003. Morals, harmony, and national identity: 'companionate feminism' in Colonial Indonesia in the 1930s. Journal of Women's History 14 (4), 38–58.

Maassen, C.C.J., 1939. Javanese agricultural migration to the outer provinces. Bulletin of the Colonial Institute of Amsterdam Volume II (3), 183–192.

McCarthy, J.F., 2004. Changing to gray: decentralization and the emergence of volatile socio-legal configurations in Central Kalimantan, Indonesia. World Development 32 (7), 1199–1223.

McCarthy, J.F., Cramb, R.A., 2009. Policy narratives, landholder engagement, and oil palm expansion in the Malaysian and Indonesian frontiers. The Geographical Journal 175 (2), 112–123.

Moore, D., 2005. Suffering for Territory: Race, Place and Power in Zimbabwe. Oxford University Press, Oxford.

Newberry, J., 2006. Backdoor Java, State Formation and the Domestic in Working Class Java. Broadview Press, Peterborough, Ontario.

Nightingale, A., 2006. The nature of gender: work, gender and environment. Environment and Planning D: Society and Space 24, 165–185.

Nightingale, A., 2011. Bounding difference: intersectionality and the material production of gender, caste, class and environment in Nepal. Geoforum 42 (2), 153–162.

Ong, A., Peletz, M., 1995. Introduction. In: Ong, A., Peletz, M. (Eds.), Bewitching Women, Pious Men. Gender and Body Politics in Southeast Asia. University of California Press, Berkeley, pp. 1–18.

Pain, M., Benoit, D., Levang, P., Sevin, O. (Eds.), 1989. Transmigration and Spontaneous Migration in Indonesia. ORSTOM, Bondy, France.

Peluso, N.L., 1992. Rich Forests, Poor People: Resource Control and Resistance in Java. University of California Press, Berkeley.

Peluso, N.L., 1996. Fruit trees and family trees in an anthropogenic forest: ethics of access, property zones and environmental change in Indonesia. Comparative Studies in Society and History 38, 510–548.

Peluso, N.L., 2009. Rubber erasures, rubber producing rights: making racialized territories in West Kalimantan, Indonesia. Development and Change 40 (1), 47– 80.

Peluso, N.L., Vandergeest, P., 2001. Genealogies of the political forest and customary rights in Indonesia, Malaysia and Thailand. The Journal of Asian Studies 60 (3), 761–812.

Pelzer, K.J., 1945. Pioneer Settlement in the Asiatic Tropics. American Geographical Society, New York.

Rankin, K., 2001. Governing development: neoliberalism, micro-credit and rational economic woman. Economy and Society 30 (1), 18–37.

Razavi, S., 2003. Introduction: agrarian change, gender and land rights. Journal of Agrarian Change 3 (1/2), 2–32.

Resurreccion, B., Elmhirst, R. (Eds.), 2008. Gender and Natural Resource Management: Livelihoods, Mobility and Interventions. Earthscan, London.

Ribot, J.C., Peluso, N.L., 2003. A theory of access. Rural Sociology 68 (2), 153–181.

Robinson, K., 2009. Gender, Islam and Democracy in Indonesia. Routledge, London.

Rose, N., 1999. Governing the Soul: The Shaping of the Private Self, second ed. Free Association Books, London.

Rutherford, S., 2007. Green governmentality: insights and opportunities in the study of nature's rule. Progress in Human Geography 31 (3), 291–307.

Schalkwijk, W.C., 1921. De kolonisatie met van Java afkomstige gezinnen in de Residentie Lampongsche Districten over het jaar 1920. Koloniale Studien 5 (1), 219–240.

Schroeder, R., 1999. Shady Practice: Agroforestry and Gender Politics in the Gambia. University of California Press, Berkeley.

Sikor, T., Lund, C., 2009. Access and property: a question of power and authority. Development and Change 40 (1), 1–22.

Sivaramakrishnan, K., 1999. Modern forests: State-Making and Environmental Change in Colonial Eastern Indonesia. Oxford University Press, Oxford.

Smyth, I., 1991. The Indonesian family planning programme: a success story for women? Development and Change 22, 781–805.

Sowerwine, J.C., 2004. Territorialisation and the politics of highland landscapes in Vietnam: negotiating property relations in policy, meaning and practice. Conservation and Society 2 (1), 97–136.

Stoler, A.L., 1985. Capitalism and Confrontation in Sumatra's Plantation Belt, 1870–1979. Yale University Press, New Haven.

Stoler, A.L., 1992. Carnal Knowledge and Imperial Power: Race and the Intimate in Colonial Rule. University of California Press, Berkeley.

Sundberg, J., 2004. Identities in the making: conservation, gender and race in the Maya Biosphere Reserve, Guatemala. Gender, Place and Culture 11 (1), 43–66.

Suryakusuma, J., 1996. The state and sexuality in New Order Indonesia. In: Sears, L. D. (Ed.), Fantasizing the Feminine in Indonesia. Duke University Press, Durham, NC, pp. 92–119.

Suyanto, S., 2004. Fire, Deforestation and Land Tenure in the North-Eastern Fringes of Bukit Barisan Selatan National Park, Lampung. <http://www. cifor. cgiar. org/ fire-project/doc/sos1>.

Ter Haar, B., 1948. In: Hoebbel, E.A., Schiller, A.A. (Eds.), Adat Law in Indonesia. Institute of Pacific Relations, New York (translated from Dutch).

Thomas-Slayer, B.P., Rocheleau, D.E., 1995. Research frontiers at the nexus of gender, environment and development: linking household, community and ecosystem. The Women and International Development Annual 4, 79–116.

Tsing, A.L., 2004. Friction: An Ethnography of Global Connection. Princeton University Press, Princeton.

Utomo, K., 1967. Villages of unplanned resettlers in the subdistrict Kaliredjo, Central Lampung. In: Koentjaraningrat (Ed.), Villages in Indonesia. Cornell University Press, Ithaca, pp. 281–298.

Van Doorn-Harder, P., 2006. Women Shaping Islam: Reading the Qur'an in Indonesia. University of Illinois Press, Urbana and Chicago.

Vandergeest, P., Peluso, N., 2006a. Empires of forestry: professional forestry and state power in Southeast Asia, part 1. Environment and History 10, 31–64.

Vandergeest, P., Peluso, N., 2006b. Empires of forestry: professional forestry and state power in Southeast Asia, part 2. Environment and History 12, 359–393.

Velayutham, S., Wise, A., 2005. Moral economies of a translocal village: obligation and sharing are important. Global Networks 5 (1), 27–47.

Whitehead, A., Tsikata, D., 2003. Policy discourses on women's land rights in Sub-Saharan Africa: the implications of the return to the customary. Journal of Agrarian Change 3 (1 and 2), 67–112.

Wieringa, S., 2002. Sexual Politics in Indonesia. Palgrave Macmillan, Basingstoke.

Witasari, A., Beilin, R., Batterbury, S., Nettle, R., 2006. How Does Social Capital Matter in Managing Protected Forest? A Case of Indonesia. Paper Presented at the 11th Biennial Conference on Common Property: 'Survival of the Commons: Mounting Challenges and New Realities', 19–23 June, 2006, Bali, Indonesia.

Yeh, E.T., 2009. Greening western China: a critical view. Geoforum 40 (5), 884–894.

Part V

Forested landscapes and farming in Latin America

The two papers in Part V are particularly good at placing gender within wider cultural and sociopolitical contexts and their constitutive power relations.

Paulson's paper (newly published), which focuses on silvopastoral systems in Bolivia, but extends across the Andes, stands out in exemplifying the more recent interest in addressing men's situations as much as women's. She begins by highlighting the importance of concepts of masculinity as they affect both genders. She also situates her analysis (political ecological in orientation) using the concept of value chains. She shows changes for people as one moves analysis across scales from the local in remote dry forest areas into towns and beyond (and the reverse). Like Agarwal and Fortmann (both this volume), she shows how inequitable conditions – for both women and men – have led to ecological degradation. She particularly highlights hierarchical regimes of masculinity that build on indigenous men's participation in and success at wage labour as a fundamental marker of manliness (like many contributions in Bannon and Correia 2006; Dolan 2002; or Kandirikirira 2002). She stresses the shortcomings of an approach that focuses on 'women' or 'men' as homogenous groups and ignores the multiple social differences that exist within each of these groups. She adopts an intersectional approach that acknowledges how different axes of social differentiation intersect to position women or men in relation to socio-ecological systems (see also in this volume: Colfer, Elias, Sijapati Basnett, and Stevens Hummel; Colfer; Norgaard). Paulson's approach demonstrates how the labour and resources of indigenous, rural men are exploited by 'a different type of men' who are more capitalized and reside in urban areas. Hence, "unequal distribution of resources and responsibilities in local socio-ecosystems is only one link in chains of asymmetrical exchange in which not only women's labor and resources, but also men's, are exploited and degraded via gendered relations of power" (p. 273)

Paulson also recounts a variety of methods she has used – including participatory methods, participant observation, and project document reviews. These will be useful to readers less familiar with gender studies as she also discusses the dilemmas she encountered in their use.

Another strength is her placement of human activity within an ecological context, much more clearly than most papers in this collection. She addresses resilience, energy, net flows, and asymmetric exchanges that adversely affect both humans and the environment; and she puts these thoughts into historical context. Like Elmhirst, Li, and Rocheleau and Ross (all this volume), Paulson shows the power of formal state (and international) systems (in this case, of land tenure) to mould and shape local systems, to the disadvantage of local men and women.

Finally, she discusses the growth of networks that strive to overcome some of the historical patterns that have so disadvantaged local communities and contributed to ecological degradation – though such networks have yet to pay sufficient attention to the gender systems that reinforce and help to maintain these patterns. She concludes with an argument for incorporating men more meaningfully – to the advantage of all – in gender analysis and related policymaking (see also Chapter 1).

In the piece by Rocheleau and Ross (1995), set in the Dominican Republic, we are again reminded of the simplistic binaries (e.g., farm versus forest, production versus preservation) and triads discussed in Chapter 1. A central and problematic differentiation in this case is among home, natural spaces, and production spaces. These authors, like Arora-Jonsson (this volume), convey the power of 'scripts' such as 'the greening of the discourse' – narratives at a broader scale – as they affect rural women and men in particular locations. They vividly convey, using the *Acacia mangium* tree as an entry point, the ways that these narratives and the international development interventions and national policies in which they take hold can both and alternatively empower and/or disenfranchise rural peoples (see also Schroeder, this volume). Their emphasis on risks and distrust between local people and formal forest managers is reminiscent of the situation in California described by Norgaard (this volume).

Rocheleau and Ross' chapter is particularly strong in its historical and multiscalar perspective, which reveals long-standing conflicts around the use and meanings of the forest and trees at various scales. A focus on vertical relations illuminates how state and local interests in forest management diverge and escalate into physical violence, corruption, and the criminalization of everyday tree-based livelihood activities. In this context, *Acacia mangium* serves as a tool for the forestry department to extend its local reach and control the terms and process of tree production and sale. This draws attention to another important aspect of forest governance: the state's social construction of the 'environment' in ways that legitimize state actions in environmental governance (e.g., Peluso 1992; Agrawal 2005). At the level of the formal organization of the Rural Federation of Zambrana-Chacuey, Rocheleau and Ross describe the processes that both include and exclude women's representation, voice and access to services (see also Rocheleau 2015 and Agarwal, this volume). At the household level, a third, intersecting arena of struggle occurs as women and men compete in a material and ideological conflict about land and tree rights. One key element of this competition involves a 'masculine mystique' about timber that is reminiscent of conditions described by Reed in the Canadian case (this volume). This masculine mystique, fomented by NGO foresters, curtails women's direct engagement with *Acacia*, yet does not apply to fruit trees, which are considered compatible with women's capabilities and which they are free to manage. The authors argue that these conflicts "articulate with conflicts of class, locality, occupation and organizational affiliation" (p. 412 in original) as contingent social relations are "embedded in the everyday practices of agriculture, forestry and conservation" (p. 424, original).

Rocheleau and Ross convey as well the value of trees as indicators of ownership. Planting *Acacia* strengthens local (mostly men's) rights to land in the eyes of the state; and in some cases, this seems to serve as a bulwark against land grabs by large-scale private industry. This is a more pleasing outcome than Colfer observed with this tree in East Kalimantan (Colfer *et al.* 1997), where it more closely resembled what Rocheleau and Ross, in a subsequent section, call a "green machete" – highlighting the deforestation that was resulting from *Acacia*'s introduction. The Federation convinced government foresters to legalize over a dozen other timber species but Acacia still threatens the bio-diverse patio and polyculture gardens that women plant, the "wealth of women" (p. 423, original).

References

Agrawal, A. 2005. Environmentality: Community, Intimate Government, and the Making of Environmental Subjects in Kumaon, India. *Current Anthropology* 46 (2): 161–190.

Bannon, I., and Correia, M. C. eds. 2006. *The Other Half of Gender: Men's Issues in Development.* Washington, DC: The World Bank.

Colfer, C. J. P., Peluso, N. L., and Chin, S. C. 1997. *Beyond Slash and Burn: Building on Indigenous Management of Borneo's Tropical Rain Forests.* Bronx, NY: New York Botanical Garden.

Dolan, C. 2002. Collapsing Masculinities and Weak States – A Case Study of Northern Uganda. In F. Cleaver, ed., *Masculinities Matter! Men, Gender and Development.* New York: Zed Books, 57–83.

Kandirikirira, N. 2002. Deconstructing Domination: Gender Disempowerment and the Legacy of Colonialism and Apartheid in Omaheke, Namibia. In F. Cleaver, ed., *Masculinities Matter! Men, Gender and Development.* New York: Zed Books, 112–137.

Peluso, N. L. 1992. *Rich Forests, Poor People: Resource Control and Resistance in Java*, Berkeley, CA: University of California Press.

Rocheleau, D., and Ross, L. 1995. Trees as Tools, Trees as Text: Struggles over Resources in Zambrana-Chacuey, Dominican Republic. *Antipode* 27 (4): 407–428.

Rocheleau, D. 2015. A Situated View of FPE from My Networks, Roots and Territories. In W. Harcourt and I. L. Nelson, eds., *Practicing Feminist Political Ecologies.* London: Zed Press, 29–66.

References

12 Changing masculinities and femininities in the (re)production of Andean silvopasture systems

Susan Paulson

Research and practice related to gender and environment have often sought to address the marginalization of women, in relation to men in their households and communities, in terms of control of natural resources and participation in environmental governance and decision-making. This chapter advances a complementary approach that is different in key ways. First, it integrates gender attention to variously positioned men and masculinities with more conventional approaches focused on women. Second, it considers interactions and influences of gender dynamics operating in various times and spaces, including along value chains that reach into communities of concern. This more holistic approach aims to help understand and strengthen ongoing roles that gender systems play in sustaining a balance between production and reproduction and in supporting resilience in the face of change.

Studies of forest management have made unique contributions to understanding gender and environment by drawing attention to diverse regimes of tenure and access, including many forms of commons, and by revealing culturally distinct practices and meanings, including symbolic and spiritual aspects of people–forest relations. Especially relevant here are those analyses that look beyond individual actors to examine historical and institutional processes through which particular forest regimes are reproduced, degraded, or transformed. Asking how women's roles in governance interact with forest health, biomass, and biodiversity, Bina Agarwal (2001, 2) suggests that "excluding women (often the principal users of community forests) from a group's decision-making bodies could have a range of negative efficiency fallouts," and she later demonstrates in parts of India and Nepal "that groups with a high proportion of women in their executive committee (EC)—the principal decision-making body—show significantly greater improvements in forest condition" (2009, 2785).

In hearty agreement that strengthening women's access and agency can—in many contexts—contribute to building healthier forests and communities,

Susan Paulson, "Changing Masculinities and Femininities in the (Re)production of Andean Silvopasture Systems," Chapter 12 in the *Earthscan Reader on Gender and Forests* (this volume). Edited by Carol J. Pierce Colfer, Marlène Elias, Bimbika Sijapati Basnett, and Susan Stevens Hummel. Published by Routledge.

I also draw attention to potential benefits of addressing gender-based constraints, challenges, and opportunities for change faced by men, and particularly by certain groups of men. The chapter begins with a brief description of socio-ecological changes in the central Andes between 1985 and 2005, a period marked by troubling developments among forested areas and women residents. The next section discusses analytic and research methods together with definitions of key concepts: forests, gender, masculinities, intersectionality, (re)production, and resilience. Then, with the goal of demonstrating that the conditions and positions of women and forests within their communities do not tell the whole gender story, the rest of the chapter builds a more comprehensive analysis by considering masculinity and by expanding the spatiotemporal frame. The case is part of a multiphased study whose methods and findings have been discussed elsewhere, including Paulson (2015, forthcoming), from which I draw materials, ideas, and text passages.

Exposed roots and exhausted women

In households across the Andes, gender has long interacted with kinship and production systems in the organization and sustenance of numerous species and varieties of plants and animals, geographically distant and ecologically dissimilar spaces, different types of economic relations, and a wide gamut of knowledge and technical skills. This chapter focuses on an arid, mountainous part of Cochabamba, Bolivia, where indigenous communities have farmed and herded for centuries in the context of evolving biophysical environments and sociocultural systems.[1]

In the communities studied, agriculture is generally considered a masculine domain and is, in practice, the principal focus of most men's labor (which does not mean that agriculture is not the principal activity of some women—it definitely is). Most women are more closely associated, both practically and symbolically, with herding (sheep, goats, some cows) and with collection in non-cultivated areas. I refer to this set of practices as "silvopasture," one among various types of forest regimes explored in this volume. Gender norms and perspectives surrounding silvopasture are communicated in many ways, such as the resistance of a teenage son who began to balk at pasturing sheep, saying that only women and children, not men, should be on the hillsides with the herds, and the needs of a widower who explained that after his wife's death, he needed to transport manure from his daughter's household to maintain the fertility of his soil. Although these and other expressions associate silvopasture with femininity and agriculture with masculinity, activities and skills in these realms are certainly not "sex segregated"; on the contrary, connections and interactions are built into nearly every aspect, and arrangements vary by season, with migration, and according to other factors.

Forests, in this context, take the form of hedgerows and borders along cultivated fields, paths, and canals; irregular rocky slopes covered with greenery; and forested ravines in steeper parts of the watershed. Residents interact with and talk about diverse species of Prosopis and Acacia, *Tipuana tipu*, *Ceiba* sp., *Jacaranda mimosifolia*, *Alnus glutinosa*, *Buddleja coriacea* Remy and *Dondonea viscosa*, among others. Gendered identities, decisions, and practices affect the density, diversity, and distribution of shrubs, trees, and grasses and also the institutional arrangements governing them. Prefiguring current global discourses about ecosystems services, local men and women identified forest features as physical barriers against wind and water erosion; shade for humans, animals, and crops; biological barriers against crop-specific pests and diseases; and nesting places for birds and animals. They also valued them for producing forage, fuelwood, green fertilizer, fruits, roots, and medicinal and culinary herbs.

During the 1980s and 1990s, this panorama was marked by troubling changes: stark erosion exposed tree roots, biomass diminished on hillsides, and torrents of rainwater carried away topsoil and created deep gullies (Calvert and Alandia 1993; Paulson and Schultz 1995). A number of governmental and nongovernmental organizations responded with programs that provided native tree seedlings, supported the planting of eucalyptus trees to reduce demand on native wood, promoted cultivation of alfalfa fodder to relieve stress on forage, and introduced improved stoves and bioenergy converters to reduce fuelwood consumption. These forest conservation strategies did not explicitly consider gender, and some didn't involve women at all.

During the same period, several rural development organizations addressed "gender issues" including perceptions that women were overworked and undervalued. These initiatives ranged from workshops aimed at bolstering women's self-esteem to projects supporting women to engage in money-making opportunities such as weaving or commercial agriculture. For the most part, these gender initiatives had little to do with silvopasture or with men.

Via involvement with the Food and Agriculture Organization's (FAO's) Forest, Trees and People Project and with Bolivia Sustainable Forest Management Project (see Paulson 1998, 2000) in the 1990s, I struggled to see and describe processes through which gender practices and meanings interact with and impact the biophysical world. Research with attention to gender and forestry pointed to interconnections among troubles affecting women residents and troubles in ecosystems they managed. In relation to men in these communities, women controlled less than an equal share of certain assets, including land and irrigation water, and exercised less than equal power in decision-making and resource allocation, particularly in formal organizations and relationships with outside actors. My analysis suggested that men-led commercial agriculture ventures were being subsidized by the provision of animal and green manure, fuelwood, food, and other uncommodified fruits of women's labor in asymmetrical processes that drained ecological energy away from silvopastoral systems, undermining work identified as feminine.

Although these observations are relevant, I do not now find them sufficient to understand or respond in sustainable ways to the gendered forest dynamics in question. A look at value chains that reach into these rural watersheds reveals that men residents were also subjected to unequal exchanges that functioned over time to draw energy of human and other natural resources away from rural socio-ecologies altogether. Importantly, these exploitative relations were shaped and justified by hierarchical regimes of masculinity and by the establishment of economic production, particularly in the cash economy, as a fundamental marker of manliness. In order to contribute to this volume's goal of advancing more balanced and holistic approaches, later sections of this chapter strive to locate my community-level research on gender and forestry within these wider-scale and longer-term considerations.

Analytic and field research methods

Political ecology analysis works together with mixed-methods research to guide my attention to a diversity of sociocultural and ecological phenomena and to prompt questions about how they relate to each other. Rather than looking exclusively at silviculture or at agriculture, for example, I explore interdependencies and mutual impacts provoked by changes in each. Rather than focusing a gender lens on women, I look at roles and identities of differently positioned men and women, asking about changing relations among them. The multi-sited frame of investigation portrayed here encompasses not only local socio-ecosystems, but also near and distant actors and landscapes, urban food markets, and national/international interests and initiatives. Finally, a focus on social difference and power draws attention to asymmetrical relations in all kinds of exchanges involving natural resources, produce, pay, respect, ideas, and development visions.

Scholars contributing to this volume richly represent conversations that have, over the years, informed my analytic and field methods. Dianne Rocheleau (2008, 716) traces political ecology's unflinching commitment to linking theory with empirical observation of biophysical and socioeconomic phenomena, with practical intent to contribute to material and social change. Andrea Nightingale (2011, 154) describes methodological steps toward understanding embodied performances of gender and caste as inseparably material and symbolic: "Attention to everyday, seemingly mundane, spatial practices gives insight into how people produce a particular relationship with 'others' including their environments, that are rarely ecologically neutral." And Melissa Leach (2007, 82) argues for deeper dialogue between academic studies and lagging policy in which "Issues of rights and resource access and control are now acknowledged, but not necessarily in relation to gender, and rarely through the relational, multilayered lens which feminist political ecologists and gender analysts of land have seen as important." Congruent with Damayanti Banerjee and Michael Mayerfeld Bell's (2007, 3) call for "eco-gender studies," this project advances "a relational and dialogic conception of gender and its intersection with other inequalities, and departs from those ecofeminist strands that sacralize an essentialist 'nature' and romanticize non-Western traditions."

Information and understandings expressed here were accumulated through numerous phases of field research that I carried out between 1985 and 2005, including 18 months residence in a Quechua farming and herding community amid nearly a decade of residence in Bolivia. Participating with local men and women in many forms of daily labor, I also observed farm-gate sales, documented transport of passengers and crops, and accompanied residents as they traveled to sell crops in urban markets and to seek migrant work.

Several short-term consultancies provided opportunities to collaborate with organizations working in the area and to access field records and studies that included information—some of it spanning decades—on trees, soils, crops, and animals; diseases and other problems that affected them; and biodiversity and biomass in grazing areas. In those records, loss of native vegetation and erosion appeared as increasingly urgent concerns. One organization fenced off sectors of hillside to measure regeneration of biomass and plant diversity in relation to areas used for pasture and collection; another collaborated with farmers to document soil erosion by planting metered stakes in the ground.

One phase of my research involved working with a team to implement methods designed to study gender and resource management. Instruments included focus groups to catalog men's and women's activities on daily hour sheets and annual calendars, participatory construction of resource and species maps, and day-long transect walks guided by diverse residents. Our efforts to apply gender-focused methodologies generated useful findings; they also led to confusions and debates that are still relevant today. Limitations in the ways our instruments construed human–environment relationships became evident in responses like that of a woman who remarked, "What do you mean, do I *control* or *use* these sheep? My duty is to assure that the herd is healthy and reproduces." Our tenure categories were especially inadequate for capturing the relationships our interlocutors' maintained with trees. While pasturing on a rocky slope, one woman observed, "I always stop here at my algarrobos [*Ceratonia siliqua*] so the sheep can forage. I've known that big tree since I was a girl, and when this smaller one sprouted, I made sure to keep the animals away from it, and pulled up nasty plants that threatened it." During a break in weeding corn, an elderly man declared: "I'm in my field all day. Here, nowhere else. This is my place, where I feel good. The house, that's for women. I even eat my lunch here, and take my siesta right here, under that molle tree [*Schinus molle*]."

Our research tools were also limited by the assumption of binary gender categories: men and women. When asked to describe women's resources and responsibilities, few residents found it easy to generalize. Several asked, "What woman? A young single woman [*sipas*]? Or a mature woman with family [*warmi*]?" Others differentiated the rights and responsibilities of widows and single mothers from those of married women, suggesting that the universalizing category "woman" does not coincide with different gendered resource-management roles at play in the community. These experiences complement those in other contexts around the world to provoke adjustments in conceptualizations and operationalization of "gender" as well as "forests," some of which we explore now.

Evolving concepts

"Forests" are complex environments shaped and changed through interactions among many biotic and abiotic elements, including trees. Scientists and conservationists classify types (such as rainforests, boreal forests, or mangrove forests) and work to understand and conserve them for purposes including biodiversity, habitat, and ecosystem services. Different scientific efforts have supported commercial forestry, aiming to maximize production and harvest of timber and other forest products, usually through the cultivation of trees (see Scott 1999, 11–52). A third approach, community forestry, highlights the roles of local actors in management and decision-making about forest resources and extends professional interest to a wider range of biophysical characteristics and management practices (see cases collected in Colfer 2005).

Forests are often conceptualized without humans, making it difficult to see or think about social identities and relations that influence their characteristics. A few scholars have explored mutual influences between commercial forestry and manliness, notably in Scandinavia (Brandth and Haugen 2005; Östman 2008); however, little attention has been paid to masculinities in relation to any kind of forestry in the global South. To date, most gender attention has focused on women in community forest management.

For me, however, gender is not (only) about women. I conceptualize "gender" as a sociocultural system that organizes practices and relationships that play out among humans, and between humans and their environment, infusing them with power and with meanings that refer symbolically to sex and sexuality. Operating through various contexts and scales, gender systems influence a range of institutions, together with the distribution and use of different assets, in ways that impact infrastructure and biophysical environment as well as human bodies, lives, and livelihoods. My thinking about gender has evolved in conversation with theorization of "territory" as a socio-ecological space that is continually produced and reproduced via power relations playing out among differentiated groups and individuals (Manzanal, Arzeno, and Nardi 2011; de Souza 2009; Schejtman and Berdegué 2007). Territoriality draws attention to dynamics that work to sustain, or to transform, the sociocultural institutions and biophysical features that interact in and on each space, including power relations that encompass conflict and synergy, competition and complementarity, inequity and opportunity.

I also expand the scale of analysis beyond the physical forest and its local managers. In contrast with longstanding tendencies to focus gender analysis on closer scales (bodies, households, communities), this approach coincides with efforts in new feminist political ecologies to "see gender as a constitutive force at all scales of analysis" (Elmhirst 2011, 131) and to ask how social hierarchies become materially manifest on the landscape (Nightingale 2011, 154). The investigation of value chains outlined later in this chapter activates synergies between ideas of gendered power relations shaping forests and ideas about unequal distribution and exchange (Escobar 2006; Walter and Martinez-Alier 2012).

Scholarship and practice that have looked at society, economy, and environment with attention to gender have challenged dominant interpretations of the world and changed actions upon it. While this work has mostly focused on women and women's issues, it has also laid groundwork for growing attention to "masculinities," which I conceptualize as constellations of qualities, behaviors, attitudes, and accomplishments that—within particular communities of interpretation—are associated with the category "man" or subcategories of men. Within the varied and rapidly evolving field of masculinities studies, I mention a few moves that can support more holistic approaches to gender–forest dynamics.

At the dawn of the twenty-first century, some voices began to identify the widespread failure to consider men as gendered actors as a barrier to the achievement of objectives of gender and development (Chant and Gutmann 2000; Cleaver 2003; Cornwall and White 2000). While subsequent work has made invaluable contributions (Cornwall, Edström, and Greig 2011; Bannon and Correia 2006), masculinities studies have become institutionalized in separate academic programs, conferences, and literature, resulting in surprisingly little integration with ongoing concerns with gender and environment.

Moves toward that integration can be supported by key moves identified by Australian sociologist Raewyn Connell (1998, 3): a conceptual shift from singular "man" to plural "masculinities," attention to hierarchies and hegemonies, and multi-scale analysis. Also key are challenges to the gender binary. A great deal of human expression observed in this heterogeneous world has been organized according to binary categories such as white versus nonwhite, economically active versus inactive, productive versus reproductive, and of course, men versus women. Summing up the complex historical/colonial processes leading to this conceptual map, María Lugones (2010, 742) writes, "Modernity organizes the word ontologically in terms of atomic, homogeneous, separable categories." The use of consistent categories in census and survey instruments clearly facilitates efforts to organize, present, and compare findings. It is also clear that these come up short in efforts to represent empirical realities observed in diverse contexts around the world. Studies showing different forest access and rights for married women versus singles, or for women of different ethnoracial or socioeconomic identities, have contributed to ongoing shifts away from addressing "women" as a monolithic group. Practical relevance of such distinctions is evident in Marlène Buchy and Bimala Rai's (2013) conclusion that women's forestry initiatives in Nepal had disproportionately served economically better-off and high-caste women, possibly exacerbating marginalization of other women.

One useful way of complicating the binary framework is by paying attention to ways in which gender intersects other differences relevant to a given issue (e.g., caste, ethnoracial, occupational, generational, religious, sexual, educational, ableness). The concept of "intersectionality" that Kimberlé Crenshaw (1991) introduced to make visible the unique challenges of those who suffer both racial and gender oppression has been adapted to study and support women marginalized in varying ways around the world (Yuval-Davis 2006). Rebecca Elmhirst (2011, 130) explores both possibilities and dangers of approaches that destabilize gender as the object of study, and "instead, emphasis is given to an exploration of multidimensional subjectivities where gender is constituted through other kinds of social differences and axes of power such as race, sexuality, class and place, and practices of 'development' themselves." Admitting that it has been much easier to identify either gender, caste, or ecology as the primary object of analysis than to address their intersection, Andrea Nightingale (2011, 153) works to "rectify this oversight by exploring the production of difference through the everyday movement of bodies in space to show how subjectivities are produced out of the multiple and intersecting exercise of power within socio-natural networks."

To date, most intersectional analysis has focused on variously marginalized or disadvantaged women. Recognition of and response to a wider range of actors and processes relevant to gendered forests requires expanding the scope to encompass attention to all positions within intersecting systems of opportunity and adversity. Ideas of hierarchical masculinities mentioned above, for example, allow us to see that indigenous-identified farmers are constituted differently from the businessmen and politicians who benefit from their labor. These differences are not reducible to economic power alone. They are experienced, embodied, and expressed through gender practices and meanings.

At the heart of sustainability for all human groups is the (re)production of human energy each day through food, rest, and care and the (re)production of human communities through the birth, socialization, and education of new generations. These are, of course, inseparable from the adaptive regeneration of trees, shrubs and associated biodiversity, soil fertility, and other aspects vital to nitrogen and hydrological cycles. "Resilience" involves finding ways to sustain the vitality of these socio-ecological processes in changing conditions. Geographer Karl Zimmerer (2014, 4) conceptualizes resilience in a way that resonates with the case examined here: "the capacity of agrobiodiversity-producing cultural landscapes, and the agroecosystems and food use they support, to respond to stressors and disturbances while maintaining structures and functional processes that include multifunctionality and some degree of socioeconomic power and autonomy."

Challenges of resilience raise ontological questions about processes through which a multiplicity of socio-natural worlds are produced and reproduced. In Dianne Rocheleau's (2011, 209) words, "We all live in emergent ecologies—complex assemblages of plants, animals, people, physical landscape features, and technologies—created through the habit-forming practices of connection in everyday life." Thinking about the co-creation of engendered forests provokes new conceptualizations of "mode of production" emphasizing that humans manufacture not only food, shelter, and clothing, but also biophysical landscapes including forests, together with regimes of production, consumption, and environmental knowledge and governance.

Space: masculine energy circulating through wider webs of power and meaning

Troubles like those described above are experienced in each household and community, but are not circumscribed there. Local socio-ecologies are impacted by gendered power relations playing out in other places. One of these is Bolivian agricultural markets where, during the 1980s and 1990s, the labor and natural capital that farmers invested in commercial crops was rarely met with returns adequate to sustain the socio-ecological conditions of production (Lagos 1994; Urioste 1987). Another is the seasonal and migrant employment through which men farmers sought to make up the deficit and where their labor was also exchanged for wages barely sufficient to sustain their own strength, let alone reproduce future generations of workers.

Asymmetrical exchanges between diversely positioned people, resources, and spaces have long been of concern in Marxist analyses and have been explored globally by dependency and world systems theorists (Frank 1967) and critical geographers (Harvey 2006). To see how these exchanges work to sustain—or to drain—specific socio-ecologies, some scholars are endeavoring to bring social science notions of value and exchange together with thermodynamic measures of the matter and energy that is transformed and transported (Bonaiuti 2014; Walter and Martinez-Alier 2012).

What thermodynamic processes are relevant for the silvopasture dynamics described above? As extracted fuelwood is burned, relatively ordered tree matter becomes relatively disordered carbon dioxide, methane, black carbon, and water vapor, representing release of potential energy and increase in entropy. As sheep and goat meat is eaten, most of the highly ordered proteins and fats work to power human bodies, where they are transformed into more disordered heat, sweat, etc., but some also function to regenerate muscle and build human bodies. For their part, manure and green fertilizer transported to agricultural fields interact with water, air, and solar energy to build well-ordered plant matter. In short, silvopasture contributes to the production of human bodies, labor, and agricultural crops via processes that continually extract and break down useful mass and energy from forest ecosystems. A basic challenge for sustainability in this and other contexts, then, is for human practice and knowledge to interact with ecological processes in ways that sufficiently regenerate biomass and energy in these areas, thus avoiding net degradation.

What about agriculture? Human energy is expended to transform soil, water, solar energy, and other inputs into well-ordered plant matter. The food farmers eat provides energy that is transformed into useful movement and material needed to regenerate blood, muscle, skin, etc. In the context discussed here, harvested crops were transported elsewhere and exchanged for money that was insufficient to regenerate all the mass and energy invested in production. When these farmers sold crops, and when they migrated to work elsewhere, part of their useful energy was channeled into economic profits enjoyed by employers and investors, a different kind of men. While remarkably low prices of food crops debilitated small farms and farmers during this period, they simultaneously bolstered the profitability of mines, factories, and agroindustries by allowing employers to maintain low wages. Displaced farmers who joined other men in exchanging their labor for those low wages contributed doubly to competitive production and export.

Without calculating the specific joules, calories, dollars exchanged here, I point to biophysical and social impacts that suggest asymmetries in the net flows. Zimmerer (1993, 1659) observes that "soil erosion in the Bolivian Andes worsened during recent decades (1953–91) due to changes in production as peasants shifted labor from conservation techniques to nonfarm employment," and cumulative impacts of degradation impelled an exodus from highland communities as people relocated to growing tropical colonies and urban satellites (Dandler 1987; Ledo 2002).

In other parts of the world as well, chains of asymmetrical exchanges have facilitated net flows of ecological mass and energy, together with economic value, away from commonly managed agroforestry and silvopastoral spaces and the people who look after them. It is often argued that rural hinterlands and indigenous women suffer the most degradation from socio-ecological dynamics of recent decades. However, if "women" and "nature" are the losers, our case certainly does not suggest that "men" and "culture" are winners, at least not those in the communities studied.

Why did (and do) so many men engage in practices and relations that seem to exploit their labor and undermine their socio-ecosystems? Among other factors to consider are norms and relations in prevailing regimes of masculinity. How have expanding associations between manliness and income influenced farmers to seek ways of earning money, even at tremendous human and environmental costs? How have balances between production and reproduction been affected by historical processes in which men have been labeled "producers" and received technical and financial support in their work with cash crops, while women have been associated with "reproduction" and spuriously associated with the modern housewife?

We also need to examine hierarchies of power and value among differently positioned masculinities. The grueling labor conditions that rural men endure and the unfair exchanges to which they submit in both product and labor markets are not simple functions of supply and demand—far from it. They are socially constituted and symbolically justified via intersections of gender, spatial, and ethnoracial systems that attribute lower value to rural indigenous masculinities relative to more privileged masculine identities. Whereas certain gender expectations and values may drive entrepreneurs to prove their business-manliness through specific forms of wealth, options for proving subordinate manliness include displays of toughness and endurance in the face of hardship, pain, and risk (as José Eduardo Calvario Parra [2011] discusses in his aptly titled book *¡Aguántate como hombre!* [Take it Like a Man]).

In sum, community-based studies carried out with the kind of gender–environment methods described at the beginning of this chapter (e.g., labor calendars, resource maps, transects) often point to inequitable gender arrangements in local culture as causes of women's exhaustion and of forest degradation. This broader frame leads to a quite different analysis in which unequal distribution of resources and responsibilities in local socio-ecosystems is only one link in chains of asymmetrical exchange in which not only women's labor and resources, but also men's, are exploited and degraded via gendered relations of power.

Time: historical development and dissemination of gender–environment models

Many discourses and approaches portray aspects of "nature" and "gender" as unchanging givens, making it difficult to even ask how gendered forests change in relation to forces and factors that start before and extend beyond them. Sometimes deep assumptions about change and its causes lead people to overlook empirical evidence that contradicts dominant narratives, as Fairhead and Leach (1996) demonstrate in their landmark study of forest-savanna mosaic in Guinea. Without pretending to provide a comprehensive history, this section raises some deeper historical questions about Bolivia's gendered forests.

Relations interconnecting Andean men, women, and forests evolved slowly over millennia, punctuated by events such as the Inca expansion (Silverblatt 1987), Spanish colonial resettlements, and mining drafts (Mangan 2005; Zulawski 1990). Archaeological and historical records show that in some periods, highland Cochabamba was more heavily forested, in others quite bare, with shifting tree populations, all of which Daniel Gade (1999, 42–74) connects to changing modes of production, forms of governance, population density, and material culture.

By the time of independence from Spain in 1825, significant portions of highland Bolivia were organized into feudal-like production units called *haciendas*. Although arrangements varied, a common feature was the appropriation of indigenous men's labor in commercial agriculture ventures designed to generate profit for Euro-identified landowners. Local gender systems adjusted as indigenous men's practices and places changed, together with meanings of masculinity, and women sought and managed a variety of peripheral resources in efforts to reproduce human and other resources (Larson 1988).

Hacienda arrangements demonstrate just one application of a historical model that has had tremendous impacts on gendered forests worldwide: the conceptual and institutional divorce of those activities identified as "productive" from those designated "reproductive," the association of the first with masculinity and the second with femininity, and the investment of greater value, resources, and power in masculine-identified production. Interacting with other aspects of expanding productivist economies (notably private property and the commodification of human labor and natural resources), this model fostered the inferior valuation and marginalization of labor and resources that would come to be known as commons, reproductive, and/or feminine, notably those supporting forms of silvopasture and agroforestry.

Bolivia's 1953 agrarian reform advanced this institutional model in bold new ways with the goal of modernizing rural production. First, former *hacienda* and community lands were redistributed via campaigns that titled land almost exclusively to men. Second, a state-supported network of peasant unions was established with the participation of only those residents identified as "male heads of households." In *Seeing Like a State: How Certain Schemes to Improve the Human Condition Have Failed*, political scientist James Scott (1999, 3) points to long-term implications of these kinds of abridged maps that leave out relevant spaces, actors, and elements.

> They did not successfully represent the actual activity of the society they depicted, nor were they intended to; they represented only that slice of it that interested the official observer. They were, moreover, not just maps. Rather, they were maps that, when allied with state power, would enable much of the reality they depicted to be remade. Thus a state cadastral map created to designate taxable property-holders does not merely describe a system of land tenure; it creates such a system through its ability to give its categories the force of law.

Indeed, impacts of these radically gendered new models for governing land-based resources were extended in following decades by countless agencies and actors who chose to work with official peasant unions and to link technical and financial support to landownership, thus bypassing older and more inclusive cultural traditions. Miguel Urioste (1987, 11) characterizes ensuing policies and programs as promoting land privatization and titling, expanding land area cultivated with commercial crops, and increasing yields per unit of land. I add that they focused almost exclusively on men, promoting commercial farming as masculine.

Following this historical trajectory, late-century alarms about the degradation of forests and the inferior social value of women who manage them should come as no surprise. However, there is far from consensus on historical causality. Powerful narratives continue to locate the causes of deforestation and of women's marginalization in traditional beliefs and practices of rural communities; from that perspective, the observed crises call for *more* modernizing development interventions. My causal reading suggests a quite different response, starting with critical assessment of assumptions about gender and environment that underlie processes of rural modernization and that have been disseminated with them throughout the world. During the second half of the twentieth century, national and international discourses, policies, and programs, together with media and scholarship, prioritized select sectors, spaces, and social groups. Agriculture was strengthened more than silvopasture; commercial production was encouraged more than subsistence; and monoculture of improved varieties replaced diverse species and varieties. Individual production on private plots garnered more attention than cooperative management of communal spaces, and activities and efforts associated with masculinity were strengthened more than those associated with femininity.

In Bolivian communities, as some areas formerly used for grazing and collecting were enclosed and others became degraded and eroded, I watched women intensify fuelwood collection and take their herds longer distances to find forage. These changes took a toll on the health and vitality of forests, livestock, and women. Some found it necessary to sell livestock or to arrange for it to be pastured in other communities. Reduced access to manure limited women's contributions to regenerating soil fertility while reduced access to milk and meat hampered efforts to sustain and regenerate family labor through nourishing food. Although many women responded to changes by investing more time and energy in their work, residents expressed concerns that women were failing to fulfill gendered expectations. One man complained, "The cost of buying fertilizers and food is constantly increasing, and we men have to earn more and more money to buy these things that everyone used to make at home." Among testimonies of women: "My husband attacks me because I don't cook well, like people used to." "My mother-in-law criticizes me because I don't make my herd multiply like she did." "My child got sick because I was off herding on a distant mountainside."

By the end of the century, growing awareness of these and other symptoms of socio-ecological imbalance helped to provoke new perspectives and initiatives. These include a range of adjusted paths under the banners of "gender and development" and "sustainable development" in addition to popular and scientific movements that question certain tenets of modern development more deeply and contest them more radically. Coexistence of these different visions is increasingly embraced as positive, and tension among them as fruitful, in growing accordance with Carol Colfer's (2005, 14) assertion that "Different individuals, groups and cultures have their own trajectories, and the global diversity this implies is a valuable resource, as well as a type of insurance for humanity."

Latin America's organizational and political landscapes include actors and organizations building alternatives to models they critique as culturally imperialist and environmentally destructive. In Perú, The Andean Project for Peasant Technologies (PRATEC) has worked for decades to recover and revitalize agrosilvopastoral and other cultural practices of Andean and Amazon communities. In Bolivia, the Andean Oral History Workshop (THOA) has been collaborating with the National Council for Ayllus and Markas of Qullasuyu (CONAMAQ) to reclaim historic territory, community, and forms of governance. The Campesino a Campesino farmers' movement that originated in Central America (Holt-Gimenez 2006) together with the worldwide Via Campesina network explore strategies for disengaging from market exploitation and input dependency in collaborative experimentation with paths to sustainable local (re)production and food sovereignty. These and other groups have connected creatively in forums like The World People's Conference on Climate

Change (Bolivia, 2010) and the People's Summit on Climate Change (Perú, 2014) (see Fabricant 2013). Networks of actors and organizations working across boundaries of local and scientific knowledge have also broadened the horizon. Their contributions include new attention to commons management, exemplified in decades of teamwork behind Elinor Ostrom's Nobel Prize, and advancement of what Miguel Altieri and Victor Manuel Toledo (2011) tout as "The Agroecological Revolution in Latin America." Even scientists focusing on nonhuman ecology are beginning to pay attention to traditional strategies for integrated management of diverse species and spaces. As Ivette Perfecto and John Vandermeer (2008, 187) summarize, "a broader landscape-level approach to biodiversity conservation in recent years has highlighted the role of silvopastoral systems as a matrix that can maintain biodiversity and facilitate the movement of organisms between patches on natural habitat."

By valuing culture-specific traditions of environmental management and promoting integral socio-ecology approaches, these and other initiatives enrich options for understanding and responding to the kinds of troubles discussed in this chapter. As a whole, however, they have not yet paid much attention to the constructive roles that gender systems have long played and can play in the (re)production and resilience of these systems.

Toward more balanced and resilient socio-ecosystems

I have tried to motivate new ways of seeing and thinking about gendered forests by exploring processes through which silvopasture in one Andean context has evolved in interaction with men's and women's practices and relations, together with meanings of femininity and masculinity playing out on various scales and places.

The chapter raises questions about why masculinities and femininities are so rarely considered together in relation to environment and development issues. Part of the response can be found in epistemological and institutional approaches that have associated gender with women, and particularly women's vulnerabilities, limiting the apparent usefulness of gender analysis to scientists and professionals who are working to improve forest or other environmental management. Also relevant are methodological approaches that have focused on individual actors in narrowly circumscribed contexts. These factors interact with ongoing resistance (exercised more and less consciously) by variously positioned men and women to integrating men and masculinities into gender research and practice.

In recent books, I challenge widespread assumptions that men have nothing to gain from gender analysis and change, and counter the fear that attention to masculinities will undermine efforts to improve the conditions and positions of women (Paulson 2013, 2015). Recent data on educational achievement, health, life expectancy, and vulnerability to violence show that men, as a category, are increasingly disadvantaged in relation to women in every country in Latin America. Additional intersectional analyses demonstrate ways in which certain types of masculinities are subordinated and exploited via hierarchical masculinities basic to predominant gender systems. Growing challenges faced by many men and boys in education and employment, together with increasingly violent regimes of masculinity, do not make women's lives better. On the contrary, they pose significant barriers to improvements in the conditions of women and of the forest and other resources on which they depend. Contrary to discourses that portray gender as a zero-sum game in which "women" struggle to wrest resources from "men," comparative data from around the world demonstrate that gender systems can evolve in ways that reduce different kinds of gender-linked discrimination experienced by men and by women and that empower both groups. In sum, I argue that incorporating masculinities into gender and environment work will benefit not only men, but also women and ecosystems.

The kind of analysis demonstrated in this chapter, one that looks at the constitution and operation of differently positioned masculinities together with femininities in specific socio-ecological processes, has political implications very different from analyses focused on women's issues. The characterization of gender initiatives as striving to empower women by increasing their share of resources and powers relative to men in their households and communities might be replaced by a vision of men and women working together to question, innovate, and change aspects of gender identities and relations that have been functioning to marginalize and exploit each in different and interconnected ways and to forge socio-ecological systems that are more satisfying and sustainable for more people.

Note

1 This analysis is based on research carried out in various parts of Cochabamba Department and draws some material and passages from earlier publications (see, e.g., Paulson 2015, forthcoming). The main focus here is on the highland part of Carrasco Province, which has an average altitude of 2,800 meters and receives annual rainfall of 60 centimeters.

References

Agarwal, Bina. 2001. *The Hidden Side of Group Behaviour: A Gender Analysis of Community Forestry Groups*. QEH Working Paper Series – QEHWPS76. Oxford: University of Oxford.

———. 2009. "Gender and Forest Conservation: The Impact of Women's Participation in Community Forest Governance." *Ecological Economics* 68 (11): 2785–99.

Altieri, Miguel A., and Victor Manuel Toledo. 2011. "The Agroecological Revolution in Latin America: Rescuing Nature, Ensuring Food Sovereignty and Empowering Peasants." *Journal of Peasant Studies* 38 (3): 587–612. doi: http://dx.doi.org/10.1080/03066150.2011.582947

Banerjee, Damayanti, and Michael Mayerfeld Bell. 2007. "Ecogender: Locating Gender in Environmental Social Science." *Society and Natural Resources* 20 (1): 3–19. doi: 10.1080/08941920600981272

Bannon, Ian, and Maria C. Correia, eds. 2006. *The Other Half of Gender: Men's Issues in Development*. Washington, DC: World Bank. doi: 10.1596/978-0-8213-6505-2

Bonaiuti, Mauro. 2014. "Bioeconomics." In Giacomo D'alisa, Federico Demaria, and Giorgos Kallis, eds., *Degrowth: A Vocabulary for a New Paradigm*, 25–38. London: Routledge/Earthscan.

Brandth, Berit, and Marit S. Haugen. 2005. "Text, Body, and Tools: Changing Mediations of Rural Masculinity." *Men and Masculinities* 8 (2): 148–63. doi: 10.1177/1097184X05277716

Buchy, Marlène, and Bimala Rai. 2013. "Do Women-Only Approaches to Natural Resource Management Help Women? The Case of Community Forestry in Nepal." In Bernadette Resurreción and Rebecca Elmhirst, eds., *Gender and Natural Resource Management: Livelihoods, Mobility and Interventions*, 127–49. London: Earthscan.

Calvario Parra, José Eduardo. 2011. *¡Aguántate como hombre!: Daños a la salud, masculinidades y riesgos en los jornaleros agrícolas*. Córdoba: Editorial Académica Española.

Calvert, Donald, and Segundo Alandia. 1993. *Evaluacíon ambiental para el proyecto de Desarrollo Regional de Cochabamba (CORDEP)*. La Paz: USAID Proyecto.

Chant, Sylvia H., and Matthew Gutmann. 2000. *Mainstreaming Men into Gender and Development: Debates, Reflections, and Experiences*. Oxford: Oxfam.

Cleaver, Francis, ed. 2003. *Masculinities Matter! Men, Gender and Development*. New York: Zed Books.

Colfer, Carol J. P., ed. 2005. *The Equitable Forest: Diversity, Community, and Resource Management*. Washington DC: Resources for the Future and CIFOR.

Connell, Raewyn W. 1998. "Masculinities and Globalization." *Men and Masculinities* 1 (1): 3–23.

Cornwall, Andrea, and Sarah C. White. 2000. "Men, Masculinities and Development: Politics, Policies and Practice." *IDS Bulletin* 31 (2): 1–6. doi: 10.1111/j.17595436.2000.mp31002001.x

Cornwall, Andrea, Jerker Edström, and Alan Greig, eds. 2011. *Men and Development: Politicising Masculinities*. New York: Zed Books.

Crenshaw, Kimberlé. 1991. "Mapping the Margins: Intersectionality, Identity Politics, and Violence against Women of Color." *Stanford Law Review* 43 (6): 1241–99.

Dandler, Jorge. 1987. "Diversificación, procesos de trabajo y movilidad espacial en los valles y serranías de Cochabamba." In Olivia Harris, Brooke Larson, and Enrique Tandeter, eds., *La participación indígena en los mercados surandinos*, 639–82. La Paz: CERES.

de Souza, Marcelo Lopes 2009. O território: sobre espaço e poder, autonomia e desenvolvimento. In Iná Elias de Castro, Paulo Cesar da Costa Gomes and Roberto Lobato Correa, eds., *Geografia: conceitos e temas*, 77–116. Rio de Janeiro: Bertrand Brasil.

Elmhirst, Rebecca. 2011. "Introducing New Feminist Political Ecologies." *Geoforum* 42 (2): 129–32. doi: 10.1016/j.geoforum.2011.01.006

Escobar, Arturo. 2006. "Difference and Conflict in the Struggle over Natural Resources: A Political Ecology Framework." *Development* 49 (3): 6–13. doi: 10.1057/palgrave.development.1100267

Fabricant, Nicole. 2013. "Good Living for Whom? Bolivia's Climate Justice Movement and the Limitations of Indigenous Cosmovisions." *Latin American and Caribbean Ethnic Studies* 8 (2): 159–78. doi: 10.1080/17442222.2013.805618

Fairhead, James, and Melissa Leach. 1996. *Misreading the African Landscape: Society and Ecology in a Forest-Savanna Mosaic*. New York: Cambridge University Press.

Frank, Andre Gunder. 1967. *Capitalism and Underdevelopment in Latin America: Historical Studies of Chile and Brazil.* New York: Monthly Review Press.
Gade, Daniel. 1999. *Nature and Culture in the Andes.* Madison, WI: University of Wisconsin Press.
Harvey, David. 2006. *Spaces of Global Capitalism: A Theory of Uneven Geographical Development.* London: Verso.
Holt-Gimenez, Eric. 2006. *Campesino a Campesino: Voices from Latin America's Farmer to Farmer Movement for Sustainable Agriculture.* Oakland, CA: Food First Books.
Lagos, Maria Laura. 1994. *Autonomy and Power: The Dynamics of Class and Culture in Rural Bolivia.* Philadelphia, PA: University of Pennsylvania Press.
Larson, Brooke. 1988. *Colonialism and Agrarian Transformation in Bolivia: Cochabamba 1550–1900.* Princeton, NJ: Princeton University Press.
Leach, Melissa. 2007. "Earth Mother Myths and Other Ecofeminist Fables: How a Strategic Notion Rose and Fell." *Development and Change* 38 (1): 67–85. doi: 10.1111/j.1467-7660.2007.00403.xf
Ledo, Carmen. 2002. "Poverty and Urbanization in the Cities of the National Economic Corridor of Bolivia." PhD dissertation, Delft University.
Lugones, María. 2010. "Toward a Decolonial Feminism." *Hypatia* 25 (4): 742–59. doi: 10.1111/j.1527-2001.2010.01137.x
Mangan, Jane E. 2005. *Trading Roles: Gender, Ethnicity, and the Urban Economy in Colonial Potosí.* Durham, NC: Duke University Press.
Manzanal, Mabel, Mariana Arzeno, and María Andrea Nardi. 2011. "Desarrollo, territorio y desigualdad en la globalización. Conflictos actuales en la agricultura familiar del nordeste de Misiones, Argentina." *Mundo Agrario* 12 (23). www.mundoagrario.unlp.edu.ar/article/view/v12n23a04
Nightingale, Andrea. 2011. "Bounding Difference: Intersectionality and the Material Production of Gender, Caste, Class and Environment in Nepal." *Geoforum* 42 (2): 153–62. doi: 10.1016/j.geoforum.2010.03.004
Östman, Ann-Catrin. 2008. "The Peasants of the Forest: Masculinity, Nation, and Landscapes in Agrarian Historiography in Finland, 1900–1930." *Agricultural History* 82 (1): 62–77.
Paulson, Susan. 1998. *Desigualdad social y degradación ambiental en América Latina.* Quito: Abya Yala/FAO FTPP.
———. 2000. "La diferencia e interdependencia social en el manejo agroforestal." *Agroforestería en las Américas* CATIE/ICRAF 7 (25): 8–14.
———. 2013. *Masculinidades en movimiento: Transformación territorial y sistemas de género.* Buenos Aires: Teseo. Available open source online: www.rimisp.org/wp-content/files_mf/1383855041978987186769lebook.pdf
———. 2015. *Masculinities and Femininities in Latin America's Uneven Development.* New York: Routledge.
———. 2016. "La (re)producción socioecológica en América Latina con masculinidades cambiantes." In *Transformaciones ambientales e igualdad de genéro en América Latina. Temas emergentes, estrategias y acciones.* Velázquez Gutiérrez, Margarita, Verónica Vázquez García, Ana de Luca Zuria, Dulce María Sosa Capistrán (coords.) México. Edición: CRIM-UNAM. 91–120.
Paulson, Susan, and Brian Schultz. 1995. *Evaluación de proyecto: Difusión de la agropecuaria alternativa, Provincia Carrasco.* La Paz: Catholic Relief Services.
Perfecto, Ivette, and John Vandermeer. 2008. "Biodiversity Conservation in Tropical Agroecosystems: A New Conservation Paradigm." *Annals of the New York Academy of Sciences* 1134: 173–200.

Rocheleau, Dianne E. 2008. "Political Ecology in the Key of Policy: From Chains of Explanation to Webs of Relation." *Geoforum* 39 (2): 716–27. doi: http://dx.doi.org/10.1016/j.geoforum.2007.02.005

———. 2011. "Rooted Networks, Webs of Relation, and the Power of Situated Science: Bringing the Models Back Down to Earth in Zambrana." In Mara J. Goldman, Paul Nadasdy, and Matthew D. Turner, eds., *Knowing Nature: Conversations at the Intersection of Political Ecology and Science Studies*, 209–26. Chicago, IL: University of Chicago Press.

Schejtman, Alexander, and Julio A. Berdegué. 2007. *La desigualdad y la pobreza como desafíos para el desarrollo territorial rural.* Working Paper 1. Santiago: Rimisp, Latin American Center for Rural Development.

Scott, James C. 1999. *Seeing Like a State: How Certain Schemes to Improve the Human Condition Have Failed.* New Haven, CT: Yale University Press.

Silverblatt, Irene. 1987. *Moon, Sun, and Witches: Gender Ideologies and Class in Inca and Colonial Perú.* Princeton, NJ: Princeton University Press.

Urioste, Miguel. 1987. *Segunda reforma agraria: Campesinos, tierra y educación popular.* La Paz: CEDLA, Centro de Estudios para el Desarrollo Laboral y Agrario.

Walter, Mariana, and Joan Martinez-Alier. 2012. *Social Metabolism, Ecologically Unequal Exchange and Resource Extraction Conflicts in Latin America.* Analytical Framework Report D.WP71, Environmental Governance in Latin America and the Caribbean.

Yuval-Davis, Nira. 2006. "Intersectionality and Feminist Politics." *European Journal of Women's Studies* 13 (3): 193–209. doi: 10.1177/1350506806065752

Zimmerer, Karl S. 1993. "Soil Erosion and Labor Shortages in the Andes with Special Reference to Bolivia, 1953–91: Implications for 'Conservations-with-Development.'" *World Development* 21 (10): 1659–75.

———. 2014. "Environmental Governance through 'Speaking like an Indigenous State' and Respatializing Resources: Ethical Livelihood Concepts in Bolivia as Versatility or Verisimilitude?" *Geoforum*, 64: 314–24.

Zulawski, Ann. 1990. "Social Differentiation, Gender and Ethnicity: Urban Indian Women in Colonial Bolivia, 1640–1725." *Latin American Research Review* 25 (2): 93–113.

13 Trees as tools, trees as text

Struggles over resources in Zambrana-Chacuey, Dominican Republic

Dianne Rocheleau[1] *and Laurie Ross*[2]

"To plant trees on the land is to put the land in chains"

Rafael Amparo, farmer, Zambrana-Chacuey[3]

"The Acacia can only improve the value of the [land] and keep our land safe from others."

Antonio Perez, farmer, Zambrana-Chacuey

Introduction

In agrarian and forest lands throughout the world sustainable development programs relegate biodiversity concerns to special spaces for special species (parks and reserves) as "pristine" habitats devoid of humans. Alternately, on the lived-in side of the map, people are exhorted to pursue sustained yield production in homogenized landscapes best described as biologically correct factory farms (or factory forests). This segmented vision rests on ecological and economic specialization of land use, clear dichotomies between "natural" places and production spaces, and the separation of both from what we have come to know as "home" (Rocheleau, 1991).[4] Croll and Parkin (1992) note that widespread and erroneous assumptions by international agencies about the boundaries between farm and forest, home and workplace have been made real through the daily practice and long term process of development efforts, including "sustainable development."

Nowhere is the facile division into preservation, and production more apparent than in the ubiquitous forest imagery (Hecht and Cockburn, 1989) and the use of forests as stages for global economic and environmental scripts. Trees and people have become the major players in what Schmink and Wood (1992) have called the "greening of the discourse" on development. Trees have become both icon and currency in the domain of sustainable development. For example, forests can be planted in agrarian landscapes in Costa Rica to atone

Dianne Rocheleau and Laurie Ross, "Trees as Tools, Trees as Text: Struggles over Resources in Zambrana-Chacuey, Dominican Republic," *Antipode* 27 (4), 407–428.

for industrial sins of emission in the United States, apparently with no penalty for the rural evictions that it may ensue. Likewise trees can be used as political force fields to buy time and create space for rural movements resisting the eviction of peasants and forest peoples by ranchers, commercial agriculture and mining, as in the case of the rubber tappers in Brazil (Mendes, 1989). Yet paying in green currency can be costly. Osmarino Rodrigues, successor to Chico Mendes as leader of the rubber tappers movement, noted that the use of green discourse opened their landscapes, livelihoods and everyday lives to the scrutiny, regulations and interventions of national and global environmentalists (Schmink and Wood, 1992).

This poses a challenge to political ecology to question the rediscovery and sequestering of biodiversity and the "recovery" of degraded lands through sustained yield forestry and agriculture (Rocheleau, 1991; Peluso, 1992; Schroeder, 1993). If trees and forests have been turned into metaphors for the green dreams of global environmentalists, the green greed of multinational corporations, and the greening of popular movements, they can also serve as a metaphorical site of investigation into the daily practice and long term process of struggle between these interests "on the ground." There is a particular need to critique and propose alternatives to the plethora of forestry and agroforestry initiatives which have been sheltered in the discursive shade of trees as symbols of green goodness.

The arguments and case study presented here take as a point of departure the recent work on discourse and development (Parajuli, 1991; Escobar, 1992; Sachs, 1992; Peet and Watts, 1993) and the writings of political ecologists, feminist environmentalists and activists on struggles over forests.[5] This paper explores the role of trees as objects, sites, symbols, and tools of material and ideological struggle within and between state agencies, NGOs, and a rural people's movement. It emphasizes the multiple roles of trees as instruments of power and as tools of empowerment deployed by a diversity of actors at national, regional, local, and household levels.

The case study presents the experience of the Rural Federation of Zambrana-Chacuey in the Dominican Republic and its collaboration with Environment and Development Alternatives Caribbean (ENDA Caribe) in the introduction of an exotic timber tree, *Acacia mangium,* into several hundred smallholdings in the region. The first section, "From Forest as Context to Trees as Text," introduces the national context, the region, the institutions, and the project, with a focus on the history of the role of trees and forests in the landscapes, livelihoods, and land tenure regimes of the region. "Trees in the Context of Contemporary Struggles" describes two distinct but intersecting arenas of struggle at national and household level in which the "miracle tree" has played a role. "Trees as tools" traces the *Acacia mangium* tree into the property relations, landscapes, and everyday lives of people in the region as it is reshaped into a variety of material and ideological tools at the intersection between culture and nature. The conclusion examines the implications for theory and practice in both forestry and political ecology.

From forest as context to trees as text

National context

The Dominican Republic has a long history of direct state intervention in the details of rural life. This has included strict regulation of farm clearing and management, forced employment in agriculture, and the direct appropriation of forest and farmlands by the state, often followed by allocation to preferred clients of the ruling regime.[6] Since the widespread clearing of valuable mahogany stands during the first third of the twentieth century (see Veras, 1984), the state has treated the forest alternately as a resource to be mined or as an obstacle to commercial agriculture to be cleared (of both trees and farmers).

In contrast, for the majority of small farmers, the forest was once the context in which agricultural production and much of rural life was nested. As plantation agriculture expanded, forest cover in agrarian landscapes (now called "conservation") became synonymous with under-utilized and unimproved land, while clearing of forest for agriculture (now called "deforestation") was legally recognized as an improvement and a "higher use." Smallholder farmers often lost their rights to land not under "productive use," which made forest lands particularly vulnerable to legal alienation by the state or local elites. Within that context land clearing and/or planting of commercial crops was essential to create and maintain land rights throughout this century.

During the early 1960's deforestation accelerated with the loss of state control over the countryside. The government reacted with the passage of Law 206 of 1967, which effectively placed not only forests, but all "trees" under the protection and regulation of the state (Veras, 1984). It prohibited the felling of any tree without the express permission of the newly formed, militarized forestry service (Direccion General Forestal – DGF). This law extended the authority of the state onto the lands of every resident and property holder, and into the daily lives of every tree user in the nation, regardless of land rights.

In practice the forestry service (DGF) selectively enforced tree cutting bans against poor smallholders, woodworkers and charcoal producers and awarded permits to commercial producers and local elites. This generated a strong climate of cynicism among the poor, as trees became green enemies and sites of "everyday peasant resistance" (see Scott, 1985). They were sabotaged through catastrophic "accidents" and through purposeful, daily mismanagement, to remove unwanted trees and to hasten the harvest of "deadwood."

Trees and forests were tenure liabilities to all rural farmers as long as trees were not-agriculture, not legally available for harvest and sale, and forests were un-improved lands or untouchable resources. Many farmers kept or established forests on their lands through intercropping of fruit trees, cocoa, coffee, and other "crops" with planted and self-sown trees for shade, fencing, soil fertility, fuelwood, and occasional harvest of poles and timber. The multi-story cash crop plots and living fences provided a shield of "productive" use in which a variety of forest trees could persist without putting the land tenure of farmers at risk.

The current national and international conditions for smallholder producers of traditional tree crops in the Dominican Republic threaten both the farmers and their tree crops as well as the forest species which they have "sheltered." Commodity prices for coffee and cocoa have remained chronically low throughout the last decade.[7] International trade and investment policies and national development policies favor assembly industries and tourism over agriculture and promote large scale commercial agriculture and contract farming for export over smallholder production (Deere, 1990; McAffee, 1991). Ideologies of environment and development have also impinged upon the livelihoods and landscapes of rural farmers (Lynch, 1994).

Environmental protection and regulation have captured the public imagination and the attention of Dominican political and technical professionals over the last five years. Mirroring international trends, forest management and enforcement of forestry laws have been among the most dramatic and politically charged of recent environmental conflicts. Forest protection has been invoked as a right of the poor, and it has also been wielded against them as a punishment, as in the eviction of 70,000 forest farmers from Los Haitises park in the northeastern part of the country in 1992. The discourse of forest protection has become a tool of national forest policy, political power and prestige (Lynch, 1994). Within that context the Zambrana-Chacuey region has been the target of a DGF forest protection campaign – complete with military helicopters – as well as the privileged site of a forest enterprise pilot project.

The forestry enterprise initiative in Zambrana-Chacuey

Zambrana-Chacuey is a hilly frontier region in the Central Valley of the Dominican Republic that is home to roughly 12,000 people. It encompasses an area of 250 km2 with elevations ranging from 100 to 600 meters above sea level and is located 100 km to the north of Santo Domingo, the country's capital. The subtropical humid and sub-humid forest which once covered the zone has given way to a complex landscape of forests, pastures, croplands, home gardens, and homesteads.

The majority of the residents are small farmers engaged in a tenuous mix of subsistence and commercial agricultural production combined with urban and local wage labor or trading in agricultural commodities. Farmers produce tobacco, citrus, root crops (cassava), other vegetable crops and livestock as well as charcoal and, more recently, timber. As in other rural regions in the country (Sharpe 1977; Vargus-Lundius 1991) the population is stratified on the basis of land size and quality, income sources, gender, family composition, and crop production, as well as their affiliation with the Federation and most recently their participation in externally-funded projects.

ENDA-Caribe (hereafter referred to as ENDA) is a regional branch of an international environment and development NGO headquartered in Senegal and the Rural Federation of Zambrana-Chacuey represents approximately 800 members from 500 households organized into 60 local farmers', women's and youth associations, and most recently the Wood Producer's Association. In 1984, ENDA and the Federation initiated a regional agriculture and forestry project (AGROSIL) that mixed commercial timber production, agriculture, soil conservation and gardening as somewhat separate but related activities with distinct "target" groups. Between 1984 and 1992 ENDA and the farmers of the Federation established experimental nurseries of timber, fruit, and soil improvement trees. They tested well over 60 tree species (mostly exotics) for ecological suitability and farmer preference. The research and extension efforts concentrated increasingly on *Acacia mangium,* the project's Australian "miracle tree" which produced timber for milling within 6–8 years of planting.

As of 1992, there were 87 community nurseries and more than 300 household based nurseries for timber and fruit trees. The project had established *A. mangium* trees on 80 percent of the farms of Federation members and in every community within the region. Overall, the project had planted 800,000 timber trees, 40,000 fruit trees and had 250,000 seedlings in nurseries (Valerio, 1992). Many of the Acacia trees that had been planted at the outset of the project were ready for harvest and scores of farmers in the Federation had legally sold round wood or small stakes and poles at farm gate.

Special accords between ENDA and the DGF allowed the farmers to harvest their "project" trees and permitted the newly formed Wood Producers' Association to transport, process, and market *A. mangium.* ENDA and the DGF developed a formal paper permit – called "tree titles" by the farmers – which granted legal permission to farmers to cut the trees certified by ENDA as "project trees," that is, trees planted by the named permit-holder under the auspices of the Forest Enterprise Project. As a result of the widespread adoption of Acacia as a cash crop and DGF permission, ENDA and the Wood Producers' Association constructed a cooperative sawmill. The introduction of the tree and the project both shaped and were defined by the ongoing struggles between the state and local communities and between men and women over control of trees, forests and land in Zambrana-Chacuey.

Trees in the context of contemporary struggles

The struggles in Zambrana-Chacuey are multivalent and complex, involving multiple poles of identity and fields of affinity, difference and power.[8] Among the crucial lines of conflict over trees and forests are those between state and local interests in resource management and between women and men as users and managers of trees at household, community and regional levels. These articulate with conflicts of class, locality, occupation and organizational affiliation. In the context of daily practice and long term institutional process these lines converge and intersect, often in surprising and unpredictable ways, based on shifting fields of power and constellations of interests from international to intra-household levels. A review of the defining struggles over resource management and trees within two very distinct arenas clarifies the broader political setting in which the miracle tree has operated as both a biological and social actor.

State versus local interests in forestry

Over the last thirty years land rights and, increasingly, resources have constituted the focus of conflicts between the state and the rural poor in Zambrana-Chacuey. Peasant struggles over both issues are rooted in a broader movement which arose in response to widespread evictions of rural smallholders from their lands and the emergence of Liberation Theology within the Catholic Church during the 1960's. The Rural Federation of Zambrana-Chacuey coalesced around the land struggle,[9] which was effectively waged by the landless and near landless farmers of the region against landed elites, over the objections – and often under the repressive counter-measures – of the State.

The greening of the discourse (Schmink and Wood, 1992) over the last decade has brought conflicts over resources in general, and forests in particular, into the limelight of public attention in the Dominican Republic. This sharpened already extant conflicts in Zambrana-Chacuey between the DGF and the members of the Federation. At present, state and local interests in resource management diverge sharply with respect to forestry and tree products. In spite of the agreement that protects tree harvesting, processing and sale under project auspices, many forest and tree based enterprises, as well as individual participants, remain outside the purview of these accords.

The prevailing policies and practice of forest regulation place charcoal producers as a group in open conflict with the DGF. As one full-time charcoal producer noted:

> It used to be that you could work with them. They would stop you and ask for one or two pesos [then about 5 percent to 10 percent of the gross earnings]. Now they want it all.

Lest we think this implied a new era of honest enforcement, he went on:

> They keep it all and sell it. They want to control it all, especially that one brute they've unleashed on us. Just a few weeks ago they killed Francisco at the edge of town on the main road. It was the middle of the night. They stopped him and he refused to give up the charcoal, so they shot him, right there.

The green halo of environmental enforcement has apparently broadened the scope for forest service corruption from petty bribery to outright theft of products and the price of resistance has escalated from harassment to death.[10] Charcoalers and timber extractors have reverted increasingly to sending their products to market with their young sons, aged 6–12, travelling by day on horses or burros with small amounts of wood or charcoal. They depend on the shredded remains of the moral economy (see Scott, 1976) that prevails between the DGF and the peasantry, only partly confident that the children will not be beaten and that the loss from any given act of "impoundment" will be but a small increment of the total inventory of charcoal for sale.

The clash between the DGF and charcoal producers is widely recognized and has been mythologized in local and national lore. This serves to remind the Federation membership of the potential for renewed conflict between the DGF and tree farmers, should the special accords fail. Many other categories of conflict over trees also remain outside the scope of project accords between the DGF and ENDA. Among the least visible and most pervasive of these is the gendered contest over trees within the household and the Federation.

Gendered struggles in the household and the rural Federation

Men and women in Zambrana-Chacuey often articulate their differences in terms of complementarity of activities, responsibilities and domains of authority (field interviews 1992). While difference does not necessarily imply dominance, the uneven relations of power between men and women have shaped the terms of the gender division of work, resources, responsibilities, and rewards in households, communities, and regional institutions from the Federation and ENDA to church organizations and the forestry service. Although many residents of the region would not identify these relations as conflictual, others have recounted the ongoing struggles of women in the region to change overall conditions as well as to advance their own interests within their homes, communities and local institutions. The introduction of the Acacia has articulated with these gendered conflicts in the form of land use practices within the household, participation in local associations and the structure of household linkages to the Federation.

Household Level Struggles over Trees

Within the household context, some women lack power over places to plant, while others lack power to protect their own spaces, plants and household lands from being overtaken by men's timber lots. In other cases women are excluded from timber production and marketing decisions. Each of these constitutes both a material and ideological contest over tree and land rights at the household level.

One type of struggle is exemplified by women whose ambitions to plant timber are thwarted by their lack of access to or control over land, and their husbands' lack of interest or active opposition. In group meetings many women stated their interest in planting Acacia and other timber trees on household lands but could not do so for lack of clear tenure rights over even the patio, much less the other household lands amenable to timber plantations. Most had already encountered direct opposition from their husbands.

Conversely, many women's husbands *have* planted Acacia, often without consulting them, at the expense of the women's patio gardens or family croplands. These women expressed skepticism about the timber production enterprise: "We have seen this before . . . peanuts, tobacco, now this Acacia, they all take over our croplands and reduce the food which we can grow for ourselves." They would far prefer, and might not resist, a tree more amenable to intercropping.

In yet another situation commonly observed in participating households, men may exclude women from any role beyond service labor in timber tree production due to a kind of "masculine mystique" associated with the Acacia tree and timber production. This may derive from a combination of the historical allocation of the timber trade to men and the current representation of the timber production project by ENDA and some Federation leaders as a technical, male-identified domain. Survey data indicate that two-thirds of the women have been planting trees since childhood and that many of them have managed trees and processed and sold the products. However, women tend to have more experience with fruit trees and cash crops such as coffee, cocoa, and their "companion species" for shade and soil fertility. Although greater intensity of care is required for fruit and cash crop trees, there is a widely shared sense among project staff and male participants that the timber trees require a man's skills, sense and strength. Ideological constructs of gendered knowledge and authority have influenced the material distribution of benefits from timber production.

One largeholder cocoa farmer conveyed the strength of gender-stereotyping about the new timber trees in a discussion about the gender division of labor and control on the farm.[11] When asked if his wife could sell the citrus fruits, coffee, or cocoa from their farm he replied: "Of course. Marta could take some to market or she could negotiate a price and sell it to a buyer who comes with a truck. Either one of us can do that."

The future sale of the recently planted Acacia was a different matter: "Ah no. This is different. Growing timber is a very technical enterprise. These trees must be measured in inches.[12] This tree is a business for men!" Marta is a well-known and specialized seamstress who designs and produces wedding dresses and silky party dresses. She was busy with customers at the time, presumably measuring them, in inches or something equally effective, to assure the good fit for which she is renowned.

While some men and women resist the Acacia and some men seek to exclude women from the commercial timber enterprise, the tree is a site of struggle over something else. Each group reacts to different dimensions of the tree. Some men fear legal intervention by the DGF. Others embrace the tree but resist women's involvement in management and marketing. The Acacia has also been strongly identified with ENDA, with men, and as a major asset, which strengthens not only its economic value, but also its symbolic value. That strength may represent a threat if not under the control of men. In contrast, women who resist the tree focus on the material threat to family subsistence and their own enterprises. Other women embrace the tree for their own purposes. The heart of the matter is who brings home the timber, and why.

The struggles over planting spaces and control of timber sales at the household level derive in part from anxiety about the future expansion of the timber enterprise within household lands by "the other spouse." This conflict is also shaped by the gendered meaning of the Acacia tree in the context of regional landscapes and livelihood strategies. In addition, the intra-household conflict over trees is rooted in the current institutional structure of the Federation and the Forestry Enterprise Project.

Gendered Struggles within the Federation

The Federation is a distinctly gendered organization, as reflected in the patterns of affiliation of men and women, and the roles that they play in the structures and activities of the organization. At the community and regional scale, men's and women's interests in Acacia and their access to the Forest Enterprise Project differ substantially due to the gendered patterns of individual and household connection to the Federation.

As of 1993, most women in Federation-affiliated households[13] were members of community Women's Associations (60 percent), and a small percentage were members of local Farmers' Associations (4 percent) and the Wood Producers' Association (4 percent). Nearly a third (32 percent) were not members of any association as individuals. While a similar number of men were non-members (38 percent), nearly half of the men (44 percent) were members of both the Farmers' Associations and the Wood Producers Association. A small percentage was affiliated with only one of these groups (Farmers, 11 percent and Wood Producers, 7 percent). Given that the timber project has been channeled through the latter two groups, this has implications for both women as individuals and for households linked to the Federation solely through women's membership.

At an individual level the gap between women and the Forest Enterprise Project is constructed in large part around the membership criteria of the Wood Producers' Association. The male-identified and commercial orientation of the original group has reproduced itself through entry requirements. A separate entry fee discourages membership for both spouses and deters women except for the most commercially oriented. Entrants must also have a minimum of 50 Acacia planted. Many women, due to constraints noted above, cannot plant more than 20 to 30 trees on the patio or the property lines. They would need to access group land or leased land to plant more,[14] which would presumably require the technical and legal assistance of the Wood Producers' Association, which is largely inaccessible to them.

Even if women are not interested in the timber project as a positive option, its progress affects their interests in land, land cover and land use options at household, community and regional levels for decades to come. Beyond women's membership *in* the Wood Producers' Association, there is a related but distinct need for women's group representation *to* the Wood Producers' Association. There is clearly a broader need for a Federation-wide forum to discuss and coordinate such projects across distinct interest groups, including non-participants, whether supportive, neutral or opposed.

As of 1993, the terms and pattern of household connection to the Wood Producers' Association left most of the individual women in Federation-affiliated households (96 percent) and the strictly women-connected households (18 percent) without representation in the emerging forestry enterprise. Yet 80 percent of all Federation households had planted timber. This implies a serious disadvantage in access to services and to decision-making by a large proportion of individuals and households.

The gendered organization of Acacia promotion, planting and commercialization has not only shaped the context for struggles over land, trees and forests. It has also affected the deployment of the "miracle tree" as a tool by men and women in a variety of circumstances. The Acacia has been used as an instrument of change to transform property rights and landscapes at the national, regional and household levels.

Trees as tools in material and ideological struggles

Like most technocratically inspired forestry programs, the ENDA-Federation project has employed trees as tools of modernization and has represented them as intrinsically "good" for the ecologies and economies of rural communities. Project staff have promoted a few exotic species ("miracle trees"), usually in monocrops or in fixed "technology packages." While the intent was to diversify existing monocrops and to "recover" degraded lands, in practice the Acacia has often replaced far more diverse stands of forest and agricultural crops or displaced resources belonging to someone other than the tree planter. International opposition to this approach has been couched largely in terms of essentialist vilification of the miracle tree at hand (often a *Eucalyptus* species or *Leucaena leucocephala)* as an enemy of the rural poor (Shiva, 1989). The debate has often degenerated into a battle over easy dichotomies: exotic vs. indigenous; forest vs. not-forest; subsistence vs. commercial; and conservation vs. development.

In contrast, the Forest Enterprise Project, with the *Acacia mangium* as its centerpiece, has both failed and succeeded, at different points in time and among different publics. Local experience and our observations suggest that trees are not always a good thing, nor are exotic timber trees always a bad thing. The state, local elites and the Federation have used the same tree and the same program as instruments of power and tools of empowerment.

The Acacia has been employed as a tool to restructure livelihoods, reconfigure landscapes and re-inscribe the relations of power between the DGF and the rural people of Zambrana-Chacuey. Using the tree as text the planters and promoters rewrite part of the social and material relations between people-in-place and between people and places. As part of the emerging landscape, the Acacia also shapes the material context for future changes in property relations, land use and land cover. The outcomes are far from predictable; they are contingent on shifting constellations of interests and on the actions of particular people situated within local, national and global webs of power, affinity, conflict, and constraint.[15] The summary below traces the workings of the Acacia as a material, textual and contextual tool in land tenure, livelihood and landscape change.

Trees and tenure transformations

Within the prevailing system of legal rights, the majority of farmers in Zambrana-Chacuey own not the land but the value of their "improvements" and investments on it. As one farmer, Wilfredo Hernandez, put it: "I don't own land . . . but the *mejora* (value added) is mine." The Acacia, with its new status as a widely recognized producer of a valuable, legal commodity, has strengthened the prior claims and tenure security of both largeholders and smallholders. It has served to protect each from encroachment by the other and both from displacement by the state or outside commercial interests. Many farmers have planted Acacia to further secure their land claims as well as to invest in the timber crop. They cite the value of Acacia and other recognized commercial trees[16] as "land improvements" that increase tenure security and land value. Wilfredo noted: "Since we are on state land, the Acacia can only improve the value of the *mejora* and keep our land safe from others."

One disadvantage emerges when the analysis scales up to a community or regional scale. Some farmers whose tenure in their own plots is more secure with Acacia are reversing the traditional place of trees and food crops in the landscape. They plant timber trees on prime land near the house and cultivate crops on peripheral plots, often on rented land. If timber were to go through a boom cycle and largeholders were to plant most of their pasture and fallow lands to timber, reduced availability and increased prices of rental land could undermine the security of access to food crop land by smallholder tree planters. In short, the scale of this practice matters and may affect smallholders and largeholders differently.

Many poor farmers in the countryside have never entertained the prospect of tree farming. They still view the Acacia and all other trees as tenure liabilities that expose them and their lands to intervention by the DGF. As one man expressed it: "To plant trees on the land is to put the land in chains." For some people those chains have been more than metaphorical. They have found themselves in police or forest service custody as prisoners accused of criminal actions in the pursuit of rural subsistence.

Ramon's experience is typical of many stories recounted by rural residents of Zambrana-Chacuey. Although his wife Rosa is an active member of a Women's Association, Ramon is not a member of the Federation. He works full time off-farm and Rosa manages their 0.5 ha plot which they recently acquired with savings from his job. As Rosa recounts:

> Last year Ramon was arrested by Foresta (DGF) and held for days for cutting two palm trees which he had purchased from his uncle to build our house. Someone reported him to Foresta and they took him away to jail.

Ramon's family and friends sawed and stacked timber for the DGF in lieu of jail time. He was released after a couple of days and was lucky not to have lost his job. He had the particular bad luck to fell a palm tree during the height of a campaign to protect that species as the preferred nesting ground of the national bird. The paradox of the situation lay in the juxtaposition of the most durable and accessible wood for construction of smallholders' homes and the tree which itself constituted the best home for a form of wildlife highly valued by the national park service.

Ramon's entrapment in that paradox also had a price for the environment, as noted by Rosa: "After Ramon's release he didn't want anything to do with trees on our own land." Based on the forestry project and her childhood training in her father's coffee nurseries she had produced 300 seedlings in a small nursery. Ramon forbade her to plant the trees on their property. She and the children eventually obtained his permission to plant three Acacia seedlings.

For many people active in the Federation the "chains" were removed from trees during a dramatic public enactment of the new forest policy at a field day convened by ENDA in 1990. The then Director of the DGF, infamous for his prosecution of peasant "offenders," came to witness the legal felling of project trees on the property of Federico Mercado and to bear witness to the efficacy of the new ENDA-DGF accords by *not* arresting the axe-wielders. He also granted "tree titles" to some of the farmers present. The group then constructed a *ramada* (a detached porch) with roundwood and poles from Federico's Acacia plot. The *ramada* stands as a commemorative monument to the new covenant between the farmers and the DGF.

For many farmers the field day was a historic moment and a personal turning point in their decision to join the project and plant timber trees on their lands. Of the farmers who had planted ENDA's "project" trees by January 1993 over 40 percent had decided to plant them on the basis of the new permits from the DGF. Ramon, however, not a member of the Federation, does not trust the Federation to protect his land from state intervention, especially from the DGF. He is still waiting for ENDA and the Federation to further prove the market and the relation with the DGF, given his own recent and bitter encounter.

Aside from the Federation and ENDA, the Acacia tree has taken on a life of its own as a kind of currency for negotiating with the DGF. Many people have fared better than Ramon by recourse to a kind of green bail, sometimes avoiding arrest for land clearing and tree felling by demonstrating their blocks and lines of planted (project) trees or by agreeing to plant one or more plots entirely to timber trees. Others have parlayed their way out of prison with similar commitments. We can only speculate about the intentions of the DGF or specific officials in promoting these plantations. What is significant is that the tree has become a negotiating tool in ongoing struggles between the public and the DGF, with uncertain outcomes and room for manoeuver.

The Acacia has become not only a tool of negotiation but a venue for an unlikely and accidental alliance between the DGF and the Federation that may protect the farmlands of the region. In a situation of land scarcity, the profit margin from timber as a cash crop may allow smallholders and largeholders alike to resist buy-outs by corporate agribusiness. Despite its bias toward monocrop block plantations, and hence, larger holdings, the value of the Acacia can constitute an economic bulwark against the encroachment of expanding agribusinesses such as Leche Rica citrus and Dole pineapple plantations. The joint efforts of the Federation, ENDA, the DGF and the farmers may have spawned an unexpected by-product in the form of a green wall of financial resistance against displacement of local farmers.

Several farmers in the area have recently purchased land specifically to plant timber for long term savings as well as regular income. The experience of Antonia and her family illustrate both the problem of exclusive investment in land and the potential of Acacia lots as an alternative investment strategy:

> A couple years ago my father-in-law became very ill. My husband and his brothers sold the family lands to finance the medical expenses. Leche Rica was ready and willing to pay a good price.

Antonia and her husband managed, through his employment as a foreman on the plantation, to save and invest in a new plot. In addition to a patio garden, tobacco and a *"conuco"* – a diverse, intercropped plot including staple foods – they planted Acacia in blocks as an investment.

Several Federation members have stated that timber plantations might allow smallholders to resist distress sales of land through sale of timber, poles and other wood products, or even through loans secured against standing timber. They also have noted that Acacia has proven to be a viable, lucrative investment crop for smallholder residents and absentees alike. It requires no major capital outlays for establishment, nor does it require constant labor and supervision. Timber plantations already provide an attractive alternative to land sales for urban migrants, part time farmers and rural/urban split households. This has major implications for the future of local people in the urban work force, since timber farms provide an independent and supplemental source of income, a long-term investment and a residential fallback site.

The commodity road to environmental stabilization (Schroeder 1993) has thus constituted a path to land tenure stabilization in an otherwise volatile land market tending toward rapid land acquisition and concentration by corporate interests. However, the eventual price of this strategy to reduce buy outs by outsiders may well be the greater polarization of landholding size between local farmers. In addition, this strategy has subjected timber farmers and their lands to the regulatory authority of the DGF.

The Acacia has created a gateway[17] into the domain of agricultural production on smallholders' lands, historically outside the purview of the foresters. The DGF has used the Acacia tree to convert smallholder lands into its territory by expanding "the forest" into areas that were "not forest."[18] The issuance of selective "tree titles" has actually served the same purpose as the original forest law of 1967; that is to increase the authority of the state in general and the forest service in particular on farmlands. It could be argued that the permits to cut, transport, process and sell trees increased the regulatory power of the DGF and the vulnerability of farmers by simultaneously setting free and legitimizing, under state control, the growth of timber as a profitable agricultural commodity.

The DGF was not the only actor to use the Acacia to expand its domain. The Federation and the farmers of the region also increased their control of productive resources and scope for activity. Through "reforestation" they gained legal access to the previously forbidden domain of forest products and forestry enterprises.

The Acacia as a gateway to an expanded resource base for both the DGF and the farmers of Zambrana-Chacuey suggests, at first glance, a win-win outcome. However, there is no guarantee that the gate will remain open and that the control over it will be equitably shared. The ability to make use of this new-found mobility between forestry and agricultural domains and between state conservation mandates and the opportunities of forest product markets will depend on the continuing ability of the DGF, the Federation and NGOs to negotiate feasible and viable agreements. A return to a ban on all tree cutting by the DGF could easily put the tree-planting farms of the region into a financial and functional state of gridlock (i.e. back in chains).

When questioned about this possibility, farmers and local leaders responded with a strong resolve to resist such policy reversals and to mount a major campaign of non-violent protests if necessary. Luis Alcantara's argument followed a moral economy line (see Scott, 1976), shored up by legal documentation and administrative precedent:

> Before they were the tree planters and we were the tree cutters. We couldn't really defend our people against them. Now we are the tree planters. We have papers which say so, which they have signed. They have already approved the cutting and have said that it is just and practical. With this we could fight them and make them keep their word.

Trees as tools of landscape change

The very characteristics that make the Acacia a prodigious producer of timber also constitute a serious threat to the ecological integrity and social fabric of the landscape. The same properties that allow the tree to thrive with little or no maintenance also result in destruction of surrounding vegetation and most of the understorey as well. The stories and the living proof in the landscape after more than a decade in Zambrana-Chacuey suggest that the Acacia competes with its neighbors for light, and probably water and nutrients, and some speculate that it may actively kill off its neighbors through chemicals emitted into the surrounding soil.[19]

This miracle tree could lead to felling and clearing of diverse forests, tree crop gardens, and *"conucos"* simply by virtue of its value as a cash crop, as in earlier booms of tobacco and peanuts. However, the Acacia actively propagates itself and creates the conditions for its own expansion as a monocrop. Farmers in the region have found that the tree's new status as a legally recognized commercial tree and its own aggressive growth habit can be used as a "green machete," to take down second growth and even mature remnant forests, as well as multistorey coffee and cocoa "forests." This occurs in full view of the DGF and perhaps with their unofficial blessing.

The tree is wielded as a machete in both a practical sense and a political and metaphorical sense, to remove "undesirable" or "non-productive" vegetation, without the smoke, the charred fields and the risk of incriminating personal presence with cutting tools in hand. The resulting maintenance or expansion of "forest" on regional maps might even draw approval from the authorities. The Acacia, with the mantle of green goodness ascribed to trees, lends environmental legitimation to both the process and the result of deforestation.

The green machete is an effective tool in struggles over land, as well as in struggles over trees as objects and commodities. One could plant Acacia in a second growth forest, let it smother the existing vegetation, harvest the Acacia poles (thinnings) in four years, and the timber in seven years. The field could be 1) kept in Acacia permanently, 2) rotated between annual crops and Acacia, or 3) simply converted to crop land after one cycle of Acacia. In the first case forest cover is maintained, commercial forest production increases, and species diversity is reduced dramatically. In the second case there is a net reduction in "forest cover" but an increase in production of commercial forest products on the land. In the last case forest cover is reduced, species diversity is reduced, but the land is "unchained" and available for a variety of uses.

In economic and social terms the green machete expands farmers' land use options in forest or shrub lands and saves both labor and litigation in the process of land clearing. It serves some farmers by cutting the chains of forest protection from their land. They may, however, simply be moving from the chains of protection and prohibition to the chains of regulation.

At household level the green machete that is the friend of the *campesino* (rural man), may become an enemy of the *campesina* (rural woman). The main danger is to the women's diverse patios and the polyculture gardens planted there, as well as the household food crop plots, particularly the diverse *conucos*. These spaces, while not consisting of large tracts of land, contain a wealth of species and may provide everything from fuel wood, medicinal plants, fruits, and green vegetables to staples such as cassava, maize, beans, sweet potatoes, yams, and pigeon peas. The Acacia, if planted in patios or shared *conucos,* would displace the wealth of women, by replacing a stand of diverse products controlled largely by women with a single cash crop controlled by men.

The paradox of the green machete is that it depends in large part on the symbolic power granted to the Acacia by the DGF through the selective recognition of this miracle tree as a legal commodity and cash crop. While the DGF has granted tree titles for some other "project trees," both indigenous and exotic, they limit these other species to domestic use. At present most farmers have no incentive, beyond local use and their own "common sense," to grow other tree species. They have a strong commercial incentive to plant the Acacia into existing forests and mixed cash crop stands, letting the green machete replace all non-project (pre-existing) trees not fully licensed by the DGF for commercial exploitation.

Conclusion

While we might be tempted to stop at exposition of the flaws and systematic injustices of existing forestry programs, followed by a prediction or promotion of their demise, the experience of the Rural Federation with ENDA Caribe and the DGF suggests otherwise. It would be easy but irresponsible to simply add *Acacia mangium* to the list of "bad trees" and social forestry to the list of "hopeless populist projects." The "miracle tree" is neither miraculous nor demonic. It is – like "the people" who use, plant, promote, uproot, resist, and sell it – multivalent and subject to change in the ebb and flow of ongoing struggles.

There is too much at stake for the people of the region to just dismiss the opportunities of commercial farm forestry. It may well constitute a long-term "ticket" to continuing connection to rural land. However, there is also too much at stake in the way of biodiversity, economic diversity, women's wealth, smallholder food production, and the local balance of power, to simply accept a regional landscape devoted to a timber monocrop under the control of the DGF. The middle ground in this case need not be the facile compromise of "sustainable-development-as-usual" but could instead constitute an embrace of ongoing struggle and participation in forestry production, with conscious use of trees as objects, sites, symbols and tools in that effort. For example, in our own policy and practical recommendations as collaborators/advisers to ENDA and the Federation, we have privileged the experience and addressed the concerns of the near landless and of women as groups still underserved or ill-served by the current forestry initiatives.[20]

The story of the Forest Enterprise Project also provides a clear indication of the murkiness and contingency of social relations embedded in the everyday practices of agriculture, forestry and conservation in this region. We see stark evidence of the use of the "project tree" in a strategy that follows the "commodity road to stabilization" identified by Schroeder (1993) in The Gambia. However, we also see equally impressive examples of that same road turned into a wall against encroachment of corporate interests into local lands. Echoing the gendered interest in agricultural development schemes in many parts of the world (Carney and Watts, 1991), the timber cash crop of one farmer's dreams is also the "green machete" of another woman farmer's nightmares.

The introduction of the Acacia has provided a meeting point between the DGF and local communities in the region, but not without a price. In contrast to its increasingly closed door policy in parks and reserves, the DGF has opened wide the gates in "not-forest" spaces to allow people to farm trees and harvest forest products and to broaden its own scope and territorial limits of regulatory authority. In this domain the DCF has shifted from controlling forests as places and trees as objects to consolidating control over the terms and process of tree production and commercialization.

From afar, we might have fancied a simple map of land struggle with clear lines between oppressed and oppressors and between good crops and bad crops. Everyday life in Zambrana-Chacuey reveals a complex articulation of shifting fields of affinity and regions of conflict to create, resist and transform uneven development and the terms of power over resources from the national to household level. This example may help to untangle the debates over which shade of green to paint the future in the far-flung corners of the planet. The interactions of the *Acacia mangium* tree with the diverse Federation membership and the complex and richly textured landscape provide an example of the cultural and natural co-construction of the tree as a social actor and its transformation into a variety of tools within ongoing struggles. The crucial question is not so much to do or not to do social forestry or which tree to plant, as it is whose decision, among what options, and under what terms, in a particular place and at a given point in history.

Acknowledgments

The authors are grateful to a number of organizations and individuals. Our work was conducted under the auspices of two Clark University initiatives, the ECOGEN Project of the International Development Program and the Gender and Environment Project of the Marsh Institute, with support from the United States Agency for International Development and the Ford Foundation respectively. We thank Louise Fortmann, Philip Steinberg, and David Edmunds for their comments and suggestions on earlier versions of this article and especially Rick Schroeder and Rod Neumann for their editorial assistance. This article draws on collaborative field research with Julio Morrobel of ISA, Mamerto Valerio and Daniel Zavallos of Environment and Development Alternatives (ENDA) as well as Ricardo Hernandez. Our jointly published works elaborate further on the methodology, resource management and land use issues in the study. Our deepest gratitude goes to the leaders and members of the Rural Federation of Zambrana-Chacuey for sharing their experiences and impressions of the forestry project with us.

Notes

1 Assistant Professor, Graduate School of Geography, Clark University, 950 Main St.Worcester, MA 01610

2 Master of Arts Candidate, International Development Program, Clark University, 950 Main St., Worcester MA 01610.

3 Pseudonyms used here and through the text to protect farmer identity. This particular saying is actually often invoked, almost as a proverb.

4 Both Merchant (1989) and Hecht (Hecht and Cockburn, 1989; Hecht, 1995) have noted the way in which the boundaries between home and workplace and environment and workplace have been re-defined and re-drawn under changes in global and national economies.

5 Key sources include: Blaikie, 1985; Richards, 1985; Blaikie and Brookfield, 1987; Rocheleau, 1987 and 1991; Fortmann and Bruce, 1988; Emel and Peet, 1989; Guha, 1989; Hecht and Cockburn, 1989; Mendes, 1989; Shiva, 1989 and 1993; Agarwal, 1992; Peluso, 1992; Schmink and Wood, 1992; Bruce, Fortmann and Nhira, 1993; Fortmann, 1993; Richards, 1993; Schroeder, 1993; Rangan, 1994.

6 Clientelism penetrated every region of the country and every sector of rural production under the dictator Rafael Leonidas Trujillo, who ruled for 30 years until his assassination in 1961.

7 Price increases in 1994 reflect periodic crop failures in large producer countries, and will most likely be temporary.

8 For a fuller discussion of feminist post-structuralist notions of power and struggle engaging multiple axes of identity see Hart, 1991 and for further discussions of affinity and coalition see Haraway, 1991 and Harding, 1991.

9 Beyond the land struggle the Federation has consistently promoted the interests of smallholders through human rights advocacy, cooperatives, social services and infrastructural development (Lernoux, 1982; field interviews, 1992–3). It currently represents a coalition of three distinct streams within the broader rural movement: liberation theology, cooperative enterprises, and the traditional Catholic Church.

10 See Lynch (1994) on the eviction of forest dwellers and farmers from national park and state lands in Los Haitises.

11 Manuel had recently spent half a day in meetings and field visits with project staff and had enthusiastically committed himself and his farmers' association to a timber tree nursery and major planting initiative. His replies seem to reflect the gendered sub-text of the project ideology as he read it.

12 Manuel referred here to the standard forestry measurement of tree diameter at breast height (DBH) used to assess the volume and in this case the value of standing timber.

13 The numbers reported here are derived from a random sample of 7 percent of Federation membership.

14 Of those surveyed 62 percent of respondents expressed interest in planting trees on group land (local Farmer's Association or Women's Group land), or group leased plots, provided that the leases were legally sound.

15 For further discussion of this sense of bounded contingency see Schmink and Wood (1992) and their analysis of the factors which have impinged on the fate of the Amazon forest and communities of forest peoples, farmers and successive waves of migrants over fifteen years in a region of Para along a section of the TransAmazon Highway.

16 Once the new role of Acacia was established, Federation members and ENDA staff sought and eventually obtained tree titles for over 20 other tree species planted by various farmers under the auspices of the forestry project. This may yet open up the possibility for future negotiation about the availability (and economic and tenurial value) of whole forest stands and previously planted or protected trees.

17 See Fortmann, 1995 for a discussion of "mythical gateways" and other discursive tools in Zimbabwe forests.

18 See Peluso, 1992 for a discussion of this process in Indonesia.

19 The first point – that Acacia effectively shades out and/or buries surrounding plants in leaf litter – is widely accepted. The second and third points are unsubstantiated speculations of forestry professional and farmers alike, based on prior experience and research on other fast-growing timber trees.

20 In that spirit – as collaborators and informal advisers to both ENDA and the Federation – we have tabled recommendations to transform (not merely reform) the forestry enterprise project or alternative programs to respond to the needs, aspirations and grievances of these groups. The recommendations, rationale and detailed implementation of these strategies are described in a joint publication with ENDA and ISA (Rocheleau, et al. 1995).

References

Agarwal, B. (1992) The gender and environment debate: Lessons from India. *Feminist Studies* 18: 119–158.

Blaikie, P. (1985) *The Political Economy of Soil Erosion in Developing Countries*. London: Longman.

Blaikie, P. and H. Brookfield (1987) *Land Degradation and Society*. London: Methuen.

Bruce, J., L. Fortmann and C. Nhira (1993) Tenures in transition, tenures in conflict: Examples from the Zimbabwe social forest. *Rural Sociology* 58: 626–42.

Carney, J. and M. Watts (1991) Disciplining women? Rice, mechanization, and the evolution of Mandinka gender relations in Senegambia. *Signs* 16: 651–81.

Croll, E. and D. Parkin (Eds.) (1992) *Bush Base: Forest Farm: Culture, Environment and Development.* New York: Routledge.

Deere, C. et al. (1990) *In the Shadows of the Sun: Caribbean Development Alternatives and U.S. Policy.* Boulder: Westview Press.

Emel, J. and R. Peet (1989) Resource management and natural hazards. In R. Peet and N. Thrift (Eds.) *New Models in Geography* Vol. 1. London: Unwin Hyman Ltd., pp 49–76.

Escobar, A. (1992) Imagining a post-development era? Critical thought, development, and social movements. *Social Text* 31/32: 20–56.

Fortmann L. (1993) "Learning from People, Learning with People, Empowering People with Research." Paper presented at the Annual Meeting of the Rural Sociological Society. Orlando, Florida, August 9.

Fortmann, L. (Forthcoming) Talking claims: discursive strategies in struggles over property. *World Development,* Volume 23, issue 6, pp. 1053–1063, 1995.

Fortmann, L. and Bruce, J. (1988) *Whose Trees? Proprietary Dimensions of Forestry.* Boulder: Westview Press.

Guha, R. (1989) *The Unquiet Woods: Ecological Change and Peasant Resistance in the Himalayas.* Delhi: Oxford University Press.

Haraway, D. (1991) *Simians, Cyborgs and Women: The Reinvention of Nature.* New York: Routledge.

Harding, S. (1991) *Whose Science? Whose Knowledge? Thinking from Women's Lives.* Ithaca: Cornell University Press.

Hart, G. (1991) Engendering everyday resistance: Gender, patronage, and production politics in rural Malaysia. *The Journal of Peasant Studies* 19: 93–121.

Hecht, S. (1995) Surplus extraction? Women, extraction and rural industrialization in Acre, Brazil. Manuscript submitted to *Journal for Peasant Studies.*

Hecht, S. and Cockburn, A. (1989) *The Fate of the Forest: Developers, Destroyers, and Defenders of the Amazon.* London: Verso.

Lernoux, P. (1982) *Cry of the People.* New York: Penguin Books.

Lynch, B. (1994) "State Formation, Public Lands and Human Rights: The Case of the Dominican Republic and Los Haitises." Unpublished paper prepared for XVIII International Congress of the Latin American Studies Association. Atlanta, Georgia, March 10–12.

McAffee, K. (1991) *Storm signals: Structural Adjustment and Development Alternatives in the Caribbean.* London: Zed Books.

Mendes, C. (1989) *Fight for the Forest.* London: Latin American Bureau.

Merchant, C. (1989) *Ecological Revolutions: Nature, Gender and Science in New England.* North Carolina: The University of North Carolina Press.

Parajuli, P. (1991) Power and knowledge in development discourse. *International Social Science Journal* 127: 173–90.

Peet, R. and M. Watts (1993) Introduction: development theory and environment in an age of market triumphalism. *Economic Geography* 69: 227–253.

Peluso, N. (1992) *Rich Forests, Poor People: Resource Control and Resistance in Java.* Berkeley: University of California Press.

Rangan, H. (1994) "Devices and Desires: Rights and Negotiation in Natural Resource Management." Paper presented to the Association of American Geographers 90th annual meeting. San Francisco, California, March 29–April 2.

Richards, P. (1985) *Indigenous Agricultural Revolution: Ecology and Food Production in West Africa*. London: Hutchinson and Co. Ltd.

Richards, P. (1993) "Fighting for the Rainforest: Debating Global Environmental Change and Local Violence on the Liberia-Sierra Leone Border." Unpublished paper presented to the Association of Social Anthropologists of the Commonwealth, Oxford: Saint Catherine's College.

Rocheleau, D. (1987) A land user perspective for agroforestry research and action. In H. Gholz (Ed.) *Agroforestry: Realities, Possibilities, and Potentials*. Dordrecht: Martinus.

Rocheleau, D. (1991) Gender, ecology, and the science of survival. *Agriculture and Human Values* 8: 156–65.

Rocheleau, D., L. Ross, J. Morrobel and R. Hernandez. (1995) *Fanning the Forests, Gardening with Trees: Landscapes and Livelihoods in Zambrana-Chacuey, Dominican Republic*. ECOGEN Project Case Study. Worcester: Clark University.

Sachs, W. (Ed.) (1992) *The Development Dictionary*. London: Zed Books.

Sachs, W. (Ed.) (1993) *Global Ecology*. London: Zed Press.

Schmink, M. and C. Wood (1992) *Contested Frontiers in Amazonia*. New York: Columbia University Press.

Schroeder, R. (1993) Shady practices: Gender and the political ecology of resource stabilization in Gambian garden/orchards. *Economic Geography* 69: 349–65.

Scott, J. (1976) *The Moral Economy of the Peasant*. New Haven and London: Yale University Press.

Scott, J, (1985) *Weapons of the Weak: Everyday Forms of Peasant Resistance*. New Haven and London: Yale University Press.

Sharpe, K (1977) *Peasant Politics: Struggle in a Dominican Village*. Baltimore: University of Maryland.

Shiva, V. (1989) *Staying Alive: Women, Ecology and Development*. London: Zed Books.

Shiva, V. (1993) The greening of the global reach. In W. Sachs (Ed.) *Global Ecology*. London: Zed Press, pp 206–18.

Valerio, M. (1992) "Agroforesteria y conservación de suelos." Unpublished AGROSIL III document. Santo Domingo: ENDA-Caribe.

Vargus-Lundius, R. (1991) *Peasants in Distress: Poverty and Unemployment in the Dominican Republic*. Boulder: Westview Press.

Veras, J. (1984) Indice cronologico de la legislacion sobre asuntos forestales en la Republica Dominicana. In *Foresta: Alternativa de Desarollo*. Santiago: UCMM.

Part VI

A glimpse of gender systems in the 'North'

The geographic area of the northern hemisphere (the 'North') is comparable in size to the southern one, yet the English-language literature on gender and forests in the North is a fraction of what is published on the topic about the 'South'. Why is gender a seemingly unimportant line of inquiry in the North? One reason is illuminated by a cross-equatorial relational analysis by Arora-Jonsson (this volume) who suggests that countries with more progress in civil rights and gender equality may actually inhibit research on discriminatory processes and practices. In meetings about community forestry in Sweden, for example, a "tension in acknowledging unequal power relations" (p. 225 in original) existed that Arora-Jonsson did not see in community forest groups in India.

Despite awareness that key themes emerging from political ecology studies in the South are applicable to understanding human–environment relations in the North, Reed and Christie (2009) found that inclusion of gender as a factor in spatial analyses of the environment remains rare across North America, despite earlier urgings to do so by Fortmann (this volume). Reasons advanced by Reed and Christie (2009) for the lack of geographic studies into how power dynamics among people of different genders affect environmental change, conflict, and management in the North include: 1) views that gender is not relevant to access concerns and decision-making about natural resources, 2) research being published in journals or by scholars outside the field of geography, and 3) equating gender with women, who remain under-represented in natural sciences. Whatever the reasons are for the comparative lack of literature on gender and forests in the North, we had both fewer articles from which to choose and less familiarity with the subject. Part VI can, therefore, provide only a glimpse into important commonalities and key differences between the hemispheres. We selected two papers for illustration, both of which are from North America. Taken together with other papers in this volume (particularly Arora-Jonsson; Leach; Paulson; Nightingale), the papers by Reed (2003) and Norgaard (2007) suggest that research and practices to integrate gender and forests in the North will gain from using mixed methods. Approaches informed by systems theory, which bridges social and ecological sciences, are also fruitful for understanding gender and forests. As Paulson (first published in this volume) notes, "gender systems influence a range of institutions, together with the distribution and use of different assets, in ways that impact infrastructure and biophysical environment as well as human bodies, lives and livelihoods" (p. 268).

Reed (2003) combined fieldwork with census data and government documents into a form of 'action research', and her results highlight how ideals of masculinity influence power relations in the lives of both men and women. Reed shows how men's 'doing gender' (see Glossary) and women's response to that both serve to reinforce women's marginality in forests of British Columbia, Canada.[1] Insofar as women became involved with work in forests, their own perceptions of men as strong, brave, and dominant were undermined. This paradoxical observation by Reed in Canada was made earlier in Pacific coast states of the USA (see Colfer 1977; Colfer and Colfer 1978; Carroll 1995).[2] Policies, regulations, and studies by the national government acknowledge women's presence in forests and forest communities primarily in relation to their husbands, reminiscent of Elmhirst's Indonesian findings (this volume). The methods Reed used in her 2003 paper can provide guidance for policymakers and managers in the North who are now interested in processes of community involvement. For example, in the USA, there are several community forest groups being formed (often collectively referred to as 'collaboratives') with various types of funding and leadership. Some collaborative groups in the USA are initiated by federal agencies, whereas others are formed by community leaders.[3] A subsequent paper by Reed (2010) builds on topics of cross-boundary interest in North America; namely, a need for community forest group processes to be mindful of 'elite capture' (see Glossary) that can directly or indirectly marginalize concerns related to gender.

Reed (2003) points out a number of similarities with countries in the South: significant landownership by national government; logging being done over vast regions by private companies under license to the government; conflicted relations between various interest groups and the state; influence on government policies and the logging industry by environmental groups; and concerns about health and safety in logging. Notably, there is also a shortage of detailed information about gender in government and industrial statistics. Indeed, it is common for the actual involvement of women in forest-related jobs to be under-represented in formal counts in the North and South, which contributes to a persistent invisibility of women in 'productive' spheres. Reed reports the same gendered discrimination in formal work settings (for similar experience from Australia, see Eriksen 2013) as is found in forestry in developing countries, where women tend also to be last hired, first fired, and have difficulties getting promotions. Both Reed and Eriksen found that women continually have to 'prove themselves' on the job and struggle to be heard in meetings (see Mallory 2006, 2010, for similar problems – including fears of sexual violence – for women among anti-logging activists in the USA).

In the North, the change from tree cutting as the primary forest-related task to the inclusion of a number of less stereotypically male professions (like silviculture, planning, inventory) has opened up opportunities for women that do not typically exist in the South. The numbers of female university graduates and formally employed women are much higher in Canada than in typical developing countries, where women have tended to be relatively undereducated and involved in *informal* work.[4] This tendency is, however, changing with educational levels improving in the South. Governmental regulations (actually *implemented* in Canada) have also made government employment more hospitable for these Canadian women than in private industry.

Norgaard's (2007) paper is about an invasive plant (spotted knapweed) and responses by different community groups to managing it in forests of northern California, USA. She cites similar biologically invasive threats globally due to increased human mobility. She notes the frequent dilemma of needing to control a species with pesticides and/or herbicides on the one hand and community perceptions of health risks on the other. Such community concerns have received less press in the forests of the South – partly through ignorance of impacts, perhaps also partly through continued use of compounds no longer legal in the North.[5] But there is long experience of such problems with pesticides and herbicides in agricultural settings.

The North and the South also share problems pertaining to ancestral land. Schroeder, Elias and Carney, Li, Elmhirst, and Rocheleau and Ross (all this volume) discuss related tenure conflicts and uncertainties with gender implications in the South. Norgaard (2007) describes the management controversies among diverse groups (Indians, non-Indians, miners, loggers, hippies, and US Forest Service personnel). These groups vie with each other for control, but also sometimes cooperate with each other. The paper provides interesting perspectives on and information about community action (in which women unusually played equal roles) in response to perceived threats.

A central question in Norgaard's research is 'who suffers from environmental degradation?' This question recurs in the South, and a common trans-equatorial theme is lack of trust between communities and governmental agencies which is often built on previous experience. Evidence exists that minority groups worldwide have greater exposure to health risks.[6] In Norgaard's study, management responses to spotted knapweed varied among community groups, with agency staff viewing herbicides as the best solution and Karuk community members seeing hand weeding as the solution to avoid herbicides and offer local employment. These kinds of top-down and bottom-up conflicts occur in forests everywhere. The fact that the spraying of herbicides has stopped suggests that valuable lessons can be learned from this experience. Creative framing of the issues, a hospitable institutional framework, networking, and alliance building all played a part in developing viable solutions. Norgaard sees the equal involvement of women as central to obtaining a full story, showing Karuk women's own embodied experience of these risks, for themselves, their elders, and their children.

In contrast to much other published literature in the North or South, Norgaard attends to intersectionality (gender, race, age, residence, and employment), health, and risk. Differential gendered perceptions of risks to health are poorly studied.[7] Norgaard found varied concerns about health risks associated with herbicides in the rural populations she studied (Karuk Indians, the non-Indian community, and US Forest Service personnel). Concerns were lowest among male US Forest Service personnel and highest among Karuk of both genders.

Looking at (and for) materials on gender and forests in the North has proven a valuable endeavour. These two authors have dealt with rarely addressed issues, like caring roles and the direct implications of health problems for people's lives and work – as called for in Chapter 1. Some parallels between the North and the South have come out clearly, providing a stimulus for further studies in both regions.

Notes

1 De Mel *et al.* (2013) also highlight women's roles in perpetuating gender stereotypes, insensitive gender attitudes, and domineering masculinities in Sri Lanka.

2 'Doing gender' as it is done in Canada does not reflect masculine ideals everywhere. In East Kalimantan, Colfer found a reluctance among everyone to wield a chainsaw, overcome only by the good wages available. Men (and women) saw it simply as dangerous work, recognizing that it required greater strength than most women or men had in that community of small-statured people. The quintessentially male act there involved circular migration, with the danger, teamwork, and adventure that involved (traits similar to those valued and associated with logging in Canada). Reed's presentation of masculine identity in the Canadian woods is very close to descriptions of 'hegemonic masculinity' (see Glossary).

3 See www.fs.fed.us/restoration/CFLRP/ for an introduction to these collaboratives. See Colfer (2005) or German *et al.* (2010) for guidance on more thorough versions of collaborative forest management.

4 Mies (2007) notes a trend of increasing informal work for both men and women.

5 In several sites in Borneo, Colfer heard complaints about chemicals applied to logs rafted down rivers and local worries about their effects on water quality and human health (though there were no public fora in which these concerns could be expressed).

6 For example, see analyses by Dounias with Colfer (2008), Froment (2008), Gomez (2008), and Persoon (2008).

7 Andersson *et al.* (2010) studied men's and women's financial risk tolerance in Sweden's forests, finding, interestingly, that men and women there have similar risk tolerance in general, but as the amounts risked move from insubstantial to substantial, women tolerate more and men tolerate less risk (contrary to previous studies).

References

Andersson, M., Kansson, C. H., and Holmgren, L. 2010. Non-Industrial Private Forest Owners' Financial Risk Taking: Does Gender Matter? *Scandinavian Journal of Forest Research* 26 (supplement 9): 6–13.

Carroll, M. S. 1995. *Community and the Northwestern Logger: Continuities and Changes in the Era of the Spotted Owl.* Boulder, CO: Westview Press.

Colfer, C. J. P. 1977. *Women's Communication and Family Planning in Rural America: The Case of Bushler Bay.* Honolulu, HI: East–West Communication Institute (Case Study #4; reprinted 1978).

Colfer, C. J. P. 2005. *The Complex Forest: Communities, Uncertainty, and Adaptive Collaborative Management.* Washington, DC, Resources for the Future/Center for International Forestry Research.

Colfer, C. J. P., with Colfer, A. M. 1978. Inside Bushler Bay: Lifeways in counterpoint. *Rural Sociology* 43 (2): 204–220.

De Mel, N., Peiris, P., and Gomez, S. 2013. *Broadening Gender: Why Masculinities Matter (Attitudes, Practices and Gender-Based Violence in Four Districts in Sri Lanka).* Colombo, Sri Lanka: Care.

Dounias, E., with Colfer, C. J. P. 2008. Sociocultural Dimensions of Diet and Health in Forest-Dwellers' Systems. In C. J. P. Colfer, ed., *Human Health and Forests: A Global Overview of Issues, Practice and Policy.* London: Earthscan, 275–292.

Eriksen, C. 2013. *Gender and Wildfire: Landscapes of Uncertainty.* New York: Routledge.

Froment, A. 2008. Biodiversity, Environment and Health among Rainforest-Dwellers: An Evolutionary Perspective. In C. J. P. Colfer, ed., *Human Health and Forests: A Global Overview of Issues, Practice and Policy.* London: Earthscan, 259–274.

German, L. A., Tiani, A.-M., Daoudi, A., Maravanyika, T. M., Chuma, E., Beaulieu, N., Lo, H., Jum, C., Nemarundwe, N., Ontita, E., Yitamben, G., and Orindi, V. 2010. *The Application of Participatory Action Research to Climate Change Adaptation: A Reference Guide.* Bogor, Indonesia: International Development Research Center/ Center for International Forestry Research.

Gomez, G. G. 2008. Forest Disturbance and Health Risks to the Yanomami. In C. J. P. Colfer, ed., *Human Health and Forests: A Global Overview of Issues, Practice and Policies.* London: Earthscan, 239–258.

Mallory, C. 2006. Ecofeminism and Forest Defense in Cascadia: Gender, Theory and Radical Activism. *Capitalism Nature Socialism* 17 (1): 32–49.

Mallory, C. 2010. The Spiritual is Political: Gender, Spirituality, and Essentialism in Forest Defense. *Journal for the Study of Religion, Nature and Culture* [Equinoxonline] 4 (1): 48–71.

Mies, M. 2007. Patriarchy and Accumulation on a World Scale – Revisited. *International Journal of Green Economics* 1 (3): 268–275.

Norgaard, K. M. 2007. The Politics of Invasive Weed Management: Gender, Race, and Risk Perception in Rural California. *Rural Sociology* 72 (3): 450–477.

Persoon, G. A. 2008. Hidden Suffering on the Island of Siberut, West Sumatra. In C. J. P. Colfer, ed., *Human Health and Forests: A Global Overview of Issues, Practice and Policy.* London: Earthscan, 333–346.

Reed, M. G. 2003. Marginality and Gender at Work in Forestry Communities of British Columbia, Canada. *Journal of Rural Studies* 19 (3): 373–389.

Reed, M. G. 2010. Guess Who's (Not) Coming for Dinner: Expanding the Terms of Public Involvement in Sustainable Forest Management. *Scandinavian Journal of Forest Research* 25 (Supplement 9): 45–54.

Reed, M. G., and Christie, S. 2009. Environmental Geography: We're Not Quite Home—Reviewing the Gender Gap. *Progress in Human Geography* 33 (2): 246–255.

14 Marginality and gender at work in forestry communities of British Columbia, Canada

Maureen G. Reed

> Let me tell you about loggers. I married one 18 years ago. I couldn't resist the smell of utter maleness in damp, sweaty black wool underwear sweetened by the heavy scent of fresh sawdust and chainsaw exhaust. . . . I still love the romance of big men, big machines and big trees . . . . loggers are the last of a dying breed of men who "work" for a living. Their work is dangerous and back breaking.
>
> Maureen Henderson, Letter to the Editor, *Chilliwack Times* 1993: 9

Introducing the paradox of paid work

How do women enter the practices and discourses of forestry work? Typically, forestry conjures up images of logging, an occupation characterised by physical labour, hard work, danger and even drama. These images are symbols of masculinity in forestry communities, symbols to be admired and even romanticised (Henderson, 1993 above). There is little room in Henderson's letter, excerpted above, to recognise women's contributions to forestry aside from their dedication to 'their' men and to forestry culture more broadly. This perspective of women's dedication has also permeated contemporary policy processes. In British Columbia Canada, during a comprehensive policy review of forestry on Vancouver Island, women were considered part of forestry communities only when they were attached as partners to male workers who were considered the dominant breadwinners. In 1994, the Commission on Resources and Environment (CORE) reported that "on Vancouver Island about 95% of resource workers are male, and about 80% of them are married. This suggests that there are about 15,000 women on Vancouver Island whose spouses work for resource industries" (Commission on Resources, Environment (CORE), 1994a, p. 205). In this statement, CORE only considered women by their conjugal status and recognised only one possible status. Furthermore, CORE neglected that women themselves

Maureen G. Reed, "Marginality and Gender at Work in Forestry Communities of British Columbia, Canada," *Journal of Rural Studies* 19 (3), 373–389.

might be forestry workers with insights and experiences relevant to such workers' changing employment conditions.

This depiction of women's solidarity with, but exclusion from, forestry also infused early feminist scholarship focused on rural resource communities.[1] Research undertaken in resource towns during the 1970s and 1980s within a labour studies framework established an image of women's unwavering devotion to male-dominated occupations, particularly during times of economic strife (e.g. Ali, 1986; West and Blumberg, 1990; Maggard, 1990). The identity of these women was constructed as conservative, dependent, myopic, yet capable of rendering unstinting support during tough economic times (Gibson-Graham, 1994). And yet, the practices and discourses of forestry work do not end there. I suggest that there is a more complex set of narratives based on marginality and desire, consistency and contradiction, associated with forestry-town women.

My own research investigated the social lives of forestry-town women living on northern Vancouver Island, British Columbia (BC) during the mid-1990s. At this time, changes were occurring in the structure of the industry, in policy regimes that governed land allocation and management, in provisions for social welfare, and in relationships with Indigenous peoples.[2] Problems of forest management in BC that prevailed then were related to the transition from old-growth to second growth timber,[3] historic over-cutting, inadequate replanting and silviculture, changes in harvesting and processing technology, land tenure, international market circumstances, corporate restructuring and a failure to resolve the land question with Indigenous peoples throughout most of the province (for more detailed discussion, see Barnes and Hayter, 1997; Clapp, 1998; Hayter, 2000; Cashore et al., 2001). International environmental campaigns against logging in the coastal rainforests that had developed over 20 years required that government act to protect forest resources and non-timber values.

In 1991, social democrats represented by the New Democratic Party (NDP) were elected and held power for almost 10 years. When first elected, the party was given an overwhelming mandate to "do good" on a number of environmental promises. Yet its power base was historically built on its labour of labour or, in its words, "the working people". Thus, it faced two historically opposing constituencies. In addition, the party (in opposition at least) was sympathetic to the rights of First Nations peoples. Corresponding to the rise of environmental activism, First Nations peoples mounted powerful political and legal campaigns to protect their lands, resources, and rights of self-determination. In 1999, the first contemporary treaty in BC, the Nisga'a Agreement, was ratified after almost 30 years of negotiations. The negotiations and the completion of this landmark agreement, however, created a great deal of debate for both Indigenous and settler British Columbians. The public debate around its signing served as a reminder of the growing importance of Indigenous rights and interests as well as the uncertainties associated with their recognition and resolution.

In an effort to balance these demands, during the 1990s, the NDP introduced several new acts or policy initiatives that affected Crown land allocation and management practices,[4] and created new policies to assist workers in making the necessary economic "transition" (Price Waterhouse, 1995; Prudham and Reed, 2001; Reed, 1999, 2003). Consequently, timber available for harvesting declined and costs for land management increased. Government policy initiatives also changed the nature of forestry work. For example, a Forest Practices Code, introduced in 1994 (and proclaimed the following year), decreased opportunities for conventional logging and production and increased opportunities for jobs in silviculture, engineering, surveillance, inventory and planning. These occupations gave opportunities for women to enter the forestry workforce. While technological and market changes were also important elements of restructuring of the forest industry,[5] forestry families, workers and companies remained largely united against government and environmental organisations to retain their hold on remaining stands (Clapp, 1998). By the mid-1990s, the public policy changes met with massive protests from men and women living in forested rural communities who argued that forestry communities and culture were now endangered species in need of protection.

In this context, I began my research about women's experiences of and perspectives about living in forestry communities. Their attachments to forestry employment revealed a paradox. The paradox lies in women's shared experiences of and opposition to their marginality from the industry and local forestry culture while simultaneously reinforcing it through both discourse and practice. I found that, consistent with earlier depictions of women's lives, the women I encountered retained a strong attachment to forestry as an occupation and as a way of life. These attachments were made, in part, from their association with, and support of, their partners and other family members in forestry. For some women, their maternal role remained an important self-referent, providing guidance to separate the appropriate attitudes and activities associated with forestry from the inappropriate ones (see also Murphy, 1995). This positive role making associated with "traditional" norms of forestry as men's work were challenged by those women who chose to work in that sector. Women who sought employment in forestry reported on outright exclusion from work opportunities or documented both sexism and marginalisation in their work environments. Yet, despite these experiences, all women retained strong support for their partners in forestry and/or the forest industry more broadly. Indeed, some women even celebrated the characteristics of masculinity that define forestry culture and effectively eliminated equal opportunities for women in forestry employment. Consequently, and perhaps inadvertently, they too reinforced their own marginality.

The purpose of this paper is to explore and explain this paradox. I argue that to understand how women enter forestry jobs and forestry discourse, it is important to locate women within the dominant employment structure as well as to interpret the meanings they grant to forestry employment—both for themselves and for other family members. I pursue this argument in two ways. First, I suggest that women's employment in forestry is located within layers of marginality and exclusion that operate both from outside and from within forestry communities. From the outside, women's experiences of forestry employment are rendered marginal by academics, government agencies and policy makers. Women's representations in forestry work are limited, in part because those who count forestry have historically overlooked types of employment where women are most likely to be found. Second, I suggest that women contribute to their own marginality by their adherence to discourses and practices that reinforce stereotypes about the industry and consequently exclude them from participating more fully. Their reinforcement of their own marginality is revealed in qualitative interviews and focus groups conducted with women on northern Vancouver Island. I have organised the paper as follows.

First, I summarise my methodology and provide a brief overview of the characteristics of the women in the study region. Next, I examine how societal norms and academic biases have worked together to reinforce the notion of women's limited involvement in forestry work. I then turn to the women interviewed for this study to obtain their interpretations of forestry work. I illustrate how women's contradictory positions simultaneously support and are repelled by dominant stereotypes of forestry culture that have marginalised women's experiences. I explain this contradiction by a theoretical discussion of social embeddedness and occupational community. These concepts help me to explain how and why women contribute to their own marginality without reverting to the more simplistic dualism of "victim" or "victor" that characterised earlier work about women in resource towns.

Research methods

My study of forestry-town women combined an analysis of policy documents and Census data to understand the "numbers" of women in forestry with methods that might be classified as feminist participatory action research. I undertook 50 interviews and three focus groups with women living in nine forestry communities on northern Vancouver Island (Figure 14.1) where, according to one government study, 51% of employment income came from forestry occupations in 1996 (Horne, 1999).

To gather the primary data, I followed methodologies outlined by Reinharz (1992) and McDowell (1993a, 1993b) and in particular, the strategies described by Gibson-Graham (1994). Initially, 32 in-depth interviews were conducted by myself and my research assistants. I deliberately sought out women with connections to the forest industry and/or who were involved in influential community organisations. From this group, 10 women were selected to be community researchers in order to discuss local issues and to be trained to conduct in-depth interviews. In total, 50 interviews were completed for analysis. Upon undertaking preliminary analysis, I conducted three focus groups with local women who had not been previously interviewed. The focus groups served a dual function; they provided an opportunity for me to provide direct feedback to the local communities about the nature and status of the research and they assisted directly in corroborating and refining emerging themes and social categories I had established. After the focus groups,

I undertook further analysis and then held a workshop with the community researchers in which I again presented my results. The community researchers offered more suggestions for refining the analysis and for undertaking extension work in their communities.

The sample was not selected randomly, nor was it statistically representative of the region at large. Table 14.1 reveals characteristics of women interviewed in 1997 in relation to those living in the Census region, the Mount Waddington Regional District (MWRD) in 1996 (see Fig. 14.1). Approximately 82% of population of the MWRD is located in the study area, so these Census data provide the basis for a general comparison of the study group with the wider population of the region. Table 14.1 illustrates that most of the women interviewed were of employment age. In fact, 98% of the interviewees were between 20 and 64 years of age. The women had higher rates of university

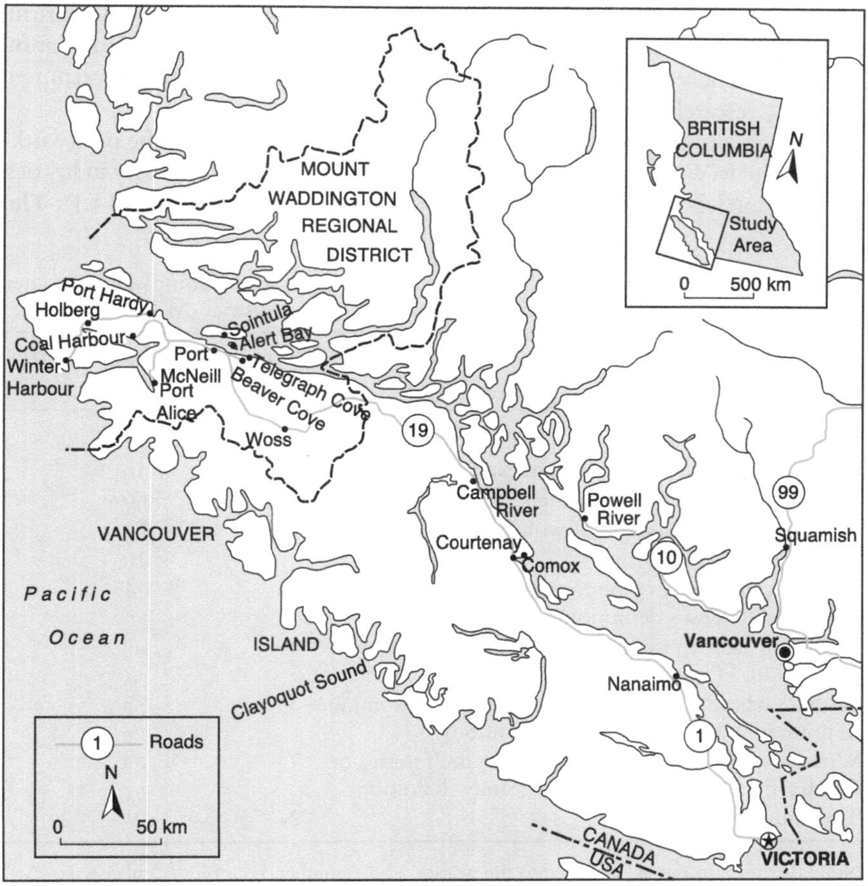

Figure 14.1 The location of the study area and boundaries of the mount Waddington Regional District (MWRD).

education and labour force participation and lower rates of unemployment than women in the MWRD. As intended by the sampling strategy, women in this study were almost four times as likely to work in primary industries as women in the MWRD more broadly.[6] The method of calculation and discussion of the employment data are described in a subsequent section.

Looking for women in the woods

Counting women "in" forestry

It is still rare to find women loggers in the forests of British Columbia. It is also hard to find them in the data. Indeed, reliable, valid, current, complete and commensurable data on gender and employment in forestry are not readily available from government agencies or private companies. Company records typically do not distinguish between women in non-traditional jobs (e.g. planers) and those employed in traditional jobs (e.g. secretaries) within the firm. Employment counts by resource sector, location and gender in combination are not routinely made available either by Statistics Canada, other government agencies, researchers or industrial employers.

According to the Census, in 1996, just under 4% of women in the paid workforce in the MWRD were employed in occupations unique to primary industries, representing less than 2% of the total workforce of the region (Table 14.1). This

Table 14.1 Selected characteristics of women in the study group compared to women (and men) in the Mount Waddington regional district (MWRD)

	% in study group (1997)	*% in MWRD (1996)*
Women aged 20–64	98	59
Women with some university education	35	16
Women living with male partners[a]	84	86
Labour force participation of women	80	71
Unemployed women	7	10
Employed women in "jobs unique to primary industries" (Census definition)[b]	15	4
Employed women in "forestry or forestry-dependent occupations" (Study definition)[b]	40	n.c.
Women whose employed partners are in "jobs unique to primary industries" (Census definition)[c]	39	25
Women whose employed partners are in "forestry or forestry-dependent occupations" (Study definition)[c]	74	n.c.
Total number	$N = 50$	$N = 7220$

Note: that proportions are rounded to the nearest whole number. n.c.: Not calculated.

a There were no known same-sex couples in the sample.

b For this calculation, only 40 women or those participating in the labour force, were included.

c For this calculation, only 39 men were included.

compared to 25% of men in the paid workforce, representing 14% of the total workforce of the region. This depiction, however, is an incomplete and inaccurate assessment of the importance of forestry as an employer in the region and as an employer for women and men, respectively. Jobs unique to primary industries include jobs in other sectors such as agriculture and mining and this category omits jobs such as truck drivers, machine operators, and registered foresters that may be specific to forestry. To obtain more details, (much) more money must be spent on special tabulations. Few academics have ordered these tabulations and yet they provide slightly different results.

A special tabulation of the 1991 Canada Census ordered by a consultant company revealed more detailed categories of employment. This tabulation, completed for BC as a whole, indicated that approximately 10% of the provincial workforce in logging industries were women and 28% of jobs related to forest services (CS/ RESORS Consulting Limited., 1997). These numbers are higher than those suggested by academic studies that have relied on basic Census data (e.g. Marchak, 1983; Grass and Hayter, 1989; Hayter and Barnes, 1992; Randall and Ironside, 1996; Halseth, 1999; Hayter, 2000). Nonetheless, as I point out later, these data are still derived from a narrow definition of forestry employment, and thereby underestimate the number of women who work in forestry occupations. For example, the Census counts include jobs in forest services such as inventory and mensuration (measurement), while jobs in management, information or administrative services are omitted.

Researchers working in other contexts have also found that Census definitions have typically not captured the extent and nature of women's participation in renewable resource sectors (e.g. Brandth and Haugen, 1998; Sachs, 1996; Wright, 2001). Academic work reinforces and is shaped by biases in the availability of data. Empirical studies that use standard Census categories continue to report the very limited participation of women in forestry occupations (e.g. Randall and Ironside, 1996; Halseth, 1999). Some researchers have attempted to address this bias by undertaking employment counts within specific firms (e.g. Grass and Hayter, 1989; Hayter and Barnes, 1992), but these data, do not provide a community-based perspective and they also have strict limits to what is counted "in forestry". The resulting research confirms and is shaped by these empirical limitations.

Academic stories

Forestry has been depicted as one of the most masculine rural occupations (Dunk, 1991; Carroll, 1995; Brandth and Haugen, 1998, 2000). According to a broad literature that addresses rural resource communities, traditional conceptions of femininity and masculinity are strong in rural resource communities where women are seen as the primary care-givers and nurturers, and men as the providers and decision makers (Gibson, 1992). These conceptions are reinforced by a dominant ideology that locates women's "rightful" place to be the home and contributes to a relative lack of employment prospects for women outside the home. The impacts of geographic and social isolation, lack of employment opportunities, financial and emotional dependence on spouses, company domination of social life, dominant ideology and limited social services have generally been viewed as limiting women's opportunities to take up paid employment in rural resource communities (Cloke and Little, 1990; Gibson, 1992; Little, 1987; Marchak, 1983; Seitz, 1995; Warren, 1992). Furthermore, "traditional" gender roles and relations have frequently been viewed as solid and unchanging.[7]

These views have shaped theoretical assumptions and empirical expectations. In contrast to a large body of labour theory to explain industrial and urban situations, there is a paucity of theory that explains gendered labour practices in rural places and resource sectors outside of agriculture (Halseth, 1999).[8] In labour segmentation theory advanced by Doeringer and Piore (1971, p. 165), the primary segment is characterised by "high wages, good working conditions, employment stability, chances of advancement, equity, and due process in the administration of work rules." In contrast, jobs in the secondary segment tend to have "low wages and fringe benefits, poor conditions, high labour turnover, little chance of advancement, and often arbitrary and capricious supervision" (Doeringer and Piore, 1971, p. 165).

According to Hayter (2000), this model is an effective representation of labour market conditions in BC's forest economy in the post-World War II era. In this context, the secondary segment is composed of non-union peripheral workers whose wage levels and employment stability are typically less structured than in the primary sector. Feminist theorists have pushed the explanation a bit further by noting that workers in the secondary sector are more likely to be non-union and female, and belong to a visible minority. According to feminist geographers Hanson and Pratt (1995, p. 6):

> [Women's exclusions from the primary segment] build on the sexist practices of male employers and employees. Male employers may be reluctant to hire women for the most prized jobs because of gender stereotypes, worries about complaints from male employees, and their more general fears about losing male advantage. . . . White, male employees also have organised through unions and professional organisations to shelter jobs for themselves.

These labour theories are important because they draw attention to ways in which local social and cultural expectations shape employment outcomes. However, they have not been systematically applied to rural places or resource industries. In this context, cultural expectations provide important explanations for employment opportunities. For example, Dunk (1991) and Carroll (1995) illustrate how male-dominated nature of occupations within resource-based communities has created and elevated the importance of a working*man*'s culture. Institutional factors, local practices, and sexist attitudes shape the culture of forestry towns, rendering less important, and in come cases, invisible, the nature and extent of women's employment within forestry communities. In this light, the classification, "primary", not only denotes occupations based on the extraction of raw materials, but also denotes occupations that hold primary importance in the local culture and economy. In short, jobs in "primary" (extraction) industries have "primary" importance to local communities and policy makers. Women's employment within these communities is viewed as secondary or tertiary, not only because they are more likely to be in manufacturing and service sectors, but also because they are seen as being of second and third-order importance to the overall workings of the forestry community (see also Marchak, 1983).

This rendering of women's paid work in forestry towns as "secondary" does not simply include women in secondary and tertiary sectors. Even women who figure within conventional counts of forestry-related occupations (e.g. professional foresters) are subject to limitations and erasure. Women employed in forestry jobs discussed restrictions in hiring and promotion practices and the lack of recognition for the contributions they made directly to the industry. The bias that focuses all attention on male wage earners is also evident in academic work that might otherwise provide accurate empirical descriptions of employment.

For example, Eric Grass and Roger Hayter (1989) collected data from a random sample of 63 plants during 1981 and 1985 to determine the employment characteristics of workers in those plants who experienced layoffs during the time of widespread recession in the industry. They noted that with the exception of women middle-managers (who were few in number), female job losses were more rapid than men's. This was particularly evident in clerical and production jobs. Yet, their data also revealed that women occupied only 3–4% of industry jobs including administration, trades and production work. More than 50% of the clerical positions were occupied by women.[9] While their analysis revealed a higher vulnerability of women workers to job loss, their data omitted employment in other sectors and locations of the industry, most notably in head offices of firms, in woods work (e.g. loggers, camp cooks), in research and development as well as in regulatory and planning positions of the Ministry of Forests. All of these sites are important components of the industry and are becoming increasingly important places of paid employment for women. Yet, the absence of research about women shapes policy debates and "available" strategies that are advanced during times of economic and social transition.

Policy debates

During times of economic transition, there is a distinct policy preference for protecting only those jobs that are directly dependent on forestry. Government programmes have focused on retaining, retraining or retirement packages aimed at the male wage earner in forestry occupations—specifically woods workers and those who worked in sawmills and pulp and paper mills. On Vancouver Island, a regional planning process undertaken by CORE was charged with determining new land use allocations that met the increasing public demands for environmental protection and with developing a transition strategy to assist forestry workers who would be displaced as a result of those allocations. As noted in the introduction, CORE only recognised women who were "attached" to male resource workers. In its final report, CORE predicted that at least as many jobs in secondary and tertiary sectors would be lost as in the primary sector, yet it made no recommendations to support those sectors. Instead, new governmental programmes were put in place for those who qualified under arrangements with (un)employment insurance.[10] Due to the structure of the industry, insurance provisions, and other issues of accessibility (e.g. lack of transportation and child care facilities), most of these programmes were primarily available to male workers.

CORE was not ignorant of the gendered impacts of its recommendations. Indeed, its report provided a brief catalogue of employment prospects and social impacts affecting women. CORE acknowledged that the disproportionate stress associated with economic dislocation would be felt by women. Furthermore, it cited a report from another province that suggested that labour adjustment was more difficult for women than for men. The report indicated that "women, in fact, take a greater cut in pay compared with men when they are displaced . . . they experience significantly more long-term unemployment and earn less when they do land a job. Because of this, there is justification for special efforts to ensure that women have full access to labour adjustment services" (Ontario Ministry of Labour (n.d.), (1994, pp. 75–76) cited by CORE, 1994a). Yet, these findings never formed a part of the transition strategies that were formulated and no adjustment programmes were created that specifically targeted women.

In sum, women's work in forestry has been mainly unexamined by researchers and policy practitioners because it intersects less obviously with male employment (on this point, see also Egan and Klassen, 1998; Brandth and Haugen, 1998; and in other resource sectors, note Porter, 1985; Wright, 2001; Sachs, 1996). Women's employment (whether on the "main stage" of forestry jobs or in "supporting roles") is viewed as secondary, both in terms of its relation to the primary industry and in terms of its importance to the overall workings of the community. Yet, alternative classification schemes that account for the gendering of work sites, as illustrated in the following section 4, reveal a different story.

Gender, work and marginality in forestry

Recent research in feminist geography has placed emphasis on diverse interpretations of gender that consider its spatial, cultural and historical manifestations. In this context, gender does not simply refer to characteristics of men and women, but also as relationships, practices and processes that give meaning to gender constructs (after Connell, 1995; Brandth and Haugen, 2000). Gender relations and identity can be studied in specific contexts such as different work sites. I adopt Brandth and Haugen's (2000, p. 344), definition of sites as "lived realities . . . [and] mental spaces constituted by the locations of family, work, occupation, class, type of industry, organisation and so forth." As Brandth and Haugen (2000) point out, forestry is not a single work site, but many. Work sites in the industry vary according to the types of work (e.g. woods work, boardrooms) and types of workers (e.g. managers and decision makers, loggers).

Gendering of work sites occurs in the structural hierarchies of power, in the practices and activities of work, and in the ways in which workers display their gendered characteristics (Brandth and Haugen, 2000). In discussing women's employment, I consider the classification offered by Brandth and Haugen (1998) where jobs were organised under headings of practical forestry work, expert work and knowledge, and organisational activities. I focus on the first two aspects here.[11] In documenting women's employment in this way, I also illustrate that women within forestry occupations have an ambiguous relationship with the industry and the culture it has engendered.

Many women I interviewed on Vancouver Island worked in "practical forestry work" as technicians, scalers, front-end loaders, enforcement officers, stream restoration workers and camp cooks. Despite their physical locations in the woods, these jobs typically did not confer the same status as men's work in logging and transportation. Women also worked in "knowledge and expert" sectors devoting their skills as administrative officers (e.g. purchasing, sales, accounting) in the Ministry of Forests or for companies. Others were engaged in planning and consultancy work related to the new regulations, so for some women, transition meant new job opportunities in silviculture, planning, engineering, enforcement, and administration. Women worked as private consultants, accountants, public educators, and administrators—all related to the expert work and knowledge of the forest industry. The Census interpretation that just under 4% of women in the region are in occupations unique to primary industries was an exceedingly narrow view to be applied to women in forestry and did not capture this range of forestry-related employment.

For example, in 1997, the Ministry of Forests District Office located in Port McNeill had a workforce of 89 in 1997, 30 of whom were women. Sixteen of 18 administrative jobs were held by women, while 11 of 53 people with occupations as technicians, foresters and planners were women (Ministry of Forests, Port McNeill District, 1998, pers. comm.). None of these jobs was classified as a forestry job by Census definitions. Women I interviewed openly challenged Census definitions, arguing that many jobs conventionally classified in other ways, are jobs that are directly reliant on forestry. As forestry occupations move beyond the physical work related to tree harvesting to include other forms of work, opportunities to employ women continue to diversify.

To "test" this discrepancy, I used the Census definition to classify the number of women I interviewed as forestry workers. In this tally, 15% of the employed women I interviewed would have been classified in forestry jobs. However, I also tallied the number of women in forestry according to the definitions provided by the women themselves. This classification revealed that 40% of employed women I interviewed might be counted as forestry workers (Table 14.1). Using a similar strategy, I classified and re-classified the jobs of the partners of interviewees and found that the proportion of their jobs classified as forestry or forestry-related rose from 39% to 74%.[12] This exercise revealed that exclusion may occur by outside researchers who determine "what counts" as forestry activity. However these numbers are tallied, these kinds of calculations have important policy implications as they determine whether employment, re-training or other social programmes are necessary and who can qualify to receive them (Fitchen, 1991). The more limited the numbers, the less likely workers will obtain assistance during times of economic and social change.

My research also revealed that marginality comes from within forestry communities as well. Women in practical forestry work as well as in fields requiring expertise and knowledge expressed their marginality from within the work force. For those working in union jobs, salaries were standardised along a scale that gave recognition to training, work experience and seniority. Union workers were well paid and women reported being able to earn additional income through overtime. Notwithstanding these benefits, sexist practices within the union and within the jobs were subtle, pervasive, and continued to exclude women from the discourses and practices of forestry work. They related to women being (in)visible for new training and promotion opportunities, having to prove over and over again that they were capable to undertake new tasks, and simply being heard in union meetings. For those working in non-union positions, women believed that hiring, wage-setting, and promotion practices were extremely irregular. Women from all job classifications experienced sexism within their daily work lives.

Table 14.2 illustrates five forms of sexism identified by women working in forestry. This situation was pervasive, in part, because of the strong attachments forestry occupations have to masculine identities. Masculine identities, in part, have been built on the notion that forestry jobs, from logging to manufacturing, require hard, dangerous, physical work, often requiring long hours, and an ability to adjust to the rough and tumble found in logging camps. Women who attempt to enter into non-traditional occupations in forestry challenge this gender ideology directly (Brandth and Haugen, 2000). Some I interviewed did so without success. One woman was told during a job interview that she would not be hired as a logger because "she could not handle the language". She viewed this excuse as a metaphor for other activities that might be expected during work in the woods and the logging camps.

In forestry manufacturing jobs, where men have "typically" held jobs on the shop floor,[13] sexist attitudes have prevented women from even being considered for particular occupations. In the gendered division of labour, secretarial positions in sawmills or paper mills paid far less with fewer benefits than positions on the production line. Yet women did not typically apply for, or gain, such positions, even in the 1990s. The presumed lack of physical strength was reason for women to be excluded from consideration for the positions requiring heavy lifting. In addition, women's success in obtaining work on the production line would challenge the deep-rooted division of labour as well as the availability of masculine spaces on the shop floor (Preston, 2002). One woman described the

Table 14.2 Forms of sexism expressed by women in forestry occupations

Form of sexism	*Quotation*	*Job classification*
Stereotyping	There's a tendency for a lot of the guys that may not really know you, especially if they're new to call you the secretary. You know, if you're a woman, and you work, then you must be the secretary.	Accountant in a private company
	. . . because there was a lot of camp work, a lot of guys. They wouldn't allow me to go into that situation. They had problems with a female sleeping out there, so they figured all females are the same.	Registered professional forester
Promotion is limited	I haven't heard of anyone from up here (scaling) ever being promoted to any position like quality control . . .	Scaler
Not being taken seriously	. . . basically, you don't get anything unless you bang your hands and feet. They won't give you a promotion or a raise because you're doing such a great jobs or you've exceeded their expectations. The only way you'll get a raise is begging or threatening to leave, but just . . . basically threatening to quit.	Registered professional forester
	I swear, if I was six-foot four, and big hairy chest, I probably could be a lot more persuasive, but as a woman, it's really hard because they look at you as not serious . . .	Scaler
Proving	Even when I first applied for the job, even though I'd worked with the guys for twenty years . . . They still have this closed mentality that they don't really want a woman in that position, you know? And I really had to prove myself, that I could do the job. And that I could learn. And that I could take the risks and that I could do it.	Front end loader
Lack of networks	Men are perceived as being more competent, in a lot of cases. And there's a lot of mentorship that goes along with men. Like men will promote men under the buddy system, but they won't do the same for women, necessarily. Like men have an edge, I'm not saying that women can't get where they want to go to, but it's usually, they have to work harder, be smarter, and they have to be lucky.	Scaler

job situation for women in her community of Port Alice, a locality where a pulp and paper plant dominated employment opportunities:

> The pulp mill doesn't hire women into the general population. The pulp mill hires women for clerical positions or cleaning positions, but the wider progression through the pulp mill is through the finishing room where you

> have these great bales of, rolls of pulp that you're heaving around and you have to be of a certain physical build to handle the job and that excludes women. . . . There's lots of jobs in there women could do, but they don't hire them. I understand that somebody's brought a suit before the Labour Relations Board, but in the past it's been you don't fight it because if you scream about not getting your job in there then that jeopardizes your husband's position.[14]

Women's employment within forestry also challenged the knowledge-based activities. In the mid-1990s, a new set of regulations, the Forest Practices Code (the Code), was established to guide forest management. When the Code was first introduced, demand for registered professional foresters in both government and private industry increased. Unable to meet demand locally, companies and government sought qualified students at forestry programmes across the country. Very gradually, the numbers of women in professional forestry began to increase across the province. In 1995, (the first year in which data were segregated by gender) 10% of registered professional forestry graduates were women; by 2000, this proportion had increased to 14% (Yochim, 2002, Manager, Forestry and communications, Registered Professional Foresters' Association of British Columbia, pers. comm.).[15]

Professional foresters were also subjected to sexism and marginalisation within their work sites. One woman who sought an engineering position to match her training said, "[The company] wouldn't hire me and I'm pretty sure it's because I was female. And they've hired people with less experience than I had, that couldn't even read a compass. You know, so it was kind of upsetting to find that they'd do that and not hire me." She believed that the province was particularly backward in its appreciation for women's abilities. She said:

> [In Ontario] I competed in woodsmen's and lumberjack competition all through university and I came second out of fifty. I came out here and it's, 'Oh my God, don't touch that, oh you might hurt yourself'. So that is quite different. So it's harder for me, like even though when I graduated I wanted sort of to be an engineer, like [in] logging, I couldn't get a job. . . . I had interviews where they wouldn't, basically, it's because there was a lot of camp work, a lot of guys. They wouldn't allow me to go into that situation. They had problems with a female sleeping out there, so they figured all females are the same. So I was sort of stuck into being something in silviculture.

Women applying to work for the provincial government in the Ministry of Forests reported the least incidence of overt sexism. Union contracts, clear job classifications, regular work hours (including flexible time) and a local administration that provided logistical and moral support for women employees had established conditions that were favourable to women's employment. For women in office jobs, the salaries paid by the Ministry were comparable or better to administrative positions in private companies. Ministry jobs also provided favourable benefits (e.g. extended maternity leave, flexible hours etc.). However, the Ministry did not match the salary possibilities provided in job categories dominated by male workers by private companies for union or professional workers. Thus, the occupational division of labour, coupled with public sector union representation and benefits,[16] likely contributed to women being over-represented in government-related forestry occupations.

New ministry positions available to women were created in areas of planning, regulation, compliance and enforcement. Existing jobs became more complex, and in some cases, women moved from clerical positions to more administrative and regulatory ones. These new positions required that women have more advanced computing skills, regulatory knowledge and "customer" service experience. The Ministry itself provided training for its employees. In part, this training was provided to women as part of general provisions to keep staff abreast of new regulations and to ensure that the local work force met rapidly changing regulatory needs. In part, the promotion of women through the administrative ranks was due to the forward thinking of the Manager of Corporate Services at the District Office. She worked very hard to ensure that women were able to improve their job skills as changes took place in the regulation of forestry and subsequently, new demands were made of ministry personnel. One woman, who began doing data entry and secretarial work, explained her experience:

> So [my job has] expanded and just grown and now I'm being trained again. We're getting a lot into the Forest Practices Code for contravention with the companies whoever they are. And that has opened a whole new area where I have to sit in on the hearings, do the minutes, prepare the packages. I'm now going down to learn this tracking system so that I can come back as a trainer to the District and train all the technical staff.

Consequently, for some women, changes in government regulation opened opportunities for employment, both in terms of numbers of positions, as well as in terms of opportunities for enrichment and advancement.

But the most pervasive and formidable challenges facing both "practical" and "knowledge-based" workers were ideological. Women from all occupational groups expressed their frustration with the constant and consistent theme that they were inadequately suited to work in forestry. This affected the perceptions men had of women's duties while on the job (e.g. where only female professional foresters were required to make the coffee and clean up after meetings), the opportunities for promotion (where women were passed over), and ultimately, to the size of the pay cheque they brought home each month. Once in jobs, women believed that they needed to constantly prove themselves. Women who had been employed for more than 20 years still believed they had to prove their worth; women who had been employed for only 2 years believed they never would be able to do so. Instead, many women stated that recognition of their abilities on the job would require a new generation of managers to replace the "dinosaurs"[17] who were currently in positions of prestige and power. Prospects for improvement appear bleak in the near future, at least. Women believed that sexism was endemic, pervasive, and enduring. Notwithstanding their marginalisation, many women expressed perspectives that reinforced their own marginality. It is to these perspectives that I now turn.

Women endorse the status quo

Reinforcing the gendering of practical forestry work

There is no doubt that the norms of practical forestry work have dominated the local culture of forestry. Practical forestry work is characterised by hard work, hard play and common-sense knowledge (Dunk, 1994; Satterfield, 2002). It is also heavy-duty, dirty and dangerous (Brandth and Haugen, 1998). It is man's work, requiring a man's strength and skill, even though much of it is now highly mechanised. Activities associated with male-based practical forestry work include felling, transportation, pruning and planting, although some of these activities (e.g. felling) have higher status as masculine activities because they register more obviously on the "hard and dangerous" criterion. Historically on Vancouver Island, young boys began their employment as teenagers, setting chokers and then moving up through the system as machine operators or other trades people. They did not have to complete high school to complete their education. Rather, boys became men in the woods. In the words of one interviewee:

> One of the hardest things I see around here is that the fathers who got jobs when they were seventeen or eighteen they walked out getting a job setting chokers now they have worked their way up and are now the machine operators with the beer bellies [with] the attitude that their sons should be able to do what they did, you know, get out there, get a job, get off your ass. They treat their sons very roughly about this, a lot of them . . . A lot of them take the hard line like their fathers did. Get out there and work like a man, be a man.

Workers were well paid for their labours. This interviewee went on to say "they had huge wages. They would go on a strike if there was no cherry pie in the cook house." In good times, there was no shortage of work. Workers could move to other companies or locations and pick up work if local conditions changed or there was a falling out with other workers. Men started in the workforce as teenagers and retired by age 55, their bodies spent, their retirement incomes secure.

If women did not participate as loggers in the early days of forestry, they certainly shared the pride in some of the masculine ideals that characterised the identity of "their" men. One woman stated her father did ". . . everything. He ran every piece of equipment in the woods. He was never out of work. He used to quit jobs left and right. And the phone would be ringing off the hook as soon as he quit because jobs were that plentiful and when you had someone as talented as my Dad." This woman entered the industry as a scaler and spoke honestly and passionately about the sexism and limitations she had encountered. She also worked from within the union in an attempt to organise changes that would support women's equal opportunities and reduce overt sexist activities she observed.

According to women interviewed, forestry occupations, particularly logging, were dangerous. One woman, who had lived all her life in a logging camp, said "It's terrible to say about your own kids, that you don't want them to be a logger, because there's nothing wrong with being a logger . . . But I've lost too many friends in logging accidents." Many women described the death of one or more friends or family members in logging, the sense of danger was not simply from acute accidents, but also daily health and safety issues.

Contemporary work standards and technological capacity has not alleviated this sense of danger. For example, hazards related to repetitive use injuries affected the necks, backs, hands of their partners and friends; injuries to body parts had led to job losses for friends and loved ones. Women described new hazards resulting from new methods of logging such as helicopter logging or new company policies. There was concern that with the downturn in industry, companies attempted to cut corners by recycling old machines, running short handed, and demanding flexibility in job tasks for workers. These dimensions of men's work had added to family stress and heightened worries about their partner's health and safety. One woman described the pressures in this way:

> they got [him] at the age of forty with a bad ankle, he's had a fused ankle for a few years now and he limps. They had him out running a hydraulic logger loader in the middle of nowhere. All winter long, all alone, in the middle of the night, he would have to walk out to the machine, which was sometimes hundreds' of feet off the road in the dark, with a flash light, and they could never have done that years ago . . .

Work in forestry manufacturing was also considered dangerous. Women described physical work in pulp and paper mill in Port Alice where men had to haul large loads and were susceptible to back injuries. Hearing loss remained a problem for those running big equipment, particularly in the mills. Women also described the hazards from exposure to chemicals used in the milling process. One woman pointed out that people with allergies were particularly susceptible or in her words, "you can't breathe that muck in day after day and have it not do anything to your body."

This sense of danger was an important component of forestry culture—it created pride in skills necessary to avert the dangers that were ever present and a bonding among those who shared those dangers. The bonds linked all members of forestry communities—men, women and children. Women who identified themselves within "traditional" feminine relationships such as help-mates and wives as well as women who self-identified as participants and partners in forestry occupations suggested that feminine identity in forestry communities were locally and mutually constituted with those of men's masculine identities. For example, some women who were interviewed chose not to risk challenging the "manliness" of their partner by taking a job. This ideological positioning was reinforced by logistical factors such as limited availability of childcare, lack of public transportation as well as institutional factors such as low pay for "women's work" and company hiring practices that precluded more than one member of the family to be employed. Men, whose jobs almost invariably accounted for the larger economic benefit, were more likely to come first. For some women, there was no need to question these norms. As one woman explained, "I understand why the logging companies aren't hiring women over the men, for the simple fact that they don't have enough jobs for the men, let alone the women".

Women spoke of "their men" with pride when they spoke of the values of hard work, danger and technical and physical skills required to perform in the woods. This shared sense of pride helped to foster a common occupational community of forestry that they projected to outsiders, including myself. In their affiliation with forestry, they shared part of their community identity with their partners and they reinforced values and norms that continued to marginalise their own contributions. While these elements dominated discourse about jobs in practical forestry work, they were also evident when women discussed expert work and knowledge-based occupations.

Reinforcing the gendering of expert work and knowledge

During the 1990s, forestry required more workers who possessed expertise and knowledge. Increased regulations affected forestry operations at all levels. Workers who previously learned how to run machinery through direct experience and mentoring from an older worker, now had to complete technical exams to certify their knowledge. Regulations required more detailed environmental plans, the advent of geographic information systems required translation of this technical expertise into planning maps that had to be approved by the Ministry of Forests. Indeed, there is now a range of experts required at all stages of forestry from planning, developing seedlings, planting, harvesting, re-planting, silviculture, enforcement and public relations (for both government and industry).

For professional foresters, changes in forestry practices and land use regulations had brought greater job opportunities and more responsibility on the job. The Forest Practices Code and a more general sensitivity in the industry to environmental issues were cited as reasons that foresters working in private companies were finally "being recognised for the training that they have". However, changes in practices and regulations required foresters to engage in more paper work, rather than work outside—a characteristic of the profession that may have first attracted them. The Code provided much less room for professional discretion, thereby downgrading some of the interpretive or professional skills of foresters and it reduced their association with outside physical labour that was an important element of forestry culture.

Government officials in the Ministry of Forests (MOF) were often considered less intelligent and less masculine than those in industry. Although many "outside" jobs in the Ministry were as demanding physically and intellectually as jobs in industry, they did not confer the same status. Ministry workers had a longstanding reputation as being "soft" and "not too bright". For example, one woman said, "Anywhere you can travel in the world doing forestry . . . . Anybody that doesn't do well in the industry always finds a job in the government, which is a standing joke. . . .when you're working for a private industry, you work very hard and fast and do it. . . . And it is a joke that people that don't make it in the industry always get a job in the government, they end up in the Forest Service. Then you're dealing with those people". Perhaps this denigration of the expertise of government workers is one reason why women were viewed as suitable to enter the Ministry more readily than in private companies. Nonetheless, changes in the regulatory climate altered the context of government work as well as the gender regime that had prevailed in earlier times.

Expansion of regulatory functions increased the number of jobs and created demands for people with new skills. Jobs as forest technicians, planners, and enforcement officers grew and opportunities for advancement through the Ministry were enhanced. During the 1980s and 1990s, a cultural shift occurred in the Ministry that was derived from at least three sources. First, the introduction of the personal computer, with a local area network, a wide area network and the Internet, resulted in significant changes in the skills of employees and in how regulations were enforced. Second, initiatives that were taken as matters of policy, discretionary procedure and indeterminate scientific or "professional" judgement were now required to meet legally enforceable standards that left much less room for error or personal judgement. Last, the Ministry was required to become much more open about its operations and decisions and to be more effective in its communications with the general public. One employee within the Ministry suggested that recent changes in the Forest Practices Code and other regulations altered the culture of the Ministry from "Smokey the Bear", the warm and fuzzy steward and protector of the forest, to a litigious character now focused on compliance and enforcement.

As the Ministry of Forests expanded to handle its increased regulatory requirements, new employees—both men and women—tried to fit within conventions upon which the masculinity of forestry jobs had been built. One woman, whose husband was an enforcement officer with the Ministry, deliberately described her husband in terms similar to those in private industry:

> one of the side effects of being a resource community is, is that when you watch your man go out into the bush every day with all the other men and he goes out there and he get bitten and he gets scratches all over his legs,

> and, you know, has to dodge the bears and the cougars . . . . these guys up here, they're men's men. They're not putting on a suit and going down to the bank. They're going out into the bush.

Yet, where Ministry officials had previously negotiated compliance with government policy, possibly drawing on their common interests and identity associated with forestry, new legal requirements meant that Ministry officials now shared with industry forestry workers the characteristics of danger, but this time as law enforcement officers. Reference to retaliation and war in the woods secured the notion of danger and positioned Ministry workers as "men" within the dominant male culture of forestry. In the words of one interviewee:

> when the Forest Practices Code came in, they gave my husband a badge, you know, in a leather pouch, just like LAPD [Los Angeles Police Department] . . . So I was really, really nervous. . . . When they originally brought in the Forest Practices Code they said that it's going to be tough enforcement and . . . I thought . . . this is the makings for a war. Like he's going to get a stick of dynamite up the tail pipe of his truck . . . they were giving these seminars . . . they were bringing in ex-cops and lawyers telling them how they should be out there, you know, instant cop training . . . And I thought, hey, this is not right. This is going to lead to absolute war in the woods.

Rhetorical reference to physicality, violence, war, and danger linked government employees directly back to elements of traditional masculinity that have been so important to defining forestry culture. This masculinity was identified as an important barrier for women—even within the Ministry—because women were unlikely to overcome these stereotypes and be taken seriously. One's physical presence influenced worker–regulator interaction. One interviewee explained:

> There's a girl (of twelve staff) who works in Timber (at MOF) and she's young, she's small, she's feminine . . . how seriously do they take her? You know, in comparison to dealing with a guy who's six foot four, and got a beer belly . . . that still hasn't changed all that much . . . they're pretty rugged dudes.

So while there were more openings for women to be employed in regulatory positions of government, they faced similar barriers and exclusions.

The quotations above, when considered in isolation from their context, may be interpreted as a commentary by forestry-town women about the marginalisation of women. Indeed, they are. For example, reference to qualities such as "girl" "young" "small" "feminine" were used by the interviewee to illustrate the challenges faced by women who entered the forestry workforce, even in government jobs. However, when these quotes are considered alongside assertions of pride in their partners—assertions punctuated by descriptions of men's physical stamina, work ethic, ability to face danger, etc.—these descriptions of women as "young", "small" and "feminine" appear to illustrate how women fall short. In subtle ways, through the course of an interview, interviewees often inadvertently reinforced the divides that they so openly opposed.

Interpreting the gendering of forestry work

Where, then, are women located within the local employment structure and local culture? How do they enter contemporary forestry practices and discourses? So far, I have suggested that gendered practices and discourses of forestry have marginalised and excluded women. Indeed, gendered identities in forestry are not just created by men in forestry occupations, they are also reinforced by women living and working in forestry communities. This situation can be understood by considering the ways in which the occupational community of forestry is socially embedded within social networks and norms that compose forestry culture.

Social embeddedness and occupational community

Scholars of sociology and geography have begun to consider social context to explain employment choices and circumstances (e.g. England, 1993; Hanson and Pratt, 1995; Carroll, 1995; Carroll et al., 2000; Tigges et al., 1998; DeBruin and Dupuis, 1999; Halseth, 1999; Smith, 1997). These studies pay particular attention to the geographic variation in female participation rates across regions and demonstrate gender typing of occupations within and between places (Massey, 1994; Parr, 1990). Like other elements of social life, paid employment illustrates how women's experiences are embedded in multiple layerings of "community". According to Tigges et al. (1998, p. 204), consideration of social embeddedness "means being alert to the influence of structural constraints on the range of choices available and to the costs and benefits of pursuing certain strategies given available resources". Social embeddedness emphasises the ways that social relationships affect choices and actions. In Granovetter's (1985, p. 487) words, "actors do not behave or decide as atoms outside a social context, nor do they adhere slavishly to a script written for them by the particular intersection of social categories they happen to occupy." Their attempts at purposive action are instead embedded in concrete, ongoing systems of social relations that link economic and non-economic goals (Granovetter, 1985). Thus, within a framework of social embeddedness, the workplace, family and community dynamics can be considered important elements in an analysis of paid work in forestry communities.

The concept of "occupational community" highlights the importance of social relations in understanding labour markets. Van Maanen and Barley (1984) and Carroll (1995) organised occupational community around four elements. People share an occupational community when they regard themselves to be involved in the same sort of work, when their identity is closely tied to that work; when they share "values, norms and perspectives" that are linked to, but extend beyond, the work setting, and when their social relationships meld the realms of work and leisure. This concept can be used within the framework of social embeddedness because it draws attention to how relations within and between the workplace, family and community affect choices and perspectives about employment. Embeddedness may also have a particular geographic resonance as occupational community works with specific sites that are locally (and historically) constituted (Hanson and Pratt, 1995).

In applying this framework to loggers' social or life world, Carroll (1995) identified occupational role identity as a prominent element in the lifeworld of the logger, contributing in large part to his (gender intentional) overall self image and central in the presentation of self to others (particularly to outsiders) (Carroll, 1995, p. 27). According to Carroll, occupational community can be characterised by common attitudes, values and norms associated with the dominant occupation. With this notion of occupational community, Carroll began a sensitive and nuanced explanation of differences in perceptions of loggers in debates of environmental protection of old growth forests of the Pacific Northwest. He explained that "wilderness" and "nature" is the "work site" for loggers, where both material and emotional well-being are exacted and where masculine identities are formed and reinforced. His work emphasised how commonalties in identity formation and the establishment of a sense of community among loggers, had repercussions for how loggers are interpreted by "outsiders" as well as their ability to adapt to changes in economic conditions and forest management regulations (see also Carroll et al., 2000). But Carroll's (1995) conceptualisation and discussion did not include women because he could not find any working in the woods.

Yet it is not necessary to restrict one's scope to "loggers" to illustrate attachment to occupational community. In my own study of forestry-town women, I found that the perspectives, values and norms of women living and working in forestry communities were also shaped by the predominant occupational identity associated with forestry. This identity, while referred to colloquially as "logging", was broader than tree felling; it included all aspects of industrial forestry—from logging, transportation, re-planting to secondary activities in saw mills and pulp mills and tertiary activities associated with public policy planning and regulation.

Two observations in this regard shaped women's interests in and attachments to the dominant occupational community. First, as a result of gendered divisions of labour within forestry communities, women's direct attachment to the industry was a contradictory one: women wanted in, but they were simultaneously repelled by structural and patriarchal norms within the industry and within the communities. Consequently, their support of the labour process was partial. Most women interviewed were quite open about the sexism they encountered in their every day lives in their places of work. This was as true for women who were active supporters of the industry as it was for those who were critical of it. Second, women's support of the labour process extended beyond material attachments. Women's identification with "logging" and forestry more broadly was key to how they presented themselves to others, regardless of whether they were directly employed or whether they felt included in or excluded from the work place. Their shared interests in, and support for, forestry embedded them within the culture of forestry communities.

Women's own work in forestry occupations has been shaped by these formations of masculinity. They have been largely unsuccessful in entering logging where "danger" and "physicality" are defining characteristics. However, where they do enter in the woods in occupations such as log scalers and foresters, they provide a direct challenge to some of the conventional elements of masculinity as well as to the emergent definition of "expertise". There is no doubt, however, that their status within the forestry labour force and within forestry communities remains marginalised. Thus, women's identity was placed at the edge of forestry culture. Women's own employment choices represented part of a complex network that embeds their understanding of paid work of women in local practices and meanings. These practices and meanings in forestry communities placed women's paid work in a marginal economic and social position in relation to the paid work of men. Despite a contradictory and ambiguous location in paid forestry work, women of this study retained a strong attachment to forestry as an occupation and a way of life.

Women's perspectives, in part, were constructed by a network of shared understandings and obligations associated with forestry. This network moved beyond material attachments associated with income. It also provided meaning to the lives of women who live within these communities. Women's choices and perspectives about employment are located within systems of social relations and cultural norms that fix their work in particular social and geographic locations. In short, the discourses and practices of women in forestry were socially embedded within local and societal norms and values. Rather, contradictory ideas about inclusion and exclusion, and appropriate feminine and masculine behaviours ran simultaneously within individual interviews and across the discussions with women of differing employment, age and life-stage status.

Women's adoption of cultural norms and values associated with forestry reflected and reinforced their own marginality. In drawing this conclusion, I do not reiterate former theoretical suppositions that characterised early feminist research—either liberal interpretations of women's victimisation or socialist interpretations of women's victory over their social, economic and political status (e.g., National Film Board, 1979; Ali, 1986; West and Blumberg, 1990; Maggard, 1990; for discussion Gibson-Graham, 1994). Rather, my approach calls for a dual focus lens, one that draws attention to the social embeddedness of women's lives in local norms and networks that reinforce their role identities and a second lens that illustrates how women's attachment to the dominant occupational community has reinforced a working*man's* culture. This dual expression of embeddedness and occupational community helps to explain the paradox of marginality that is both externally imposed by researchers, policy makers, employers, and partners, and internally reinforced by women living and working within forestry communities. Using this two-part theoretical lens, women are viewed neither as victims of nor as victors over their circumstances. The binary of victim/victor does not grant sufficient attention to the complexity and contradiction that characterise women's lives and perspectives. Rather, I suggest that through discourse and practice, women are *co*-creators of the forestry culture and communities that provide openings and closures for women in the paid work of forestry. Greater attention to women's participation in forestry—in practice and in discourse—provides more nuanced theoretical explanations and more accurate empirical descriptions to inform policy choices about forestry employment.

Acknowledgements

Funding for this research was provided by Forest Renewal BC. In addition, the seeds of this research were sown by virtue of a Scholar-in-Residence position I held at the Centre for Research in Women's Studies and Gender Relations at the University of British Columbia in 1996. Technical assistance was provided by Paul Jance (the University of British Columbia), Keith Bigelow and Elise Pietroniro (University of Saskatchewan). Maija Heimo, Janice May, Mary Pullen, Rhonda Koster and Sharmalene Mendis were students at different stages of the project. Two reviewers gave this paper careful consideration, while the editor provided constructive feedback and support. Of course, the greatest thanks go to the women who participated in this study as interviewees, collaborators, and friends.

Notes

1 I use the term, "forestry communities" to denote a multi-layered concept of community including territory, interest and attachment (after Crowe and Allan, 1994). This definition is consistent with more recent discussions of Silk (1999); and Liepins (2000a, b). I consider forestry communities within the broader rubric of rural communities, partly because this term has been chosen by people who live in these places during policy and planning debates (CORE, 1994b) and partly because people living in forestry communities share common elements and concerns with residents of other rural places who rely on extraction and/or processing of natural resources for their livelihoods (e.g. mining, fishing and agricultural communities). When I refer solely to a territorial definition of forestry communities, I use the more restricted term, "forestry towns".

2 In British Columbia (BC), 95% of the forest land base is Crown land, meaning it is owned and managed by the provincial government. Treaties were never signed (with small exceptions) with Indigenous peoples and there are still outstanding issues to be resolved regarding jurisdiction. Since World War II, the forest industry has been mostly composed of a small number of large-scale, integrated multi-national companies who log the property under different lease and licence arrangements with the Province. Small independent forest operators are few in number, particularly on Vancouver Island, although there is a growing public insistence to open up the industry to allow them entry. Despite differences in ownership patterns, elements of work culture are remarkably

similar to those described by researchers in other westernised contexts (e.g. Brandth and Haugen, 1998, 2000; Carroll, 1995; Carroll et al., 2000; White, 1995).

3 This has been known as the falldown effect. Industry uses the term falldown to refer to lower volumes available from second growth timber stands in which large (older aged) trees have been replaced by younger trees with smaller volumes of wood. Sometimes falldown has also referred to the decline in timber harvests resulting from over-harvesting of timber at rates that exceed its ability to replace itself within reasonable investment time frames (Clapp, 1998).

4 In BC, 93% of the land base and 95% of forested lands are Crown lands, owned and managed by the Province.

5 Whether government, industry, environmentalists, First Nations or others are primarily responsible for changes to the industry is a matter of considerable debate, much of which is ideologically driven (for diverse opinions, see Barnes and Hayter, 1997; Binkley, 1997; Clapp, 1998; Hayter, 2000; Marchak et al., 1999; Cashore et al., 2001). The point here is that those living in forestry communities blame changing government policies and environmental interests.

6 For a richer discussion of the interviewees, please see Reed (2000, 2003).

7 Part of the reason for this is that alternative family forms outside the nuclear family were rarely discussed, although there has been some recent effort to redress this imbalance (e.g. Brown, 1995; Cloke and Little, 1997). Part of the reason lies in the way in which particular social relations have been classified as traditional by researchers themselves. For discussion on this point, see Gibson-Graham (1996) and Reed (2003).

8 There is more research about these topics in agriculture (Whatmore, 1991; Little, 2002) and fisheries (Porter, 1985; Wright, 2001), but very little in forestry. For example, Egan and Klausen (1998) surveyed research dealing with gender and the restructuring of BC's forest industry. They noted that while some investigations used a gender-sensitive approach (e.g. MacKenzie, 1987; Grass, 1987; Grass and Hayter, 1989; Stanton, 1989; Hayter and Barnes, 1992; Hay, 1993), "the bulk of recent research . . . neglects gender as a central category of analysis (e.g., Drushka, 1985; Ettlinger, 1990; Hayter et al., 1993; Drushka et al., 1993; Barnes and Hayter, 1994; Hayter and Barnes, 1997), . . . and overlooks the marginalised position of women in the paid labour force and forest-sector unions and, moreover, ignores the broader issue of the sexual division of paid and unpaid labour in forest-dependent communities" (Egan and Klausen, 1998, p. 9). Given the limitations of research, it is not surprising that policy debates reflect and reinforce the same biases.

9 In a review of the US Forest Service, Thomas and Mohai (1995) also found that women were over-represented in clerical and administrative jobs and under-represented in professional and technical positions.

10 During the course of this study, the federal government changed the terminology of worker insurance from unemployment insurance (UI) to employment insurance (EI). This insurance programme is managed by the federal government, but is paid by employers and employees. Upon job loss (temporary or permanent), insurance benefits are generally available to only those workers who receive income from an employer. This plan omits many self-employed and/or contract workers who are increasingly important players in the forest industry.

11 On Vancouver Island, women who undertook organisational activities (ranging from lobbying government, organising demonstrations, doing public 'education' in the schools, providing forestry tours, working on salmon enhancement projects, and organising forestry days or logger sports events) did so as part of their volunteer community activities. These are discussed in some detail in Reed (2000, 2003).

12 Unemployment rates are typically calculated from the population of people who are participating in the labour force, not the total population. Therefore, I calculated those in forestry from within the sample of those in the paid workforce.

13 I say that men have "typically" held these jobs because if a longer time frame were adopted, we would find women working in these plants during the Second World War.

14 The woman referred to in this interview was successful in her claim raised with the Labour Relations Board. After applying to work in the mill, she was told that no hiring had taken place. Discovering that younger, less experienced men had indeed been hired to work in the finishing room, she again approached the company and was told that only men were hired for the shop work while women were hired for secretarial positions. She would be notified when such a position became vacant. She took her case to the Labour Relations Board. After 2 years of debate, and a demonstration of her physical proficiency, she was hired in the finishing room of the plant. Her efforts paved the way for other women to be hired on the shop floor.

15 Only 1% was 40 years of age or older.

16 Public sector unions typically have higher proportions of women than do private sector unions and have worked systematically to promote parity in income and to obtain benefits that might be more congenial to women in the labour force (e.g. extended maternity benefits, flexible work hours).

17 The term "dinosaurs" was used by several interviewees.

References

Ali, M., 1986. The coals war: Women's struggle during the miners' strike. In: Ridd, R., Callaway, H. (Eds.), Caught up in Conflict: Women's Responses to Political Strife. Macmillan, London, pp. 84–105.

Barnes, T.J., Hayter, R., 1994. Economic restructuring, local development and resource towns: forest communities in coastal British Columbia. Canadian Journal of Regional Science 17 (3), 289–310.

Barnes, T.J., Hayter, R. (Eds.) 1997. Troubles in the Rainforest: British Columbia's Forest Economy in Transition. Canadian Western Geographical Series No. 33. Western Geographical Press, Victoria BC.

Binkley, C.S., 1997. A crossroad in the forest: The path to a sustainable forest sector in British Columbia. In: Barnes, T., Hayter, R. (Eds.), Troubles in the Rainforest: British Colombia's Forest Economy in Transition. Canadian Western Geographical Series No. 33. Western Geographical Press, Victoria, BC, pp. 16–35.

Brandth, B., Haugen, M., 1998. Breaking into a masculine discourse: women and farm forestry. Sociologia Ruralis 38, 427–442.

Brandth, B., Haugen, M., 2000. From lumberjack to business manager: masculinity in the Norwegian forestry press. Journal of Rural Studies 16, 343–355.

Brown, B.A., 1995. In Timber Country: Working People's Stories of Environmental Conflict and Urban Flight. Temple University Press, Philadelphia.

Carroll, M., 1995. Community and the Northwestern Logger. University of Colorado Press, Boulder, CO.

Carroll, M.S., Daniels, S.E., Kusel, J., 2000. Employment and displacement among northwestern forest products workers. Society and Natural Resources 13, 151–156.

Cashore, B., Hoberg, G., Howlett, M., Rayner, J., Wilson, J., 2001. In Search of Sustainability: British Columbia Forest Policy in the 1990's. UBC Press, Vancouver, BC.

Clapp, A.R., 1998. The resource cycle in forestry and fishing. The Canadian Geographer 42, 129–144.

Cloke, P., Little, J., 1990. The Rural State? Limits to Planning in Rural Society. Clarendon Press, Oxford.

Cloke, P., Little, J. (Eds.), 1997. Contested Countryside Cultures: Otherness, Marginalisation and Rurality. Routledge, London and New York.

Connell, R.W., 1995. Masculinities. Polity Press, Cambridge.

CORE [Commission on Resources, Environment.] 1994a. Vancouver Island Land Use Plan, Vol. I. Commission on Resources and Environment, Victoria, BC.

CORE [Commission on Resources, Environment.] 1994b. Vancouver Island Land Use Plan, Vol. III. Vancouver Island Table Report. Commission on Resources and Environment, Victoria, BC.

Crowe, G., Allan, G., 1994. Community Life: An Introduction to Local Social Relations. Harvester Wheatsheaf, New York.

CS/RESORS Consulting Limited, 1997. Women and the Forest Industry. Report to the Policy Development Division, Ministry of Employment and Investment. Ministry of Employment, Investment, Victoria, BC.

DeBruin, A., Dupuis, A., 1999. Towards a synthesis of transaction cost economics and a feminist oriented network analysis: An application to women's street commerce. American Journal of Economics and Sociology 58, 807–827.

Doeringer, P., Piore, M., 1971. Internal Labour Markets and Manpower Analysis. D.C. Heath, Lexington.

Drushka, K., 1985. Stumped: The Forest Industry in Transition. Douglas and McIntyre, Vancouver.

Drushka, K., Nixon, B., Travers, R. (Eds.), 1993. Touch Wood: BC Forests at the Crossroads. Harbour Publishing, Madeira Park, BC.

Dunk, T., 1991. It's a Working Man's Town: Male Working-class Culture in Northwestern Ontario. McGill-Queen's University Press, Montreal, QU.

Dunk, T., 1994. Talking about trees: environment and society in forest workers' culture. Canadian Review of Sociology and Anthropology 31, 14–34.

Egan, B., Klausen, S., 1998. Female in a forest town: the marginalization of women in Port Alberni's economy. BC Studies 118, 5–40.

England, K., 1993. Suburban pink collar ghettos: the spatial entrapment of women? Annals of the Association of American Geographers 83, 225–242.

Ettlinger, N., 1990. Worker displacement and corporate restructuring: a policy-conscious appraisal. Economic Geography 66 (1), 67–82.

Fitchen, J., 1991. Endangered Spaces, Enduring Places: Change, Identity and Survival in Rural America. Westview Press, Boulder, San Francisco and Oxford.

Gibson, K., 1992. Hewers of cake and drawers of tea: women, industrial restructuring, and class processes on the coalfields of central Queensland. Rethinking Marxism 5 (4), 29–56.

Gibson-Graham, J.K., 1994. Stuffed if I know!: reflections on postmodern feminist social research. Gender, Place and Culture 1, 205–224.

Gibson-Graham, J.-K., 1996. The End of Capitalism (As We Knew it): A Feminist Critique of Political Economy. Blackwell, Oxford.

Granovetter, M., 1985. Economic action and social structure: the problem of embeddedness. American Journal of Sociology 91, 481–510.

Grass, E., 1987. Employment Changes During Recession: the case of the British Columbia forest products manufacturing industries. MA thesis, Simon Fraser University.

Grass, E., Hayter, R., 1989. Employment change during recession: the experience of forest product manufacturing plants in British Columbia, 1981–1985. Canadian Geographer 33 (3), 240–252.

Halseth, G., 1999. We came for the work: situating employment migration in BC's small, resource-based communities. Canadian Geographer 43, 363–381.

Hanson, S., Pratt, G., 1995. Gender, Work, and Space. Routledge, London and New York.

Hay, E., 1993. Recession and restructuring in Port Alberni: Corporate, community and household coping strategies. MA Thesis, Simon Fraser University.

Hayter, R., 2000. Flexible Crossroads: The Restructuring of British Columbia's Forest economy. UBC Press, Vancouver.

Hayter, R., Barnes, T.J., 1992. Labour market segmentation, flexibility, and recession: a British Columbian case study. Environment and Planning C 10, 333–353.

Hayter, R., Barnes, T.J., 1997. The restructuring of British Columbia's coastal forest sector: Flexibility perspectives. In: Barnes, T., Hayter, R. (Eds.), Troubles in the Rainforest: British Columbia's Forest Economy in Transition. Canadian Western Geographical Series No. 33. Western Geographical Press, Victoria BC, pp. 181–202.

Hayter, R., Barnes, R.J., Grass, E., 1993. Single Industry Towns and Local Development: Three Coastal British Columbia Forest Product Communities. Lakehead University Centre for Northern Studies, Research Report No. 34, ISSN 0843-9885. Lakehead University Centre for Northern Studies.

Henderson, M., 1993. Letter to the editor. Chilliwack Times 9.

Horne, G., 1999. British Columbia Local Area Economic Dependencies and Impact Ratios 1996. Business and Economic Statistics, Victoria, BC.

Liepins, R., 2000a. New energies for an old idea: reworking approaches to 'community' in contemporary rural studies. Journal of Rural Studies 16, 23–45.

Liepins, R., 2000b. Exploring rurality through 'community' discourses, practices and spaces shaping Australian and New Zealand rural communities. Journal of Rural Studies 16, 325–341.

Little, J., 1987. Gender relations in Rural areas: the importance of women's domestic role. Journal of Rural Studies 3, 335–342.

Little, J., 2002. Gender and rural Geography: Identity, Sexuality and Power in the Countryside. Pearson Educational, Essex.

Mackenzie, S., 1987. Neglected spaces in peripheral places: homeworkers and the creation of a new economic centre. Cahiers de Geographie du Quebec 31 (83), 247–260.

Maggard, S.W., 1990. Gender contested: women's participation in the Brookside coal strike. In: West, G., Blumberg, R.L. (Eds.), Women and Social Protest. Oxford University Press, New York, pp. 75–98.

Marchak, P., 1983. Green Gold: The Forest Industry in British Columbia. UBC Press, Vancouver, BC.

Marchak, M.P., Aycock, S.L., Herbert, D.M., 1999. Falldown: Forest Policy in British Columbia. David Suzuki Foundation and Ecotrust Canada, Vancouver, BC.

Massey, D., 1994. Space, Place and Gender. Polity Press, Cambridge.

McDowell, L., 1993a. Space, place and gender relations: Part 1: feminist empiricism and the geography of social relations. Progress in Human Geography 17, 157–179.

McDowell, L., 1993b. Space, place and gender relations: Part II: identity, difference, feminist geometries and geographies. Progress in Human Geography 17, 305–318.

Murphy, K., 1995. Women's political strategies in a logging town. In: Felt, L.F., Sinclair, P.R. (Eds.), Living on the Edge: The Great Northern Peninsula of Newfoundland. Memorial University of Newfoundland, Institute of Social and Economic Research, St. John, pp. 164–182.

National Film Board, 1979. No Life for a Woman, Video. Ottawa, National Film Board.

Ontario Ministry of Labour (n.d.) The Displaced Workers of Ontario: How do they fare? Unpublished report prepared by Economic and Labour Market Research, Ontario Ministry of Labour reprinted in Commission on Resources and Environment 1994 Vancouver Island Land Use Plan, Volume II, Appendices. CORE, Victoria, BC.

Parr, J., 1990. The Gender of Breadwinners: Women, Men, and Change in Two Industrial Towns: 1880–1950. University of Toronto Press, Toronto, ON.

Porter, M., 1985. She was skipper of the shore-crew: notes on the history of the sexual division of labour in Newfoundland. Labour? Le Travailleur 15, 105–123.

Preston, V., 2002. Making paper, constructing workers: A case study of women in the newsprint industry. Presented at the Annual meeting of the Canadian Association of Geographers, May 28–June 02, Toronto, Canada.

Price Waterhouse. 1995. Analysis of Recent British Columbia Government Forest Policy and Land Use Initiatives, Report to the Forest Alliance of British Columbia. Price Waterhouse, Vancouver, BC.

Prudham, S., Reed, M., 2001. Looking to Oregon: comparative challenges to forest policy reform and sustainability in British Columbia and the US Pacific Northwest. BC Studies 130, 5–40.

Randall, J.E., Ironside, R.G., 1996. Communities on the edge: an economic geography of resource-dependent communities in Canada. The Canadian Geographer 40, 17–35.

Reed, M.G., 1999. Jobs talk: retreating from the social sustainability of forestry communities. Journal of Forestry 75, 755–763.

Reed, M.G., 2000. Taking stands: a feminist perspective on 'other' women's activism in forestry communities of northern Vancouver Island. Gender, Place and Culture 7 (4), 363–387.

Reed, M.G., 2003. Taking Stands: Gender and the Sustainability of Rural Communities. UBC Press, Vancouver (in Press).

Reinharz, S., 1992. Feminist Methods in Social Research. Oxford University Press, New York and Oxford.

Sachs, C., 1996. Gendered Fields: Rural Women, Agriculture and Environment. Westview Press, Boulder, CO.

Satterfield, T., 2002. The Anatomy of a Conflict: Identity, knowledge and emotion in old-growth forests. UBC Press, Vancouver, BC (in press).

Seitz, V.R., 1995. Women, Development, and Communities for Empowerment in Appalachia. State University of New York Press, Albany, NY.

Silk, J., 1999. Guest editorial: the dynamics of community, place and identity. Environment and Planning A 31, 5–17.

Smith, Y., 1997. The household, women's employment and social exclusion. Urban Studies 34, 1159–1178.

Stanton, M., 1989. Social and Economic restructuring in the forest industry: a case study of Chemainus. Unpublished MA Thesis, The University of British Columbia.

Thomas, J.C., Mohai, P., 1995. Racial, gender and professional diversification in the forest service from 1983 to 1992. Policy Studies Journal 23, 296–309.

Tigges, L.M., Ziebarth, A., Farnham, J., 1998. Social relationships in locality and livelihood: the embeddedness of rural economic restructuring. Journal of Rural Studies 14, 203–219.

Van Maanen, J., Barley, S.R., 1984. Occupational communities: culture and control in organisations. Research in Organisational Behaviour 6, 287–365.

Warren, K., 1992. Role-making and coping strategies among women in timber-dependent communities. Unpublished M.Sc. Thesis, University of Washington.

West, G., Blumberg, R.L. (Eds.), 1990. Women and Social Protest. Oxford University Press, New York.

Whatmore, S., 1991. Farming Women: Gender, Work and Family Enterprise. Macmillan, London.

White, R., 1995. Are you an environmentalist or do you work for a living ?: work and nature. In: Cronon, W. (Ed.), Uncommon Ground: Toward Reinventing Nature. W.W. Norton & Company, New York, London, pp. 171–185.

Wright, M., 2001. Women in the Newfoundland fishery. In: Cook, S.A., McLean, L.R., O'Rourke, K. (Eds.), Framing Our Past: Canadian Women's History in the Twentieth Century. Oxford University Press, Toronto, pp. 343–346.

15 The politics of invasive weed management

Gender, race, and risk perception in rural California[1]

Kari Marie Norgaard

"Biological invasions" are now recognized as the cause of significant ecological and economic damage: zebra mussels clog plumbing in the Great Lakes region, and tamarisk overtakes native willows all over the Colorado Plateau (Bright 2001; Pimentel et al 2000; Pimentel, Zuinga, and Morrison 2005; Schmidt and Simberloff 1997). Although humans have always moved organisms from one place to another as we travel, and participated in the shaping of so-called "natural ecosystems," the rates of human travel and trade, and hence new species introductions, have increased rapidly with the advent of free trade—the latest phase of globalization. Living organisms are moving around the globe at an unprecedented rate through direct importation and also as "hitch-hikers" on freighters, packaging, and equipment. Some of these species take hold and spread rapidly in their new environment. When this occurs, native organisms, and ecosystem relationships may be quickly altered, as with the arrival of Dutch elm disease in the eastern United States, or the brown tree snake in Hawaii.

Biological invasions have been characterized as among the "most dangerous and least visible forms of environmental decline" (Bright 2001) and the second leading cause of biodiversity loss. Of all the impacts of these events, ecological and economic consequences have been the most readily identified and measured and are consequently the best understood, for example, Pimentel's oft-cited figure of $120 billion per year (Pimentel et al. 2000, 2005).

Contention over herbicides is one of the most common sources of social controversy regarding invasive species, and such controversy is increasing alongside the rising numbers of rural and urban acreage that are sprayed. Spraying of Malathion has provoked recent urban controversies in Southern California over the control of fruit flies and in Sacramento over the control of West Nile virus vectors. Similarly, in 2001 the possibility of spraying for the glassy-winged sharpshooter in rural Sonoma County, California, led to county-wide opposition and threats of direct action. One of the key issues this study highlights is the increased public concern over herbicide use and the bases for differences in the perception of risk regarding herbicide use. Agencies from the National Park Service to the U.S. Forest Service are mandated to control invasive organisms. Yet, management is often a political process: Who gets to decide which invasives are a priority for management? What methods are to be used? And who may benefit or be impacted by different management processes?

In 1997, spotted knapweed was found along the Clearwater River (this is a pseudonym) in a remote and rural region of northern California. Spotted knapweed is a serious concern of the California State Department of Agriculture due to its impacts on rangeland quality. The proposed use of herbicide treatment by the Forest Service sparked an immediate, intense, and ongoing controversy. Ninety percent of community members in the region oppose the Forest Service's plan to apply herbicides. The Karuk Tribe passed a resolution against the use of herbicides in their ancestral territory. Other community members threatened direct action. The Forest Service has received more comment letters on the current Noxious Weeds Environmental Impact Statement than any other recent action—including timber sales, which are usually considered the most controversial of Forest Service activities. As of this writing, the weeds project has been placed on hold "due to lack of funding." Forest Service employees have also alluded to community controversy as a factor.

What accounts for the different views of Tribal members, the general community, and the Forest Service on the safety or appropriateness of herbicides? To what extent do different risk perceptions reflect unequal patterns of exposure by race and gender? Despite agreement that spotted knapweed is a problem, members of the local community— the Karuk Tribe of California, the non-Indian community, and the U.S. Forest Service—are each affected differently by and have different notions of the best way to respond to the presence of spotted knapweed. This study provides a comparative analysis of how the three groups within the region have come to hold very different perceptions of invasive weed management. In so doing it highlights how race and gender inform risk, the relationship of risk perception to environmental justice and who pays the price for environmental degradation in the form of species invasions.

Perceptions of environmental risks

In the Clearwater River area, the proposed use of herbicides on invasive species led to immediate controversy. Public concern about potential health hazards of pesticides' use is widespread and crosses many demographic categories (Chimpan and Kendall 1995; van Tassell et al. 1999). There are also documented differences in perception of environmental risks by gender (e.g., Bord and O'Connor 1997) and among race and ethnic groups (Finucane et al. 2000; Flyn, Slovic, and Mertz 1994; Marshall 2004; Palmer, Carlstrom, and Woodward 2001). For many

Native Americans in California, herbicides are strongly opposed and referred to as "poison." Opposition to herbicide use was a central factor in the formation of the California Indian Basketweavers Association. This organization developed in 1992 to promote safe gathering conditions for Indian basketweavers, many of whom live in Northern California.

Overall, literature on gender, race, and risk perception describes what Flyn, Slovic, and Mertz (1994) call a "white male effect," that is, that white men differ from members of all other groups in perceiving risks as smaller and more acceptable and in being more willing to impose environmental risks on others. Gender differences in environmental concern have been greatest with local problems and with issues that pose health and safety concerns (Davidson and Freudenburg 1996). Explanations for this gender gap have focused on social roles of women as family nurturers and caregivers. Less research has been conducted on the basis of racial differences in risk perception. Social and economic conditions of racial minorities, including high poverty, low wages, and inadequate access to information and health care, contribute to greater risk of exposure, more significant consequences of exposure, and may also lead to perception of greater risk. Yet, aside from aspects of social class, what factors underlie racial differences in risk perception? Why are Native people in particular so strongly opposed to the use of herbicides? Are there racialized aspects of the life experiences of these individuals that lead them to different interpretations of safety and risk? In addition, many of the social factors discussed below, such as institutional trust, have specifically gendered and racialized dimensions. Voices from the Clearwater River provide insight into the understudied basis for racialized differences in risk perceptions.

Risk research over the past decade highlights the importance of social context in understanding the public's sense of risk. This may be due in part to gaps between theoretical explanations of risk and the realities of risks as they are experienced by people "on the ground." Beamish notes that "economic modeling of environmental trade-offs, contingency based probability assessments and psychological work on risk perception decontextualizes and oversimplifies the scenarios and experiences that inform lay-public interpretations" (2001:11). Similarly, Clarke and Freudenburg observe that "experts tend to concentrate on the details whereas the public focuses on the bigger picture" (1993:71). Furthermore, much research outlines how perceptions of herbicides are intertwined with people's trust in institutions of government and industry (Freudenburg 1997; Murdock, Krannich and Leistritz 1999; Wulfhorst 2000). The notion of "institutional trust" is widely studied in risk literature. For example, Beamish (2001:6) describes how

> In the case of the Guadalupe dunes, the public's sense of risk emerged in part from perceptions of the threat posed by the immediate hazard, but more importantly from the sense of institutional failure on the part of both industry and government agencies.

He also notes that,

> When a complete breakdown in trust of this kind occurs, community members' perceptions of and reactions to risk can be seen as 'rational' but cannot be understood as merely calculative responses to the physical hazards associated only with the immediate, discrete event (p. 5).

Finally, in contrast to conventional risk evaluation models that tend to privilege the risk perceptions of experts over the public, sociological research indicates that even professionals may underestimate the crisis potential of the systems they operate (e.g., Perrow 1984; Schrader-Freschette 1991). Research by Slovic et al. (1995) found that occupational affiliation with chemicals lowers an individual's perceived risk of chemicals. Similarly, Hawkes and Stiles (1986) note that individuals with "pesticide connections" perceive the smallest amount of risk compared with scientists, government employees, the general public or elected leaders. From another angle, Dunlap and Buess' (1992) public opinion survey in the Pacific Northwest uncovered an interesting result: the belief that pesticides are necessary is one of the most important predictors of their acceptability. Research by Winston (1997) also found that the belief that pesticides are necessary is related to the beliefs that there are no alternative methods to remove pests, and that the benefits of pesticide use outweigh their risks. Institutional trust, proximity to exposure, gender, and race were each clearly visible dimensions of risk evaluation that in turn led to controversy in the Clearwater case.

Environmental justice

When differences in risk result in part from relative chances of exposure, and when those who face exposure are members of poor or racial minorities, issues of risk perception fall into the broader issues of environmental justice. International, national, and regional trade is responsible for the transport of invasive weed species such as spotted knapweed across the nation or region. Yet, rather than instituting tighter regulations or fines on the movement of species, citizens and land managers combat these ecological and economic problems in specific locations on the ground. As a result, the use of herbicides for the management of invasive species is increasing across the United States. This increase is furthered by the intensification of weed invasions, the increased visibility of invasives as a social problem, and the direct marketing of herbicides for use in invasive management by chemical companies such as Monsanto.

Chronic exposure to low levels of pesticides or their residues on plants has been linked with serious human health problems including cancer, birth defects, and infertility (Baldi et al. 2001; Garry et al. 2002; Oliva, Spira, and Multigner 2001; Savitz et al. 1997; Zahm and Ward 1998; Zheng et al. 2001). Racial minorities experience disproportionate exposure to a variety of environmental problems in rural communities, including toxic waste (e.g., Bullard 2000;

Pastor, Sadd, and Morello-Frosch 2004) and pesticides (Arcury and Quandt 1998). Furthermore, poorer people and members of racial minorities generally have less access to information and decision making options than whites. In their recent review of the effectiveness of the environmental justice Executive Order, Murphy-Green and Leip (2002) note the widespread lack of information on pesticides and pesticide laws among American farmers, about 90 percent of whom are Hispanic. The authors conclude that farm workers in the United States are "one of the least statutorily and constitutionally protected occupational groups in America today . . . they are provided unequal protection under the law, which leads to environmental injustice" (p. 685). Few investigations have been conducted into the experiences, herbicide exposure levels and access to information of other racial minorities in land management activities such as forestry, mushroom picking, tree planting, and highway maintenance. Even less is known about the experiences of Native Americans in particular. Most scholarly work on Native exposure to environmental problems has been in connection to energy use, mining activities, and radioactive disposal. Little has been done on exposure to herbicides resulting from cultural and subsistence ties to the land. This paper contributes to this understudied area of social experience.

Executive Order 12898, signed in 1994, requires that federal agencies identify and address adverse effects of their actions on human health or the environment of minorities and low-income populations, as well as the equity of the benefits and risks of their decisions across populations. Despite the Executive Order and minorities' increased risk of exposure, potential effects on the poor and racial minorities are rarely or never discussed in planning documents regarding species invasions. One of the broader sociological questions this paper highlights is the issue of who in society pays the price for environmental degradation.

Gender, race, and cultural opportunity structures

Whether protest is successful depends on a range of factors. In addition to work on resource mobilization and political opportunity structures, Johnston and Klandermans (1995:5) note that culture may function to channel or constrain the development and success of social movements. The notion of "cultural opportunity structures" refers to the kinds of actors and possibilities for action that exist based upon the distribution of means within society. Taken-for-granted notions of who and what counts in society may serve as "frames" available to social actors (Snow et al. 1986; Swidler 1986). For example, since the 1960s, national-level cultural emphases on racial and gender equality have created a new cultural opportunity structure that elevates the voices of women and people of color in a variety of social movements. In the Clearwater case, gender and race led not only to different perceptions of risk, they also created opportunities for mobilization. These aspects of social experience formed the basis of "cultural opportunity structures" that were used in organizing opposition to spraying.

Data and methods

The results presented here are part of a larger ongoing investigation into the social impacts of and responses to invasive weeds in a rural northern California community. Data for this paper were drawn from 15 months of ethnographic field work, including participant observation, interviews, and archival analysis. The Clearwater River community was selected because it appeared to provide an example of how a community could successfully mobilize and eradicate a serious invasive weed on the watershed level (this topic is the subject of another paper). In addition, the diverse racial and political make-up of the community lent itself to a comparative analysis of the various ways that invasive species may be viewed and the effects on different social groups. Finally, the regional presence and concerns of the three largest indigenous tribes in California suggested an important but understudied environmental justice issue. I began with broad research questions: How were people being affected by invasive weeds (if at all)? How did different groups within the community view invasive weeds? What compelled community members to spend so much energy on volunteer hand-removal tactics? And, in terms of the specific focus of this paper, what was the basis for the vastly different perspectives on whether herbicides were safe?

As a participant observer, I spent time in numerous community meetings, dug spotted knapweed, and attended regional and statewide meetings of land managers concerned with invasive weeds. In each of these settings I learned much about the perspectives, assumptions, concerns, and daily struggles of members of the Forest Service, Karuk Tribe, and non-Indian community. I also spent a great deal of time living "as a community member" in the Clearwater River watershed. This was necessary to develop trust and gather data in this remote area.

In addition to participant observation and extensive informal conversations with community members, I conducted in-depth interviews with 42 individuals: 5 members of the Forest Service (plus 3 other relevant land managers), 7 members of the Karuk Tribe (and 3 additional people connected to the California Indian Basketweavers Association who were not in the Karuk Tribe), 9 members of the Clearwater River Restoration Council, 8 general community members and 6 outside "experts" (including other professionals working with invasive species and a physician involved in the collection of herbicide exposure depositions). Informants were selected to cover the widest possible range of viewpoints in the community. Interviewees were also selected based on their ability to serve as key informants on specific issues. In-depth interviews were directed towards expanding understanding of current or past events and clarifying, confirming, or denying suspicions about social dynamics that I had developed from participant observation activities. All interviews were transcribed and coded.

Cultural and political diversity 'on the river:' Indians, hippies, loggers, miners, and the Forest Service

The Clearwater River watershed is remote, pristine, and biologically significant. The watershed, which is largely free of invasive species, boasts exceptionally high water quality and is considered a key refugium for a number of fish at risk of extinction, including summer and winter runs of wild Klamath Mountains Province steelhead, spring and fall Chinook salmon, green sturgeon, Pacific lamprey, and Coho salmon. The human community along the Clearwater River is politically and culturally diverse. Residents of this relatively isolated watershed include the Karuk Tribe, loggers, "back-to-the-land hippies," small scale farmers, miners and Forest Service employees. Poverty and unemployment are high among the approximately 250 residents.

The region is the ancestral territory of the Karuk Tribe of California. About one quarter of the community is Karuk, and the presence of the Tribe is a significant cultural and political feature of the region. The Karuk people have endured demographic and cultural decline since contact with white settlers as a result of systematic genocide, forced assimilation through boarding schools, and other aspects of legal discrimination (Bell 2002; Lowry 1999; Norton 1979). Since 1979, when they gained federal recognition, the Karuk Tribe has experienced a political, economic and ethnic renewal (Bell 2002; Nagel 1996). Members are actively recovering cultural traditions, including fishing techniques, language use, ceremonial practices, and traditional basketweaving. The Karuk Tribe has a Department of Natural Resources and is involved in land management, although the Forest Service is the dominant land manager in the region (98.7% of the land area is managed by the Forest Service).

Non-Indian settlers entered the area as miners during the 1850s (Bell 2002), and some remained. Logging also brought non-Indians to the region in several waves. Timber has been a significant source of income for short periods in the watershed, especially in the 1970s and 80s when up to three quarters of the community were employed in forestry or related activities. A number of urban whites entered the river region during the 1960s as part of the "back to the land" movement, exerting their own cultural influence on the watershed (Salter 1981). Besides the Tribe, one of the important community organizations involved in land management is the Clearwater River Restoration Council (hereafter simply "the Restoration Council") formed in the early 1990s. The Restoration Council consists of a unique blend of miners, loggers, and environmentalists all working towards watershed health in the community. The Restoration Council employs a dozen or so community members and coordinates a much larger network of volunteers. This group of mostly non-Indians keeps track of community perspectives on land management and educates the community on upcoming management issues.

The local National Forest was established in 1905, and the U.S. Forest Service became the primary land managers along the Clearwater River by the 1930s. Forest Service management efforts over the last eighty years have followed notions of ecology and forest health derived from European models and have been shaped by national and regional mandates to provide timber (Davies and Frank 1992; Hirt 1994). Since the 1970s increased attention has been directed towards forest ecology. During the height of timber output from the region, the Forest Service had several offices located along the river and housed many employees in the region. More recently, the Forest Service has scaled back its activities and presence in the watershed, closing offices along the river. Most Forest Service employees now live and work in areas up to two hours drive from the River.

In this small community, Indians, non-Indians, and the Forest Service are not always distinct groups and sometimes overlap. Members of both the Tribe and the non-Indian community have worked for the Forest Service, and members of the Tribe are on the board of the Restoration Council. And the three groups have interacted differently at different times: all three have worked on cooperative projects, and all three have held conflicting perspectives.

"No poisons in our watershed!" controversies over weed management and the safety of herbicides

While there was general agreement that invasive species such as spotted knapweed were not welcome in the watershed, the possibility of herbicide use as a strategy to control weeds was a topic of serious contention. Forest Service employees working on the issue considered herbicides to be either "safe" or a "necessary risk that could be adequately managed." Community members in the Restoration Council and the Karuk Tribe, on the other hand, were almost universally opposed to the use of herbicides because of potential risks. Although both Indian and non-Indian community members were opposed to herbicide use, the basis for their opposition was somewhat different. How did each group develop such different perceptions of safety and risk? Here I will describe how social context, including the possibility of direct exposure, lack of institutional trust, gender, and race, shaped the differing views of those involved about the safety, meaning, and significance of herbicide use. Although both the Indian and the non-Indian community were opposed to the use of herbicides, there were differences between these groups that highlight significant racial dimensions to risk perception. The different bases for evaluation between groups led to different conclusions regarding herbicides as an appropriate management strategy. Gendered and racialized experiences in turn formed the basis of mobilization against herbicide use. I further describe how women and members of the Kaurk Tribe each used aspects of their experiences as the basis for generating opposition.

Local history and institutional mistrust

Sociological literature on risk perception describes the social bases of risk (Slovic 2000) and the significance of institutional trust (Beamish 2001; Freudenburg

1997)—two factors that are clearly relevant on the Clearwater. Probably the most significant defining event in the relationships of these groups was the aerial spraying of a mixture of 2,4-D and 2,4,5,-T on the forest and surrounding communities as part of the Forest Service's timber management practices in the 1970s and early 1980s. The spraying was part of what is known as a "conifer-release program," in which herbicides were used on logged areas to prevent the growth of broadleaf trees and brush, species perceived to compete with newly planted conifer seedlings after clear cutting. This program was highly controversial in rural communities locally and throughout California and Oregon (Ortiz 1993; Van Strum 1983). Dioxin, the active ingredient of 2,4,5-T and Agent Orange, has been linked to hormonal and endocrine disruptions in Vietnam veterans and their wives and children (Le and Johansson 2001). Incidents of water supply contamination, late-term miscarriages, and unusual cancers and birth defects were documented in the community. It was this event more than any other that galvanized and united this otherwise politically and culturally diverse community. Indians, non-Indians, miners, loggers, and hippies all joined efforts to stop the use of herbicides by the Forest Service. They did so successfully with a court injunction in the mid-1980s.

It is clear that current perceptions of herbicides and questions about the Forest Service's intentions are influenced by the earlier history of herbicide spraying. In addition to specific health problems such as cancers and birth defects, residents described experiences with the Forest Service that led to significant mistrust, including the spraying of a spring that was a family's water supply:

> [We had a tarp covering our spring] . . . And we had it tested. Our side had it tested and it was covered with whatever they sprayed and Atrazine . . . there was red dye still up in [our spring] and they [the Forest Service] had told us they were going to leave these buffers . . . But they hadn't.

This woman went on to say that, although they left their home for several days while the spraying occurred, when they returned she and her daughter became very ill with feverish flu-like symptoms that lasted several months and recurred for years afterwards. Another resident described the attitude of the Forest Service as disrespectful and generally mean:

> There were really a lot of people in the Forest Service that were nasty and mean. You couldn't really trust their motives or their interpretation of science, because they didn't really care . . . [T]o trust the government who was pretty freely using this stuff to be watching out for the welfare of people, it just wasn't happening. You have to watch out for your own welfare and health, because the Forest Service truly did not care.

And for some, distrust in the Forest Service in particular was linked to the growing general distrust of government and science in the 1980s and 1990s (Beck 1992). For example, one community resident described his successive experiences of being told that first DDT and then Agent Orange were "safe," only to see both compounds classified as dangerous substances some years later. From his experience, it only made sense that many chemicals presently considered to be "safe," given their minimal testing, would be recognized as dangerous in the future:

> What we keep hearing was that the herbicides are safe. Don't worry about it, we used bad chemicals, but they're safe now. My personal experience about that: I was sprayed with DDT heavily when I was a kid, and they felt it was safe. We used to ride behind the fog trucks and hide in the fog. We had also had this experience earlier in Clearwater River where Agent Orange was supposed to be totally safe, and that had been around for a while. Now these chemicals that are now told to be safe; there is just a lot of distrust, that in 5 or 10 years, these will be banned too, and then there will be a lot for residues. So, there is a nervousness about the herbicides.

In contrast, Forest Service ecologists and range scientists believed the use of herbicides for invasive plant control was very different from the aerial spraying of the past. People in the Forest Service emphasized that they had learned from past mistakes. Not only were herbicides safer now, but application techniques had improved:

> I knew the history, but wasn't directly involved with any of that, and the herbicides that were used in the seventies were aerial herbicides used for reforestation . . . I thought, "Oh wait a minute, this is completely different context—we're talking about spot spraying" And 20 acres was the maximum of the infestation. It's actually small little patches, you know, the size of this room.

Finally, not all experiences that generated mistrust were far back in history. Members of the Restoration Council described how, at a time when a cooperative hand-eradication program was in place, the Forest Service illegally sprayed knapweed infestations in a nearby area and failed to inform the community until much later (sprayed areas should be indicated with signs to minimize community exposure). One man described how this incident heightened his sense of mistrust:

> They didn't register that in the state, so it's illegal for them not to do that; they broke the law there . . . There's about seven places they broke the law. They violated the labels, sprayed closer to the creeks than they're supposed to. They used Tordon, which is not registered for use in California, so that wasn't good. There are all these things that they did that were really weird. So we really made an effort to document. After they had sprayed, I was quite upset and I didn't trust them and I wanted to make sure it was documented well.

While river residents seemed to agree that the application methods for proposed herbicides were not as extreme, and even that the chemicals were potentially "safer" than what had been used in the past, the risk remained above their thresholds. As mentioned in an earlier interview passage, concerns existed about human exposure and the exposure of Salmonid populations in creeks and rivers. This situation echoes Beamish's (2001:5) scenario, in which

> Perceptions of present and future risk associated with the massive contamination of Guadalupe Dunes grew more from impressions of the way corporate and government institutions in the area mishandled this and previous oil-related hazards than from fear of health risks associated with the discrete Guadalupe Dunes event.

Social context of risk evaluation and proximity to exposure: institutional versus local, abstract versus direct

Another factor contributing to the conflict was that people's perceptions of the risk, significance, and meaning of herbicides were constructed within three very different social contexts. Indeed, the reactions of members of the Tribe, the general community, and Forest Service staff provide a window into the power dynamics of the local social structure. Forest Service employees were accountable to the agency at regional and national levels for funding, an organizational structure based outside the immediate area. The Forest Service made judgments and choices about strategies within an institutional framework that prioritized the importance of weed control at a national level and favored the use of herbicides to achieve this. Furthermore, Forest Service employees—most of whom did not live in proposed spray areas—viewed risk and safety in an abstract sense, referring to scientific literature and risk assessment studies. In contrast, the issues and concerns raised by community members in the Restoration Council and the Karuk Tribe derived from their attachment to place and residence in the community as well as long time observations of the area and the possibility of their direct exposure. Community members viewed weeds as a problem that should be managed, but also as an issue that could provide long term potential for employment to the region. And while Forest Service employees evaluated risks using scientific literature, community members' sense of risk was developed not only through the use of science, but also within the context of local social, historical, and political forces.

Institutional vs. local risk evaluation

Forest Service employees perceive herbicides to be the appropriate choice within a context of management direction (they receive mandates on the federal level) and existing resources (there is a set amount of funding for invasive weeds). Yet despite the overall institutional emphasis, Forest Service ecologists and weed managers are frustrated by too little funding to complete the task on the ground. The wildland area in the region of concern is extremely large, plant populations are located in remote areas, and staffing for the task is minimal. From the Forest Service perspective, herbicides are viewed as the primary strategy because they are seen as the most (and in some cases only) effective eradication tool and as the only cost-effective strategy to approach such a widespread problem. Additionally, the structure of funding itself favors the use of herbicides—funding is allotted on a per-acre basis, favoring the use of what community members see as the "quick fix" solution. As one Forest Service botanist explained:

> We get dollars relative to acres we treat. When you can't use herbicides, you can't treat that many acres, so forests that were able and counties and areas that, Lassen, they do spraying, and all those kind of places, they tend to get a higher budget, because they can treat more acres. So there is a disadvantage of doing it manual, because you're never going to be able to compete with hundreds of acres when you're doing manual treatment.

Furthermore, although they may perceive themselves as neutral, Forest Service employees are also evaluating risks within social and institutional context. Forest Service employees, like other land managers working on invasives, receive information about available "treatment options" provided by chemical companies at trainings on invasive weed management at county and statewide events. For example, the California Invasive Plant Pest Council—the annual conference devoted to weeds in wildland areas of the state—receives major funding from Monsanto (producers of Round-Up, one of the more commonly used herbicides on weeds). At these meetings, agency staff is exposed to information that normalizes the use of these materials. Chemical companies use displays and brochures to promote the safety of their merchandise and make presentations about their latest available products. Although some information about non-chemical approaches is present, the dominance of chemical options in these training settings creates the sense that chemicals are the primary effective strategy. Furthermore, one manager I spoke with at this meeting, whose research showed that Round-Up was less effective than mulching on a particular plant species, described instances of intimidation by chemical company representatives.

Whereas agency people almost universally lived and worked outside the watershed and evaluated management in the context of regional and national strategies, members of the Restoration Council and the Tribe evaluated appropriate management strategies from a localized context. Although they, too, used outside information sources; attachment to place for Indian and non-Indian residents led to questions about impacts on human health and fish and other species. Fish are an important species to both Indian and non-Indian river residents. People expressed concern that proposed herbicides have been reported as acutely toxic to anadromous Salmonids. Here one community member describes their concern about how herbicides might impact salmon:

> We're worried about the whole ecosystem really, but we tend to be fish-centric. So, noxious weeds are a potential problem to the water quality, as I said, fishery and watershed health. Anyways, we're really concerned about the impacts of spotted knapweed and the pesticide approach . . . It may impact spring Chinook . . . That set off the buzz because we have the only run of spring Chinook left.

Community members also expressed concern about human health effects and the possibility that chemicals could not or would not be used in accordance with guidelines. People asked questions about the likelihood of chemical spills in the river, as sprayers with backpacks inevitably slipped in the course of many necessary river crossings. In addition to concerns about the safety of the herbicides, residents also brought up the issue of what might go wrong in their application. When I asked one resident whether she believed that the present chemicals were safer than those in the past, she replied:

> No. No. I don't believe that for a minute. There is no way. And people are still people, and people still aren't perfect, so there's still going to be accidents, and there's still going to be mistakes, and they say they have training. But I've heard that training is a big fluke anyway. There is just no way. There is too much possibility for something going wrong and we don't know enough about it, and there's just—No.

Abstract vs. "embodied" risk evaluation

Most significantly, what underlies these differences in the social context of decision making is the issue of direct versus abstract evaluation of risks. One key issue that came up again and again was the notion of "who pays the price." As one community member put it:

> My general feeling is that they are really isolated from the consequences of their decision. So, it's fine for them to say that it's safe, but it doesn't really matter if it's safe or not, because they don't live in this community, and they're not a part of the river the way that people who live around here are. It's easy for them to say that.

This difference between embodied or direct and abstract risk evaluation is evident in the narratives given by these groups about the issue of safety. For example, when the issue of safety came up in an interview, this Forest Service employee referred to the standards and scientific procedures of testing that were used to evaluate herbicide effects:

> The Forest Service contracted with this group: Syracuse Environmental Research Associates . . . And they have a full staff of, you know, toxicologists and biologists, etc. etc. that have done these risk assessments on how these herbicides impact human health, wildlife, soil, water, fish . . . How they do their risk assessment is they take the dose . . . the EPA gives a reference dose, which is the amount that it takes for any effect on a test population . . . The reference dose is the amount that you do—how much do you give until there's No Observable Effect—well the reference dose is that cutoff where there is an effect. Some observable effect. And then that's divided by the dose that the application rate that we're proposing, and that comes up with a hazard quotient . . . So, you know, we have to disclose the effects and we have to be responsible with science. I believe that the science is valid.

Facing potential direct exposure, community members evaluated risks in a direct, embodied way, describing incidents of birth defects and cancers from personal histories or from people they knew and the concerns these events raised for everyone in the community. One woman described how her own child's birth defect caused her to question the safety of herbicides used in her area at the time. In this personal testimony, this woman explains the different criteria for risk she uses when trying to understand the impacts on her child:

> Then I moved back to San Francisco some time in 1974 and gave birth to a baby, who was born with two holes in his heart . . . I felt particularly sensitive to the issue of human health hazard, and I felt as if, I had no way of knowing, but since I didn't know what caused my baby to be born with this life threatening problem, and I knew that herbicides had been used in a place where I lived, *I had no way to know that I wasn't—that my events weren't contributed to by that activity. I had no way to prove that it was and I had no way to know that it wasn't.* A few years later, my friend, Mary, as well as Edith, had a baby who also had a hole in his heart that wasn't as severe (emphasis added).

This woman's statement that, "I had no way to prove that it was and I had no way to know that it wasn't" illustrates the use of a precautionary principle. When it came to thinking about what had happened to her child, she used different criteria than Forest Service managers. Rather than looking for proof that there had been a connection, she wanted proof that there had not been one.

Gendered exposure and meanings of herbicides

Along with the overarching mistrust of the Forest Service, questions about the process of science and differences in abstract versus direct risk evaluations, gender also shaped the experience of past herbicide exposure and the meaning of potential herbicide use in the future among both Indian and non-Indian community residents. Although nearly all community members were opposed to the use of herbicides, a number of women in the community held unique experiences of past herbicide exposure, in their bodies and through their social roles in the community as evidenced in the last quote. Gendered bodily impacts included late-term miscarriages, birth defects, and alteration of menstrual cycles. In 1976, one third of the pregnancies in the immediate area ended in miscarriages after the third month, one deformed child was born, and there were three molar pregnancies. Dramatic physical experiences such as miscarriages and menstrual-cycle disruption exemplify the difference between an abstract and embodied sense of risk. The local nurse described how the miscarriages were later in pregnancies than usual, and the fact that she was convinced they were due to herbicide exposure: ". . . later, after they sprayed around them, we had, the rest of the next two years, we had miscarriages, and they're always around the third month, where normal miscarriages usually are six weeks to eight weeks." Of course, not all women in the community had actually experienced such events, yet the stories of those who did made it clear to all women that very serious, personal consequences of herbicide exposure could affect them.

In addition to concerns regarding direct bodily exposure, both Indian and non-Indian women have particular worries about herbicide safety related to their social roles as both mothers and caregivers. Incidences of birth defects impact women as mothers of children, while cancers in adult people affect women as caregivers of the elderly. Karuk women in particular raised health issues facing children, such as exposure by teething on woven baby rattles and by food served in handmade bowls:

> For me, I worry about myself, but I like to make rattles, and the first thing you do is you give it to a baby and they're going to put it in their mouth. My kids teeth on their rattles, and I think that most kids do. So, it spends a lot of time in their mouth. Then, you also have your bowls that you eat out of. So, your food sits right in there.

The added risk of exposure, due to child's small body size, was also raised as a concern: "After I had kids, I thought about it even more, because I make them rattles, or they eat the food, and I think about how little they are and how things affect them differently than they would your average person." Karuk women also expressed concern for seniors in the community, including elders who themselves held an important role in carrying on culture. "Not that I shouldn't worry about me, too, it's just that I think about their little, fragile bodies, and the elders in health states where just little things like that can really affect them." Finally, there is the above mentioned potential of added exposure for weavers, who are predominantly women. Weavers in particular are exposed when cleaning materials in the field, when they put materials into their mouths, and are out in the woods visiting and tending gathering sites throughout the year:

> With willow and even hazel, when you go to clean them, I put the willow stick in my mouth and I peel off the bark. So that hasn't been soaked, it hasn't been washed. Nothing has happened to it; you clip it right off the plant, put it in your mouth, and strip off the bark. With stuff when you're making your basket, it's been soaked probably a couple times, but when you go to clean them, it's right there. I think most of all my concern is that you're out there. You're walking through everything even if they're not targeting the plant that you're gathering. You walk through whatever plant it was that they were targeting, to get to where you're, especially on the river sites.

Racialized exposure and meanings of herbicides

While there exists an extensive literature on gender and risk perception, speculating, for example, on the relative importance of gender norms versus social roles of motherhood, little sociological work addresses the basis for observed racial differences in the perception of risk. Even less work examines the basis for widespread opposition to herbicides among Native Americans. For Karuk people, the meaning of herbicide exposure must be understood within at least three different contexts. First, traditional Karuk people have additional threats of exposure based on specific cultural practices, as for the basketweavers mentioned above. Second, control over land management is an issue of cultural sovereignty. And finally, the threat of poison in the watershed is viewed by at least some as one more event in a series of acts of genocide over the past 150 years.

Probably the aspect of the Clearwater story that makes herbicides most clearly an environmental justice issue is the fact that Native people would experience additional threats of exposure as a result of subsistence and cultural practices, including gathering, tending, weaving, and eating foods from the forest. In addition to the practices affecting weavers, Karuk people eat foods from the land, including plants and animals. There are numerous stories of deer killed for meat during earlier spraying whose livers and internal organs were deformed and abnormal. The fact that a common herbicide for use in forestry in the region is not registered for use on food shows the implicit cultural assumption that people do not get their food from the forest. This was explained to me by a staff member of the California Indian Basketweaver's Association:

> Garlon,[2] this is the most frequently used chemical in the county. It has pretty long persistence . . . What's interesting and shocking about this chemical is that it's not registered for use on food crops at all . . . and there is no drinking water safety limit either. They're spraying about 100,000 pounds of this in the County every year, and there is no drinking water safety limit. It is just totally under the radar for the Safe Drinking Water Act. The issue about it not being registered for use on food crops, when

> there are people getting food plants out of the forest is pretty disturbing, too. That is just one way that the traditional lifestyles aren't being taken into account when they register these chemicals.

The assumption that food and water supplies do not come from the forest, and thus that forestry herbicides need not be tested for use on foods, puts traditional Native American people at greater risk. For the Forest Service and others, such problems may seem manageable. The species weavers use (e.g., willow, hazel) are not themselves targeted for spraying. The Forest Service has made efforts to accommodate weavers, yet for many these efforts are themselves intrusive, requiring basketweavers to identify their personal gathering areas.

This leads to a second racialized dimension of the decision-making process. Above and beyond the evaluations of hazards of individual chemicals, many Karuk viewed invasive weed management in the larger context of tribal sovereignty. The physical and cultural survival of Karuk people has depended upon their relationships with the land for tens of thousands of years. These relationships have been disrupted through different processes over the course of Indian–non-Indian relations during the last century and a half. The Karuk do not have a reservation, instead the Forest Service is the legitimate land manager of most of their ancestral territory. Part of sovereignty is not having to argue over details of spray areas or provide sensitive information to a non-Indian federal agency. That is to say, "the bottom line is, we don't want herbicides in the watershed. End of discussion." Shortly after the Forest Service proposed spraying of spotted knapweed, the tribe passed a resolution against the use of herbicides. This document refers to both tribal sovereignty and ecological concerns:

> WHEREAS The Karuk Tribe of California is a historic sovereign aboriginal People, that have lived on their own land since long before the European influence of white men came to this continent; . . . and WHEREAS the Karuk Tribe is dedicated to the preservation and ecological integrity of the Clearwater River and WHEREAS the application of pesticides/herbicides greatly threatens the fragile ecosystem of the Clearwater River watershed, THEREFORE BE IT RESOLVED; that the Tribal Council opposes application of any type of pesticides/herbicides by the United States Forest Service or any other agency in the Clearwater River Basin (Karuk Tribe of California, 1999).

Thus to understand the significance of the herbicide issue, it is necessary to view it as many people did, in the context of 150 years of struggle over access to and control over resources and cultural survival.

Finally, as illustrated in this letter written by a Karuk community member to the Forest Service in protest of herbicide spraying in 1981, at least some Indian people experienced past herbicide spraying and high rates of miscarriages as one more event in a series of acts of genocide.

> I guess it's easy for such a large organization to ignore such a small group of people. But it is not as easy for us to ignore the Forest Service when its actions cause such terrible damage to us. There are only 800 of us left. When we lose one baby, it is the same proportion as if you lost 275,000 babies. The herbicide spraying is clearly threatening our very survival as a people. Our cultural group is already endangered enough as it is.

Furthermore, cultural practices such as the act of weaving have been threatened through direct genocide and forced assimilation over the past 150 years. And the present moment is critical. Weaving, language, and other cultural traditions that were nearly lost are being actively recovered by new generations. One weaver in her forties described how her mother was discouraged from weaving. She herself did not learn to weave until she was older:

> When I was young . . . the elders didn't feel it was in the young one's best interest. They were pushing us more in the direction of basically leaving our culture behind, because they went through so many struggles, really severe struggles. Both my parents were forcibly removed from their home, their parents, and taken to what's talked about as Indian school. . . .

The perceived threat to weavers and the cultural practice of weaving carry a heavy weight of cumulative effects to physical and cultural survival. Thus, while Forest Service employees may be frustrated with the inability to "get to the facts" and discuss the details of particular chemicals, for the Karuk people, what is at stake is the ability to carry out traditional relationships with the land and, hence, ensure their cultural and physical survival without fear of exposure to herbicides, without having to ask permission from a foreign government to access sites, and without having to depend upon the risk assessments and cultural frameworks of non-Indians to determine the safety of their activities. For Forest Service employees, the question at hand in determining appropriateness of herbicides as a management tool was the numerical values of No Observable Effect Levels calculated in laboratories. The proposed use of herbicides is understood by locals in a large context. For community members and native people in particular, the questions were about who makes decisions and how, and who experiences the consequences of such decisions.

Gender, race and cultural opportunities for mobilization

Despite the fact that spotted knapweed is a Class A pest in California, as of this writing no herbicides have been used on invasive weeds in the Clearwater area. Furthermore, this resource-poor community succeeded in stopping the widespread use of aerial herbicide spraying in the 1980s. How has this politically and culturally diverse community succeeded in achieving their vision of resource management in the face of a large federal agency like the Forest

Service? To completely address this would be the subject of another paper. However, it is worth noting that gender and race, the same variables that led to differences in risk perception, were themselves used to mobilize against the use of herbicides.

Narratives about unequal exposure, particularly when voiced by women and Karuk tribal members, served as mobilization techniques because they provided frames which in turn resonated with dominant cultural beliefs about equality. These provided localized examples of "cultural opportunity structure" (see Jenkins and Klandermans 1995; McAdam, McCarthy, and Zald 1996) that work alongside "political opportunity structures." More particularly, the sharing of testimony about racialized and gendered aspects of experience provided cultural elements of opportunity structures. For example, women in the community wrote comment letters and spoke at public meetings regarding their unique concerns. Social equality between women and men has also reduced the barrier between the public and the private spheres, which has probably helped to constitute a claims structure in which stories about care work and concern for the exposure of children in the home held greater potency. One of the women whose son was born with a hole in his heart describes her involvement: "I organized a press conference for the first time, got the press to come, and got an article written in the San Francisco Chronicle. We made our presence known to the Forest Service regional office in San Francisco." In the earlier struggle against aerial spraying, midwives and female medical practitioners played key roles in disseminating information:

> Because I was doing midwifery and I was spending time with pregnant women and babies, I was really concerned that people not be drinking contaminated water. We began educational stuff in our community. People talking to each other about the effects; for me it was learning some from the Native Americans who started to have miscarriages. We became aware of that, and then I started to hear about a lot of different possible effects down river where there had been a lot of spray use. People started passing that information word of mouth in a way in the beginning.

Similarly, although the material resources of Karuk Tribe are meager when measured alongside the U.S. Forest Service, the unique experiences of Karuk tribal members formed cultural opportunity structures that facilitated work against spraying. Karuk tribal members used their legitimacy as a sovereign nation as framing in their cause. The resolution against spraying in their territory and the organizing with other Native people across the state to form the California Indian Basketweavers Association—both of which were described earlier—are examples of these efforts. In addition, tribal members worked locally to educate land managers regarding specific uses that put Indian people at particular risk. Here, a basketweaver describes spending a day with people from the state highway department and a major user of herbicide to keep highways clear in order to show him the plants they used along roadsides.

> They were doing roadside spraying, and we invited CalTrans to come up and go on a field trip with us. We would stop along the way and there would be so many plants that—I think we met at nine in the morning, and it was one by the time we got just eight miles down the road. He was really astounded by how many plants we used.

The highway department stopped using herbicides in the county shortly after this visit.

Discussion and conclusions: herbicides—a new environmental justice issue for rural communities?

As trade continues to increase and more and more species move around the globe, land managers, including the Forest Service, Bureau of Land Management, Park Service, State Parks, and private groups will face increased weed invasions. For some groups, herbicides may be a useful and welcome tool. For others, herbicide use is problematic and even represents a threat unto itself. The most visible impact of biological invasions on the Clearwater community resulted from the proposed use of herbicides to manage unwanted weeds. While there was general agreement between members of the Karuk Tribe, non-Indian community, and the Forest Service that invasive species were not welcome in the region, there were different reactions to the issue of potential herbicide use. I have described how four interconnected factors led to different perceptions of the risk of herbicide use in response to spotted knapweed.

First, the history of Forest Service herbicide spraying was a significant factor in shaping the perceptions of both Indian and non-Indian community members. Moreover, in the Clearwater region, the generalized erosion of scientific and institutional credibility carries with it the rural anti-government flavor of many western communities and is accompanied by the revitalization of and support for tribal land management.

Second, a variety of other social context factors clearly differentiated the formation of risk by both the Indian and non-Indian community from that of the Forest Service. Whereas employees of the Forest Service made decisions within an institutional context that favored the use of herbicides, many of the concerns and perspectives of community members derived from their attachment to place, historical observations of the area, and the possibility of their direct exposure. A sizeable body of research points to the importance of social and historical factors in shaping the public's sense of risk. Less attention has been given to the development of risk perceptions among scientists and land managers who also develop their understandings of risk within social and historical contexts. The present study extends the limited research in this area, providing reflections on institutional organization that may have led Forest Service employees to view herbicides as more acceptable than did members of the other two groups. Winston's (1997) observation that the belief that pesticides are necessary is related to both the beliefs that there are no alterative methods to remove pests and that the benefits of pesticide use outweigh their risks is highly useful in understanding the different orientations of groups towards herbicides in the Clearwater River. In this case, both Indian and non-Indian community members viewed weeds as a problem that should be managed, but also as an issue that could provide long term employment for the region. Furthermore, given that community residents initiated an intensive volunteer hand-eradication effort, they clearly did not see herbicides as the only way to defeat the knapweed invasion.

Third, gender differences in the risk evaluation of community members, many of which intersected with racial differences, concerned not only the documented importance of women's social roles, but also the less studied potential for direct bodily impacts such as miscarriages. Gender did not mark a divide between those who did and did not support herbicide use. Both women and men in the Indian and non-Indian community were opposed to the use of herbicides. And many women in the Forest Service were supportive of their use. Rather, in this case, gender provided an added dimension to the meaning and experience of herbicides for both tribal and non-tribal women.

Finally, voices from the Clearwater River give insight into the understudied basis for racialized differences in risk perceptions. Here the meaning and significance of herbicide exposure for members of the Karuk Tribe extended far beyond a simple calculation of probability of exposure risk. Social context features—including institutional mistrust, history of genocide, current land management struggles, and awareness of the unique Karuk uses of the forest, as well as missing scientific information regarding these uses—all contributed to the conclusion of tribal members I spoke with that herbicides were not "safe." In summary, some meanings and perceptions of risk are a function of distinct cultural uses. Other meanings and perceptions are themselves an outgrowth of the process of racial formation. In particular, the desire for autonomy and institutional mistrust are outgrowths of historical experiences of colonialism.

This case study suggests that as herbicide use increases in rural areas, there may be new kinds of community concerns and new dimensions to ongoing struggles over land management. Furthermore, because: (1) forestry workers and others at the intersection of land and land management practices are often racial minorities, (2) people of color are both politically and economically disenfranchised and perceive greater environmental risk, and (3) Indians use the land for subsistence and cultural purposes, many of these new impacts may be on people of color. Pesticide exposure already makes agriculture one of the most hazardous industries in the United States (Arcury and Quandt 1998; Quintero-Somaini and Quiridongo 2004). Like farm workers, both forestry workers and Native American people are disenfranchised and medically underserved populations. Factors including economic dependence, lack of control over work and living environments, high poverty, low wages, inadequate health insurance, inadequate access to information, and cultural barriers to political participation all contribute to increased risk of exposure and consequences for these groups.

Given the increasing prevalence of both invasive species and their management with herbicides, this paper reflects upon the broader social implications of differences in risk perception between communities and land managers. One of the larger sociological questions this paper highlights is who within society pays the price for species invasions. The use of herbicides for the management of invasive species is increasing across the United States. Many of those who face increased exposure to herbicides used for invasive species are members of racial minorities. When groups of citizens evaluate risk differently from land managers who have decision making authority, when differences in risk result in part from relative chances of exposure, and when those who face disproportionate exposure are members of poor or racial minorities, differences in risk perception become matters of environmental justice.

Notes

1 I wish to thank all those whom I interviewed for this research for taking the time to share their valuable perspectives; the NSF funded IGERT Program on Biological Invasions at U.C. Davis for research funding; Shawn Bourque, Adrieene Reed Storey Harling, Holly Hays, and Laura Smith for research assistance. I thank James Cramer, Megan Kemple, Lyn Lofland, Salm Stroich, Savanna Fergusen, and the editor and anonymous reviewers for helpful comments in the preparation of this manuscript.

2 Note that Garlon is not a proposed chemical in the current Weeds Environmental Impact Statement.

References

Arcury, T.A. and S.A. Quandt. 1998. "Occupation and Environmental Health Risks in Farm Labor." *Human Organization* 57:331–34.

Baldi, I., L. Filleul, B. Mohammed-Brahim, C. Fabrigoule, J. Dartigues, S. Schwall, J. Drevet, R. Salamon, and P. Brochard. 2001. "Neuropsychologic Effects of Long-Term Exposure to Pesticides: Results from the French Phytoner Study." *Environmental Health Perspectives* 109:839–44.

Beamish, T.D. 2001. "Environmental Hazard and Institutional Betrayal: Lay-Public Perceptions of Risk in the San Luis Obispo County Oil Spill." *Organization & Environment* 14:5–33.

Beck, U. 1992. *The Risk Society*. London: Sage.

Bell, M. 2002. *Karuk: The Upriver People*. Happy Camp, CA: Naturegraph Publishers.

Bord, R.J. and R.E. O'Connor. 1997. "The Gender Gap in Environmental Attitudes: The Case of Perceived Vulnerability to Risk." *Social Science Quarterly* 78:830–40.

Bright, C. 2001. "Biological Adversity: The Hidden Costs of Trade and Economic Globalization." *Harvard International Review* 22:4.

Bullard, R.D. 2000. *Dumping in Dixie: Race, Class, and Environmental Quality*. Boulder, CO: Westview Press.

Chipman, H. and P. Kendall. 1995. "Consumer Reaction to Risk/Benefit/Option Messages About Agricultural Chemicals in the Food Supply." *Journal of Consumer Affairs* 29:144–65.

Clarke, L. and W.R. Freudenburg. 1993. "Rhetoric, Reform and Risk" *Society* 30:78–81.

Davidson, D. and W.R. Freudenburg. 1996. "Gender and Environmental Risk Concerns: A Review and Analysis of Available Research." *Environment and Behavior* 28:302–39.

Davis, G.W. and F.M. Frank. 1992. *Stories of the Klamath National Forest: The First Fifty Years*. Hat Creek, CA: History Ink Books.

Dunlap, R.E. and C.E. Buess. 1992. "Understanding Public Concerns About Pesticides: An Empirical Examination." *Journal of Consumer Affairs* 26:418–38.

Finucane, M.L., P. Slovic, C.K. Mertz, J. Flynn, and T.A. Satterfield. 2000. "Gender, Race, and Perceived Risk: The 'White Male' Effect." *Health, Risk & Society* 2:159–72.

Freudenburg, W.R. 1997. "Contamination, Corrosion and the Social Order." *Current Sociology* 45:19–39.

Flynn, J., P. Slovic, and C.K. Mertz. 1994. "Gender, Race, and Perception of Environmental Health Risks." *Risk Analysis* 14:1101–8.

Garry, V.F., M.E. Harkins, L.L. Erickson, L.K. Long-Simpson, S.E. Holland, and B.L. Burroughs. 2002. "Birth Defects, Season of Conception, and Sex of Children Born to Pesticide Applicators Living in the Red River Valley of Minnesota, USA." *Environmental Health Perspectives* 110:441–49.

Gendar, J. 1991–2. *"Report on Panel Discussions and Workshops in California Indian Basketweavers Gathering."* Special Supplement *News from Native California*.

Hawkes, G.R. and M.C. Stiles. 1986. "Attitudes About Pesticide Safety." *California Agriculture* 40:19–22.

Hirt, P. 1994. *A Conspiracy of Optimism: Management of the National Forests Since World War Two*. Lincoln, NE: University of Nebraska Press.

Jenkins, J.C., and B. Klandermans (eds.) 1995. *The Politics of Social Protest: Comparative Perspectives on the State and Social Movements*. Minneapolis, MN: University of Minnesota Press.

Johnston, H. and B. Klandermans. 1995. "The Cultural Analysis of Social Movements" Pp. 3–24 in *Culture and Social Movements*, edited by H. Johnston and B. Klandermans. Minneapolis, MN: University of Minnesota Press.

Karuk Tribe of California. 1999. Resolution on Use of Herbicides in Ancestral Territory Passed Karuk Tribal Council. Happy Camp, CA.

Le, T.N. and A. Johansson. 2001. "Impact of Chemical Warfare with Agent Orange on Women's Reproductive Lives in Vietnam: A Pilot Study." *Reproductive Health Matters* 9(18):156–64.

Lowry, C. 1999. *Northwest Indigenous Gold Rush History: the Indian Survivors of California's Holocaust*. Arcata, CA: Indian Teacher and Educational Personnel Program.

Marshall, B.K. 2004. "Gender, Race, and Perceived Environmental Risk: The 'White Male' Effect in Cancer Alley, LA." *Sociological Spectrum* 24:453–78.

McAdam, D., J.D. McCarthy, and M. Zald. 1996. "Culture, Ideology and Strategic Framing" Pp. 261–74 in *Comparative Perspectives on Social Movements*, edited by D. McAdam, J.D. McCarthy, and M. Zald. New York: Cambridge University Press.

McNeely, J. (ed.) 2001. *The Great Reshuffling: Human Dimensions of Invasive Alien Species*. International Union for the Conservation of Nature. Washington D.C.: Island Press.

Murdock, S.H., R.S. Krannich, and F.L. Leistritz. 1999. *Hazardous Wastes in Rural America*. Boulder, CO: Rowan and Littlefield.

Murphy-Greene, C. and L.A. Leip. 2002. "Assessing the Effectiveness of Executive Order 12898: Environmental Justice for All." *Public Administration Review* 62: 679–87.

Nagel, J. 1996. *American Indian Ethnic Renewal: Red Power and the Resurgence of Identity and Culture.* New York: Oxford University Press.

Norton, J. 1979. *When Our Worlds Cried: Genocide in Northwestern California.* San Francisco, CA: The Indian Historian Press.

Oliva, A., A. Spira, and L. Multigner. 2001. "Contribution of Environmental Factors to the Risk of Male Infertility." *Human Reproduction* 16:1768–76.

Ortiz, B. 1993a. "Contemporary California Indian Basketweavers and the Environment." Pp. 195–211 in *Before the Wilderness: Environmental Management by Native Californians*, edited by T.C. Blackburn and K. Anderson. Banning, CA: Ballena Press.

Palmer, C.G., L.K. Carlstrom, and J.A. Woodward. 2001. "Risk Perception and Ethnicity." *Risk Decision and Policy* 6:187–206.

Pastor, M., J.L. Sadd, and R. Morello-Frosch. 2004. "Waiting to Inhale: The Demographics of Toxic Air Release Facilities in 21st-Century California." *Social Science Quarterly* 85:420–40.

Perrow, C. 1984. *Normal Accidents: Living With High-Risk Technologies.* New York: Basic Books.

Pimentel, D., L. Lach, R. Zuniga, and D. Morrison. 2000. "Environmental and Economic Costs of Nonindigenous Species in the United States." *Bioscience* 50:53–64.

Pimentel, D., R. Zuniga, and D. Morrison. 2005. "Update on the Environmental and Economic Costs Associated with Alien-Invasive Species in the United States." *Ecological Economics* 52:273–88.

Quintero-Somaini, A. and M. Quiridongo. 2004. *Hidden Danger: Environmental Health Threats in the Latin Community.* Natural Resources Defense Council.

Salter, J. 1981. *Shadow Forks: A Small Community's Relationship To Ecology And Regulation.* Ph.D. dissertation, Department of Sociology, University of California Santa Cruz, CA.

Savitz, D.A., T. Arbuckle, D. Kaczor, and K.M. Curtis. 1997. "Male Pesticide Exposure and Pregnancy Outcome." *American Journal of Epidemiology* 146:1025–36.

Schmidt, D. and D. Simberloff. 1997. "Biological Invasions: A Growing Threat" *Issues in Science and Technology.* Washington, DC: National Academy of Sciences. Retrieved February 17, 2007 (http://www.issues.org/13.4/schmit.htm).

Schrader-Frechette, K.S. 1991. *Risk and Rationality: Philosophical Foundations for Populist Reforms.* Berkeley, CA: University of California Press.

Slovic, P. 2000. *The Perception of Risk.* London: Earthscan.

Slovic, P., T. Malmfors, D. Krewski, C.K. Mertz, N. Neil, and S. Bartlett. 1995. "Intuitive Toxicology II. Expert and Lay Judgments of Chemical Risks in Canada," *Risk Analysis* 15:661–75.

Snow, D.A., E.B. Rochford, Jr., S.K. Worden, and R.D. Benford. 1986. "Frame Alignment Processes, Micromobilization, and Movement Participation." *American Sociological Review* 51:464–81.

Swidler, Ann. 1986. "Culture in Action: Symbols and Strategies" *American Sociological Review* 51:273–86.

Van Strum, C. 1983. *A Bitter Fog: Herbicides and Human Rights.* San Francisco: Sierra Club Books.

van Tassell, L.W., M.A. Ferrell, B. Lang, D.E. Legg, and J.E. Lloyd. 1999. "Pesticide Practices and Perceptions of Wyoming Farmers and Ranchers." *Journal of Soil and Water Conservation* 54:410–15.

Winston, M. 1997. *Nature Wars: People vs. Pests.* Cambridge, MA: Harvard University Press.

Wulforst, J.D. 2000. "Collective Identity and Hazardous Waste Management" *Rural Sociology* 65:275–94.

Zahm, S.H. and M.H. Ward. 1998. "Pesticides and Childhood Cancer." *Environmental Health Perspectives* 106:893–908.

Zheng, T., H. Zahm, K.P. Cantor, D.D. Weisenburger, Y. Zhang, and A. Blair. 2001. "Agricultural Exposure to Carbamate Pesticides and Risk of Non-Hodgkin Lymphoma." *Journal of Occupational and Environmental Medicine* 43:641–49.

Part VII

Learning and building on the classics

In Part VII, we provide some concluding remarks and an analysis of the themes addressed (and not addressed) in this corpus of material on gender and forests.

16 Concluding reflections for the future

Bimbika Sijapati Basnett, Marlène Elias, Susan Stevens Hummel, and Carol J. Pierce Colfer

Melissa Leach (this volume) has cogently noted that environment and development policies and practices have increasingly been framed in terms of property rights, resource access, and power. At the time of writing (in 2007), these were welcomed as challenges to ecofeminist and Women, Environment and Development (WED) fables about 'women's closeness to the environment'. But Leach warned that in the apparent turn away from ecofeminism and WED, scholars and development practitioners (still) undervalued gender and how property, access, and power relate unevenly across gender lines. Consequently, "gender-blind development work seem[ed] to be on the rise, and there is rather little evidence of a more politicized, relational perspective on gender and environment taking root" (p. 68 in original). One decade later, concerns over gender equality and women's empowerment are re-emerging as part of the 2030 global development agenda.

The Sustainable Development Goals (SDGs), which were adopted by 193 countries of the UN General Assembly on 25 September, 2015, epitomize the renewed focus on gender equality in more significant ways than their predecessor – the Millennium Development Goals (MDGs; Kabeer 2015). The fifth goal (SDG 5) calls on governments in developing and developed countries alike to "adopt and strengthen sound policies and enforceable legislation for the promotion of gender equality", and measures such as guaranteeing the full and effective participation of women in decision-making, undertaking reforms to give women equal rights to productive resources, and recognizing the value of unpaid care work are highlighted (UN 2015). SDG 5 also renews and reinforces commitments to women's rights and gender equality made in other international conventions such as the Convention on the Elimination of all forms of Discrimination Against Women (CEDAW), Agenda 21, the Beijing Platform for Action, and relevant national-level policies and legislation. While critics argue that the SDGs do not go far enough (Kabeer 2015), what is nevertheless noteworthy about SDG 5 is that it is premised on a strong human rights-based approach. While women's role in promoting sustainable development is not ignored, the granting of women's rights is seen as an end in itself and not just a means for effectively and efficiently contributing to such development. This is a welcome change from other contemporary approaches that view gender equality as requiring defence on the grounds that it is "smart economics" and leads to the realization of other seemingly more pressing goals (Asher and Sijapati Basnett 2016).

Such a global level consensus to promote gender equality and women's empowerment presents an unprecedented opportunity to firmly root gender issues in the forestry and landscape management agenda in the 'North' and 'South' alike. And yet, gender and feminist scholars caution against uncritical optimism and compel us to engage with and learn from the 40 years of feminist research on environmental and developmental policy. In this way, gender is not just 'added in', 'technocratized', and reduced to stereotypes about women and men, thereby ending up doing harm rather than resulting in inclusive changes on the ground (Arora-Jonsson 2014; Bee and Sijapati Basnett 2016). Hence, a multi- and inter-disciplinary volume that enables scholars to build on classical works while encouraging practitioners to apply lessons from the 40 years of critical feminist research on the topic could not be more timely.

A common thread across the chapters in this volume is the need to unpack the 'household' and see it as composed of members who are unequally situated with their respective needs, claims, goals, and bargaining power mediated by gender in interaction with a range of other social relations (such as age and generation, ethnicity, income, amongst others). Intra-household inequalities are manifest in the forestry arena in multiple contrasting ways across diverging geographies (Agarwal; Li; Rocheleau and Ross; all this volume). Many of the chapters in this volume respond to the need for attention to interconnecting processes and spaces, from the very intimate to the national and transnational (see Elias and Carney; Elmhirst; Li; Schroeder; all this volume). Questioning the assumption that gender issues in the North and South are either distinct or only concentrated in the South, the chapters on the 'developed world' shed light on gender inequalities and constraints that travel across geographies (Arora-Jonsson; Paulson; Norgaard; Colfer; all this volume). These chapters, therefore, provide a glimpse into why the justification for a divide between North and South should be examined more carefully. Chapters in this book also engage with the inevitable partiality of truth or knowledge in a range of ways – by advocating for greater attention to women's knowledge, which may be distinct, shared, and complementary to men's (Fortmann; Rocheleau and Ross; both this volume), and by illustrating the value of mixed methods in undertaking feminist research on forests (Nightingale; Arora-Jonsson; both this volume). On that point, several chapters point to the importance of 'politicizing the field' by situating researchers and practitioners within it – engaging, interpreting, negotiating, and reconfiguring gender relations through their own positions in social and institutional hierarchies and critically challenging (un)acknowledged assumptions and epistemological commitments (Arora-Jonsson; Nightingale; Colfer; all this volume).

What gaps remain in the literature?

As we did in *Gender and Forests* (Colfer *et al.* 2016c), we approach this question by examining the "Gender Box" (Colfer and Minarchek 2013; Colfer 2013). The Gender Box (Figure 16.1) is a conceptual framework developed to help forestry professionals recognize important and recurrent elements and topics in gender analyses as guidance for conducting or facilitating studies pertinent to their own work.[1] The framework is based on several extensive literature reviews (focused on gender and forests, but also tapping literature with other foci), and it is intended to evolve as we learn more about women, men, their interrelations, and forests. It identifies 11 topics (see Table 16.1), three levels (micro, meso, and macro), and a time dimension.

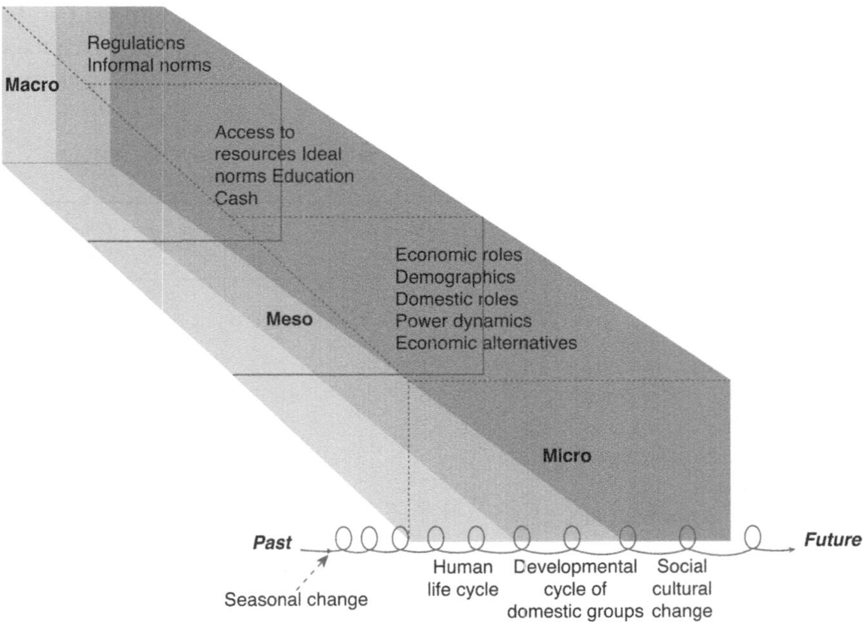

Figure 16.1 The Gender Box

These are all likely to affect relevant social processes in forests (including intersectionality, production–reproduction issues, and population/health, as suggested in Chapter 1). Since the Gender Box was first published in 2013, we have identified five additional topics of importance, so the Gender Box now totals 16 topics (Colfer *et al.* 2016a). These additions are:

- gendered knowledge, of an informal and/or traditional nature;
- local involvement in management processes, such as participation and decision-making;
- local leadership;
- networks/groups; and
- violence against women.

Table 16.1 provides a Gender Box analysis, including these new topics.[2] The second column in Table 16.1, labelled 'Classics', refers to the issues addressed in this collection. We also compare these 14 classics to the 16 articles published in our 2016 *Gender and Forests* book (column 3 in the table). Although the number of articles analyzed is small, none were explicitly selected with Gender Box issues in mind. We thus assume there is minimal inherent bias in our sample; but the analysis remains illustrative rather than predictive. The population of articles based in the 'Global North' from which the two studies were selected is far smaller than the population of articles in the 'Global South'.

We first consider the level of attention paid to topics within the 14 'classic' articles in this collection. A paper is counted as having dealt with an issue (and thus included in the percentages in Table 16.1) if a substantive discussion of the issue occurs in the paper. A mere mention of the topic is insufficient to be included. We consider a topic is generally well attended when the percentage of papers dealing with it substantively is between 75 per cent and 100 per cent;

Table 16.1 Gender Box analysis of issues addressed in these classic contributions (compared with analysis of current, 2016, contributions)

	Classics %	*2016 Collection* %	*Change* %
Part A: Original Issues			
Formal laws/policies	86	63	−23
Cultural/religious trends	71	63	−8
Access to natural resources	86	100	14
Norms of behaviour	57	88	31
Access to education*	36	63	27
Access to cash	57	75	18
Day to day economic roles	93	81	−12
Demographic issues	29	38	9
[Migration]**	21	25	4
Domestic roles	64	38	−26
Intra-household power dynamics	64	50	−14
Available economic alternatives	50	69	19
*Part B: Additions****			
Knowledge*	50	38	−12
Management process involvement	79	94	15
Leadership	29	31	2
Networks/groups	79	63	−16
Violence against women	21	25	4

Notes: * 'Education' refers to formal education; 'Knowledge' refers to informal or 'traditional' knowledge.

** 'Migration' has been extracted from 'Demographic issues' to show its prominence vis-à-vis other important and neglected demographic issues.

*** Part B Additions are those issues not originally identified in the 2013 Gender Box, but included in the 2016 analysis.

intermediate levels of attention range from 50 per cent to 74 per cent; and minimal attention ranges from 0 to 49 per cent. We then compare these patterns with those in the recent papers from the 2016 collection to make a rough assessment of changing interests in gender analysis. We conclude this section with a discussion of remaining gaps and issues that might well be added.

Topical emphases in this collection, with commentary

We turn first to the issues prominently addressed vis-à-vis those comparatively neglected in this group of 'classic' articles, published between 1983 and the present. As a whole, the papers are consistent with a common trend that began in the 1970s with Boserup's (1970) groundbreaking work on African women's roles in agriculture: an emphasis on women's 'productive' activity (most typically, the work that explicitly and directly produces food or goods – see Chapter 1 for a critique of this concept). The most consistently addressed issue is day-to-day economic roles. The authors in this volume have tended to explain what women (and sometimes men) did for subsistence or cash in the forests they studied. Other commonly addressed issues include formal laws/policies, access to forest resources, forest management process involvement, and networks/groups. Formal laws and policies often disenfranchise women in forests; and even when they support women's rights of access, there is no guarantee women will have access to such resources in practice. Traditional systems, often implemented more effectively, may still disenfranchise women (see in this volume: Arora-Jonsson; Elmhirst; Li). Women's involvements in forest management processes and networks are often discussed as tools for empowering them (see in this volume: Agarwal; Schroeder). The emphasis in the early days of gender studies was on Women in Development (WID) rather than gender, with a strong economic focus – a common goal was to prove that women also contributed to the dominant version of 'development'. This was an important first step, but the field has since progressed beyond this single-minded focus.

Issues that were addressed at an intermediate level of attention include cultural/religious trends, domestic roles and intra-household power dynamics, norms of behavior, and access to cash. Only half of the authors addressed available economic alternatives (options) or informal kinds of knowledge (see in this volume: Elias and Carney; Fortmann; Nightingale). Leach's chapter (this volume) on ecofeminism discusses one of the broad trends that has significantly influenced studies of women in forests. Domestic roles and intra-household power dynamics are increasingly of interest to gender specialists, and are addressed in Colfer, Elmhirst, Li, and Schroeder (all this volume). Cash, whether from forest products or other sources, is almost always less accessible to women than to men. This topic comes into sharp focus in Li, Paulson, Reed, and Rocheleau and Ross (all this volume).

Interestingly, some of those issues that are least addressed in this classic body of literature would seem to be particularly important for the empowerment of women – a current emphasis within gender studies. Few examined access to formal education (see in this volume: Agarwal; Norgaard; Reed). Our broader background surveys suggested that demographic issues would include attention to women's control over their reproductive health, including access to birth control – typically unavailable in remote forested areas of the world, yet important for the long-term health of both women and forests. Although clearly important in opening up empowering avenues for women (Colfer *et al.* 2008), the glaring lack of attention to reproduction is not specific to this collection, but common to all those on forests that we have examined. Instead the emphasis under 'demographic issues' is on migration. The least addressed of the issues in the expanded Gender Box is violence against women, also important in relation to women's empowerment and capacity to make their voices heard (see in this volume: Agarwal; Fortmann).

These two issues – reproductive health and violence against women – were represented in the literature reviews that led to the development of the Gender Box mainly because gender-related materials were drawn from a wider body of literature that included health-related gender studies in forested areas. The bodies of literature we have analyzed using the Gender Box have all been focused on forests or agroforests. The relevance of health to forests and people is discussed in Colfer (2008) and Colfer *et al.* (2006). Identification of this kind of gap is precisely the advantage of moving outside our disciplinary silos.

Topical changes over time

We realized that a brief comparison between this analysis and that done for our 2016 collection of new articles might provide some useful insights about changing directions in gender analysis.[3] We have structured our analysis below according to large topical changes (over 20 per cent), intermediate changes (10 per cent to 19 per cent), and small changes (under 10 per cent) to give a sense of these directions.

The biggest change was an increase of interest in the topic of norms of behaviour.[4] Among the classics, 57 per cent of the authors addressed this issue, whereas in the newer collection, 88 per cent did. The second largest change was in access to education, increasing from 36 per cent among the classics to 63 per cent in the new collection. Interestingly, and perhaps somewhat discouragingly (see Chapter 1), attention to domestic roles fell substantially. Attention to formal laws and policies also fell.

The rise in interest in available economic alternatives and access to cash represents an intermediate level of change. Interest in access to natural resources and in forest management processes also increased. Coverage of networks/groups fell, as did interest in intra-household power dynamics, economic roles, and informal knowledge.

The smallest increase was in attention to leadership; a topic that would seem, like violence against women, to deserve more attention, given the general interest in empowering women. Migration received slightly more attention in our sample of recent literature; this is a reasonable concern given the accelerating pace at which movements into and out of forest areas are occurring. Attention to demographic issues more generally rose, with more attention paid to issues beyond migration (though still not in access to contraception), whereas attention to cultural/religious trends fell.

What are the lessons for the future?

As the introductory chapter in this volume points out, future studies on the intersection between gender and forests need to engage with three key issues – intersectionality, the pseudo divide between production and reproduction, and population and reproductive health. The 'giants' whose works are showcased in this volume provide considerable food for thought. 'Women' are portrayed as constrained and enabled by structures (gender norms that pervade sociopolitical and economic systems at all scales) of varying flexibility that are subject to change through political struggles, negotiations, and social change. Gender is seen as relational and context specific, always interacting with other social relations, even as some chapters privilege class/livelihood trajectory, while others refuse to single out any particular feature. The issue of women's invisible and unremunerated work is equally at the heart of the book, with chapters highlighting spaces (including room to manoeuvre within existing inequalities) that women occupy and negotiations over women's labour that foment economic and social change.

Apart from deepening and nuancing gender analysis along the lines we suggest, future research on gender and forests needs to carefully examine the implications of the key globalizing tendencies that are fundamentally altering forested landscapes and people–forest relationships. Some of the changes are being triggered by: governments transferring tenure rights to indigenous peoples through assistance from international organizations such as the World Bank or the Food and Agriculture Organization; tropical forests being revalorized, monetized, and commodified in the name of food, fuel, and climate change (Sikor 2010); and demographic shifts, migration, urbanization, and changes in consumption patterns that are altering forested landscapes in the North and South alike (Hecht *et al.* 2015; Vira *et al.* 2015). The emerging research on these issues find that although the growing rush for land in forested landscapes is boosting rural productivity, generating employment, and in some cases addressing food insecurity, it is simultaneously undermining the rights and livelihoods of workers, indigenous communities, and smallholders – all with serious gender implications (Elmhirst *et al.* 2015; Li 2015; Chung 2016). Even seemingly well-meaning reforms aimed at granting greater tenure rights to indigenous peoples may serve to exclude women or exacerbate pre-existing gender inequalities within these communities (Bose 2011; Mairena *et al.* 2012). Migration and demographic shifts can have contradictory implications for women who are migrants and left behind – by enabling them to assume greater responsibilities and control over household and community affairs and/or increasing their workload and limiting their voice and agency in institutions governing forests and trees (Resurreccion and Elmhirst 2008; Rigg 2012; Sijapati Basnett 2013). These processes underscore the pressing need to include gender explicitly in scholarly analyses of forest dynamics and to recognize that gender is a critical aspect to the understanding of the social processes of differentiation in general.

As Sikor (2010) notes, these developments are prompting a "new agenda for research and practice" on forest justice. Are these globalizing tenure reforms leading to the creation of new opportunities for previously marginalized actors, or are they instead contributing to enclosures, exclusion, and dispossessions? Are local level concerns being sidelined in favour of global concerns with monitoring carbon? Whatever the answers, there is a need to return to the classical literature on gender and forests to unpack who is likely to benefit, who to lose, why, and what can be done to put gender equity and human rights – *including both women's and men's equally* – at the forefront of this emerging research agenda. We need to stimulate additional up-to-date studies by others who can continue to build on the insights, methods, and findings represented in these pages. To do that, we must work closely with women and men at various levels, including communities if we really want to seize the opportunities accorded by SDG 5.

Notes

1 These topics are general and do not provide ethnographic details about how women and men use forests (this is nearly infinitely variable) but, rather, what topics have a bearing on people's use of forests and therefore warrant research attention.

2 Gender Box analyses have also been done regarding a corpus of material on agroforestry (Colfer *et al.* 2015) and on most of the material that was found on gender and forests in the global North (presented in 2016 at the Society for Applied Anthropology Meetings, Vancouver, BC, Canada; Colfer *et al.* 2016b).

3 We do recognize that this collection is not representative of all gender and forests literature – the findings, as noted earlier, are illustrative only.

4 These are expectations regarding normal or acceptable behaviour within a given group. Norms can be tacit or overt, oppressive or supportive to specific groups.

References

Arora-Jonsson, S. 2014. Forty Years of Gender Research and Environmental Policy: Where Do We Stand? *Women's Studies International Forum* 47 (Part B): 295–308.

Asher, K., and Sijapati Basnett, B. 2016. Gender Equality as an Entitlement: An Assessment of the UN Women's Report on Gender Equality and Sustainable Development 2014. *Development and Change* 47 (4): 952–964.

Bee, B., and Sijapati Basnett, B. 2016 (online). Engendering social and environmental safeguards in REDD+: Lessons from feminist and development research. *Third World Quarterly*. http://dx.doi.org/10.1080/01436597.2016.1191342

Bose, P. 2011. Forest Tenure Reform: Exclusion of Tribal Women's Rights in Semi-Arid Rajasthan, India. *International Forestry Review* 13 (2): 220–232.

Boserup, E. 1970. *Woman's Role in Economic Development*. London: George Allen & Unwin.

Chung, Y. 2016. *Gendered Impacts of Large-Scale Land Acquisitions, with a Focus on Women's Land and Forest Rights: A Literature Review*. Washington, DC: Rights and Resources Initiative.

Colfer, C. J. P., ed. 2008. *Human Health and Forests: A Global Overview of Issues, Practice and Policy*. People and Plants International Conservation Series. London: Earthscan.

Colfer, C. J. P. 2013. *The Gender Box: A Framework for Analysing Gender Roles in Forest Management*. CIFOR Occasional Paper 82. Bogor, Indonesia: Center for International Forestry Research.

Colfer, C. J. P. and Minarchek, R. D. 2013. Introducing 'the Gender Box': A Framework for Analysing Gender Roles in Forest Management. *International Forestry Review* 15 (4): 1–16.

Colfer, C. J. P., Sheil, D., and Kishi, M. 2006. *Forests and Human Health: Assessing the Evidence*. CIFOR Occasional Paper 45. Jakarta, Indonesia: Center for International Forestry Research.

Colfer, C. J. P., Dudley, R. G., and Gardner, R. 2008. Forest Women, Health and Childbearing. In C. J. P. Colfer, ed., *Human Health and Forests: A Global Overview of Issues, Practice and Policy*. London: Earthscan, 113–134.

Colfer, C. J. P., Catacutan, D., Naz, F., and Pottinger, A. 2015. Introduction: Contributions and Gaps in Gender and Agroforestry. *International Forestry Review* 17 (S4): 1–10.

Colfer, C. J. P., Elias, M., and Sijapati Basnett, B. 2016a. A Gender Box Analysis of Forest Management and Conservation. In C. J. P. Colfer, B. Sijapati Basnett, and M. Elias, eds, *Gender and Forests: Climate Change, Tenure, Value Chains and Emerging Issues*. London: Earthscan/Routledge, 3–16.

Colfer, C. J. P., Hummel, S. S., Sijapati Basnett, B., and Elias, M. 2016b. Gender and Forests – Bringing 'Southern' Issues to the 'North'? Paper presented at the *Society for Applied Anthropology, 76th Annual Meeting*, Vancouver, British Columbia, Canada, 31 March.

Colfer, C. J. P., Sijapati Basnett, B., and Elias, M. eds, 2016c. *Gender and Forests: Climate Change, Tenure, Value Chains and Emerging Issues*. London: Earthscan/Routledge.

Elmhirst, R., Siscawati, M., and Sijapati Basnett, B. 2015. Navigating Investment and Dispossession: Gendered Impacts of Oil Palm 'Land Rush' in East Kalimantan, Indonesia. Paper presented at *Land Grabbing, Conflict and Agrarian-Environmental Transformations: Perspectives from East Asia and Southeast Asia*, Chiang Mai University, Thailand, 5–6 June. Conference Paper No. 69.

Hecht, S., Yang, A. L., Sijapati Basnett, B., Padoch, C., and Peluso, N. L. 2015. *People in Motion, Forests in Transition: Trends in Migration, Urbanisation, Remittances and their Effects on Tropical Forests and Forest-Dependent Communities*. Bogor, Indonesia: Center for International Forestry Research.

Kabeer, N. 2015. Gender equality, the MDGs and the SDGs: Achievements, Lessons and Concerns. *London School of Economics and Political Science: Africa at LSE*, 6 October. http://blogs.lse.ac.uk/africaatlse/2015/10/06/gender-equality-the-mdgs-and-the-sdgs-achievements-lessons-and-concerns/.

Li, T. 2015. *Social Impacts of Oil Palm in Indonesia: A Gendered Perspective from West Kalimantan*. Bogor, Indonesia: Center for International Forestry Research.

Mairena, E., Lorio, G., Hernández, X., Wilson, C., Müller, P., and Larson, A. M. 2012. *Gender and Forests in Nicaragua's Indigenous Territories: From National Policy to Local Practice*. Bogor, Indonesia: Center for International Forestry Research.

Resurreccion, B. P. and Elmhirst, R., eds. 2008. *Gender and Natural Resource Management: Livelihoods, Mobility and Interventions*. London: Earthscan.

Rigg, J. A. V., ed. 2012. *Revisiting Rural Places: Pathways to Poverty and Prosperity in South East Asia*. Singapore and Honolulu: National University of Singapore and Hawaii University Press.

Sijapati Basnett, B. 2013. *Taking Migration Seriously: What are the Implications for Gender and Community Forestry?* CIFOR Info Brief 65. Bogor, Indonesia: Center for International Forestry Research.

Sikor, T. 2010. Forest Justice: Towards a New Agenda for Research and Practice? *Journal of Integrative Environmental Sciences* 7 (4): 245–250.

UN (United Nations). 2015. Sustainable Development Goal 5: Achieve Gender Equality and Empower All Women and Girls. *UN Sustainable Development Goals*. Retrieved 10 August, 2006, from www.un.org/sustainabledevelopment/gender-equality/

Vira, B., Wildburger, C., and Mansourian, S., eds. 2015. *Forests, Trees and Landscapes for Food Security and Nutrition: A Global Assessment Report*. Vienna: International Union of Forest Research Organizations.

Index

Page numbers in *italics* denote an illustration, **bold** indicates a table, n indicates an endnote